INTERNATIONAL CENTRE FOR MECHANICAL SCIENCES

COURSES AND LECTURES - No. 329

ROTATING FLUIDS IN GEOPHYSICAL AND INDUSTRIAL APPLICATIONS

EDITED BY

E.J. HOPFINGER
UNIVERSITY OF GRENOBLE AND CNRS

Springer-Verlag Wien GmbH

Le spese di stampa di questo volume sono in parte coperte da
contributi del Consiglio Nazionale delle Ricerche.

This volume contains 176 illustrations.

Originally published by CISM, Udine in 1992.

In order to make this volume available as economically and as rapidly as possible the authors' typescripts have been reproduced in their original forms. This method unfortunately has its typographical limitations but it is hoped that they in no way distract the reader.

ISBN 978-3-211-82393-4 ISBN 978-3-7091-2602-8 (eBook)
DOI 10.1007/978-3-7091-2602-8

PREFACE

Most of the research on rotating fluids has been motivated by geophysical applications because in these situations rotation effects are of primary importance. Understanding atmospheric and oceanic motions is of vital interest for climate prediction and we are also increasingly concerned with the dispersion of pollution by biochemical and chemical substances released in the atmosphere and the oceans. The release of CO_2 *for instance is of great concern in the general question of global warming.*

Rotation is also plays a key role in rotating machines where, for example, it effects the boundary layers structure on the runner blades of turbines and can give rise to intense vortex formation. Vortex flows or swirling flows are widely encountered and are made use of in engineering problems related with cyclone separators, the aerodynamics of combustion chambers and vortex valves. There is renewed interest in the fundamental understanding of these complex engineering flows. Although the scales of motions in rotating industrial devices are very different from those encountered on planets, the fundamental concepts of rotating fluids presented in this volume have general value; much can, therefore, be gained by interdisciplinary research.

The main effects of rotation are the tendency toward two-dimensionality of fluid motions, the rapid spin-up or spin-down of fluid columns due to Ekman pumping and the possibility of sustained wave motions. A number of flow instabilities are a result of rotation or the combined effect of rotation and stratification. Some of these are well known while others are at the forefront of research. Considerable progress has also been made recently on turbulence and shear flows in rotating fluids. The mechanisms of energy and anstrophy transfers in turbulence with rotation is better understood, and exciting new phenomena like the transition to a rotationally dominated turbulence state and the possibility of vorticity concentration have recently been demonstrated. This raises new questions concerning vortex waves, vortex breakdown and vortex stability in general (in two and three dimensions).

The topics covered in this volume include general concepts and the fundamentals of rotating fluids, a comprehensive treatment of Ekman layer and Stewartson layer theories, aspects of instabilities (including baroclinic instability and vortex instabilities), different types of wave motions, vortex flows, interactions of geostrophic vortices, geostrophic and rotationally dominated turbulence and shear flows with rotation. Emphasis is placed on both, theory and experiments, either physical or numerical. The book is intended for graduate students and research workers interested in geophysical fluid dynamics or those concerned with engineering problems of rotating fluids.

The volume comprises the lecture notes of the course on rotating fluids given in Udine in July 1991. All the lectures were ready to revise and amend their notes for publication in the form of this book and I owe them special thanks. The willingness to make this additional effort is partly due to the warm welcome by the staff of the International Centre for Mechanical Sciences in Udine and to the stimulating atmosphere of the center which were greatly appreciated by the lecturers and the students. I think the lecturers will join me in expressing our gratitude for the support received from the CISM and in particular to Professor G. Bianchi, S. Kaliszky and the administrative staff of the center for their constant readiness to help.

E.J. Hopfinger

CONTENTS

PART I

PRELIMINARIES

Chapter I.1

GENERAL CONCEPTS AND EXAMPLES OF ROTATING FLUIDS

E. J. Hopfinger
University J.F. and CNRS, Grenoble Cedex, France

1. INTRODUCTION

Planets and stars are rotating objects and large scale fluid motions relative to the planets solid body rotation are constrained by rotation. Theoretical treatment of these fluid motions must, therefore, include rotation effects and it is for this reason that most of the advances in the understanding of rotating fluids have been made in the context of Geophysical Fluid Dynamics (GFD) and Astrophysical Fluid Dynamics (AFD). GFD can be considered a special case of AFD and many approximations made in GFD (see for instance Pedlosky [1]) are not valid in AFD [2]. For example, in GFD horizontal scales of motion L are much less than the radius of the planet, allowing the use of a constant or linearly varying Coriolis frequency $f=2\Omega$. The fluid layer depth H is much less than L so that only the component of Ω perpendicular to the planets surface needs to be considered. The Boussinesq approximation has also only limited validity in AFD. All the lectures in this course use approximations current in GFD when problems of geophysical interest are considered.

The main difference of GFD problems compared with problems arising in Industrial Rotating Fluid Dynamics (IRFD) is that Rossby numbers in IRFD are generally of order one and that solid boundaries are present. Depending on the orientation of the rotation vector with respect to the boundary plane and the mean shear vorticity, the effects of rotation can be radically different. The flow in industrial rotating machines is often 3D and the vorticity vector may change its orientation with respect to the system rotation vector. Nevertheless, there is much to be learned from GFD when trying to solve IRFD problems.

2. PHENOMENA PARTICULAR TO ROTATION

In Greenspan's book [3] it is mentioned that there are three primary phenomena brought about by rotation. One is the tendency for fluid motion to become two-dimensional when rotation effects are important or, more precisely, when the Rossby number, expressing the relative importance of inertia and Coriolis force,

$$Ro = U/L f \ ,$$

is low (neglecting viscous effects); U is the fluid velocity relative to the rotating frame, L the horizontal length scale of the motion and $f=2\Omega$. This rotational constraint is a consequence of geostrophic equilibrium and is known as Taylor-Proudman theorem. It states that fluid motions are invarient along a direction parallel to the rotation axis.

The Taylor-Proudman column is closely related with inertial waves which are another remarkable manifestation of rotation effects [4]. Inertial waves exist in a rotating fluid when a body or a fluid element oscillates at a frequency $\omega < 2\Omega$. The energy is radiated along characteristics which form a cross well known from internal waves. For larger excitation frequencies $\omega > 2\Omega$, the flow field created by the oscillating object is the same as in non-rotating fluids. In the other limit of $\omega \to 0$, the characteristic surfaces form a column. In this limiting case the information is passed vertically through the depth of the fluid layer at the inertial wave group velocity of order ΩL. Therefore, for 2D flow to be realized it is necessary that the characteristic time scale L/U of any eddy motion is small compared with the travelling time of inertial waves of order $H/\Omega L$. The condition for two-dimensionalization is then $H \ll L\, Ro^{-1}$.

The third important and general property of rotating fluids is the spin-up or spin-down of fluid columns in a time $t_s \sim (H^2/\nu\, \Omega)^{1/2} \ll L^2/\nu$. When there is a differential rotation between the fluid column and the end boundaries the viscous boundary layer (Ekman layer, see Read [5]) will pump fluid outward or inward setting up a secondary circulation until spin-up or spin-down is completed.

More subtle phenomena in rotating fluids are Rossby and baroclinic waves and baroclinic instability [7]. Rossby and baroclinic waves are low frequency inertial waves which require a variation in Coriolis frequency with latitude, a so-called β effect, or an equivalent stratification effect [6]. The combination of rotation and stratification can give rise (because of angular momentum conservation) to baroclinicity and a vertical shear which are necessary conditions for the growth of baroclinic instability (see Klein [8] for a recent review).

3. SOME ILLUSTRATIONS OF ROTATION EFFECTS.

The atmospheric cyclones and anticyclones, made visual by clouds, illustrate most strikingly the effects of rotation on fluid motions. In the atmosphere potential energy is released by baroclinic instability and converted into kinetic energy. The most unstable mode gives rise to an eddy scale of about 1000 km which is of the order of the internal Rossby radius of deformation

$$\Lambda = NH/f ,$$

where $N = (-g\, \partial\rho/\partial z)/\rho_0)^{1/2}$ is the buoyancy frequency and H is a pressure scale height. The motion is quasi geostrophic (nearly 2D) and the interaction between vortices results in an inverse energy cascade via vortex merging or other mechanisms, hence in a growth of the vortices. There is an upper limit to this growth where eddies are carried away as Rossby waves. This crossover scale is about 3000 km and corresponds to the scale where the eddy turnover time L/U is greater than the Rossby wave period. This scale is known as the Rhines radius

$$L_\beta = \pi(2\,U/\beta)^{1/2} ,$$

where $\beta = \partial f/\partial y$ (y is the longitude coordinate) or

$$\beta = 2\Omega \cos\theta\, /R \quad ,$$

with R the radius of the planet and θ the latitude angle.

It is interesting to note that in the Jovian atmosphere L_β has about the same value as on the earth. In the oceans, similar processes take place as in the atmosphere with the corresponding scales being one order of magnitude less. For many stars, the Rossby radius of deformation Λ at mid latitudes is of the order of the radius of the star and the whole spherical geometry has to be included in the model (Zahn [9]).

An extraordinary phenomenon is Jupiter's GRS [10] and when talking about rotating fluids it is difficult not to make mention of it. It is a large scale, about 10^4 km radius, high velocity, about 10^2 m/s, 300 years old vortex. It is a geostrophic vortex; because of the large scale the Rossby number is low. Its scale is however such that it should degenerate into Rossby waves ($L_\beta < L$). Maxworthy et al [11] have in fact proposed a solitary Rossby wave model for the GRS. More recently, models of vortex merging on a β plane have shown that a persistent large scale isolated vortex patch of uniform potential vorticity can form [12]. This vortex is fed by smaller vortices which arise from shear instability of the zonal flow. Because of the uniform potential vorticity, energy radiation by Rossby waves, which would require a potential vorticity gradient, is inhibited. The zonal flow must adjust to cancel the β effect. There is evidence that the potential vorticity of Jupiter's GRS is in fact not uniform and that the topography of the lower layer must play an important role [13]. A knowledge of the different conditions which lead to stable isolated vortices having uniform potential vorticity or not is of value in the conext of the GRS and other similar phenomena like Neptune's GDS.

Vortex merging of like-sign, barotropic vortices is now well understood and the existence of a critical merger distance between vortices is well documented. Merger of baroclinic vortices, where an additional length scale Λ must be considered, raises many questions to which we have no answer. Experiments and numerical simulations are necessary and are complementary means of investigation to resolve the questions raised [14], [15]. This problem will be discussed in more detail in a later lecture. Vortex stability is a related problem to which numerical and experimental work has much contributed [16], [17]. Much research has also been done on geostrophic turbulence and the related energy and enstrophy cascade processes. However, the response of the flow structure to perturbations either by 3D turbulence or by topographic effects is a subject of present day research.

Concerning shear flows with rotation, three basic configurations can be distinguished depending on whether the rotation vector is:

i) perpendicular to the mean flow vorticity vector and also to the boundary plane,

ii) perpendicular to the mean flow vorticity vector and parallel to the boundary,

iii) parallel or antiparallel to the mean flow vorticity vector and the boundary plane.

Configuration i) is the Ekman layer case, Fig. 1, which is of primary importance in rotating fluids and has, therefore, received considerable attention in the past. The stability of Ekman layers has fundamental fluid dynamical interest because of the interaction between two classes of waves, waves A and B [3], where transition to turbulence seems to originate. Spin-up and spin-down of fluid columns is related with Ekman layers and there is still work in progress on particular cases of spin-up where free surface, geometric or non-linear effects must be taken into account. The flow near rotating discs is a non-linear counterpart of the linear Ekman layer problem. The latter has widespread engineering applications, in particular the situation where there are two parallel closely spaced discs of finite dimensions, representative of turbine rotors or computer disc drives. This problem has also received attention in the context of disc pumps.

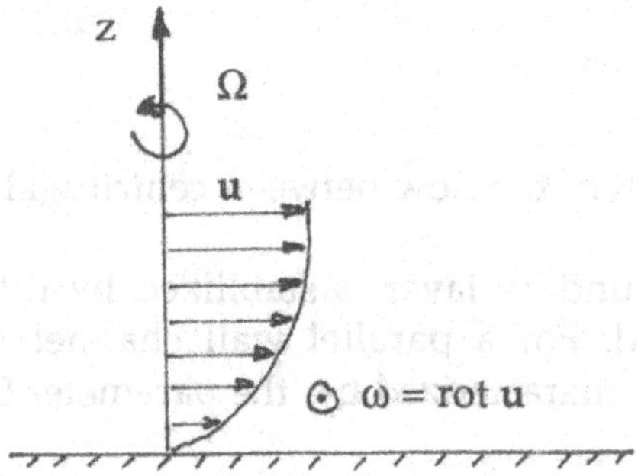

Fig.1 Ekman layer configuration

Configuration ii) is the rotating pipe flow problem, the stability of which was studied experimentally by Nagib [18] and theoretically by Maslowe [19] for instance. Intuitively, it would be expected that rotation would stabilize the flow but theory and experiments show the contrary. The critical Reynolds number for instability due to finite amplitude perturbations decreases from 2500 for zero rotation to about 900 when the swirl number

$$M = \Omega\, r_0 / U_m$$

is about 4, where r_0 is the pipe radius and U_m the mean velocity. A related problem is the swirling jet which is destabilized by rotation for swirl numbers less than 1.5 [20].

Configuration iii) corresponds for instance to the flow in a channel which rotates about the spanwise axis [21]. It is a fundamental problem in IRFD because it is representative of the flow between blades of centrifugal impellors. Schematically this is shown in Fig. 2.

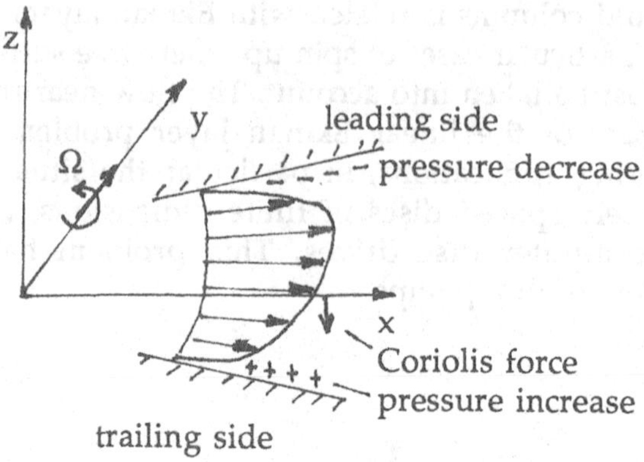

Fig. 2 Sketch of the flow between centrifugal impellor blades

The leading side boundary layer is stabilized by rotation and the trailing side can be destabilized. For a parallel wall channel the range over which destabilization occurs is characterized by the parameter S given by

$$S = -2\,\Omega/(\partial U/\partial y)\ .$$

The range of destabilization in the parameter range S can be obtained from a stability analysis. It is found that destabilization occurs when $S(1+S) < 0$. For $S(S+1) > 0$ rotation stabilizes the flow. It must be realized however that in a boundary layer the value of $\partial U/\partial y$ is a function of rotation and is not a given quantity. We will see this discussed in more detail in the lecture on rotating shear flows (Maxworthy, [22]). It should be mentioned that rotating channel flow has analogies with curved wall channel flow and this was demonstrated by Bradshaw [23]. The number R_Ω or $B = S(S+1)$ is often referred to as the Bradshaw number. In rotating curved channel flow the two effects may act in the same or opposing sense and this has interesting consequences [24].

Vortex flows are most widely encountered in IRFD applications. Here we mean large Rossby number vortex flows compared with the low Rossby number geostrophic vortices mentioned above. Well known are the trailing edge vortices. The formation of such vortices and in particular the breakdown conditions have stimulated much theoretical and experimental research (see for instance Leibovich [25]) and the lectures by D. Moore [26]). Vortex flows are also used in combustion chambers to stabilize combustion as discussed by Escudier [27]. Discussions of the physical principles of other applications of vortex flow are also found in Escudier. I would like to draw the attention of the reader to two or three of these interesting flows and their usefulness. One of these is the vortex valve which has widespread applications. The principle is to regulate the flow by controling the swirl in the valve chamber with a concentration of the vorticity in the exit shaft [28]. All the pressure head can be lost in the vortex and the flow may be completely blocked. Operating conditions depend on the ratio of the control flow rate to through flow rate Q_c/Q_s and on the ratio of the control pressure to supply pressure p_c/p_s [27].

In cylone separators vorticity concentration can lead to large head losses and there we want to prevent it from occuring. The simple method for prevention is the use of radially oriented barriers. This was already suggested long ago and guidevanes are generally used in the standard, cone shaped, cyclone separators widely used for separating large (>10 μm) heavy particles. More unusual are rotating wall hydrocyclones which have a higher efficiency for smaller and also light particles. These are used for separating water from oil and for cleaning recycled paper pulp in the paper industry. A special feature of the later use is that turbulence is required to defloculate the paper fibers and this is achieved by creating an annular flow relative to the rotating walls. The conditions of the maintenance of turbulence need to be known to determine the necessary excess velocity relative to the boundaries. Something in this direction can be learned from the stability conditions for laminar curved rotating channel flow investigated by Matsson & Alfredsson [24].

A new concept in hydrocyclones is to make use of the rotating analogue of the Boycott effect in sediment basins proposed by Greenspan and Ungarish [29]. The idea is to replace the gravity by the centrifugal force and the effect of the inclined container wall by conically shaped cylinder walls. Because of the Coriolis force, conical walls do not enhance separation performances unless a radial barrier is introduced to prevent conservation of angular momentum. The flow structure is however complicated by the presence of such barriers and further work is needed to understand fully the intersecting boundary layer flows.

An interesting effect is that of spacers used in fast rotating centrifuges constructed by Alfa-Laval. These centrifuges contain in the central part a stack of closely spaced conical discs and spacers are used for structural reasons. Without the spacers the flow is essentially confined to thin Ekman layers. The wakes caused by the spacers destroy the Ekman layers and distribute the flow over a larger section [30]. The consequence is a lower pressure drop.

A well known example of nuisible vortex flow is the draft tube vortex in turbines shown in Fig.3 .It occurs when there is residual rotation in the draft tube as is the case when the turbine is operated at partial load or overload. The vortex core is a source of cavitation which makes the vortex visible. Cavitation can cause damage through erosion. Under some conditions the vortex is unstable and this gives rise to pressure fluctuations thoughout the turbine. This unsteadiness is probably related with vortex breakdown which would occur above a critical swirl number G R/Q (G is angular momentum, Q the volum flow rate and R the radius of the draft tube) which is a function of the

Reynolds number [31].

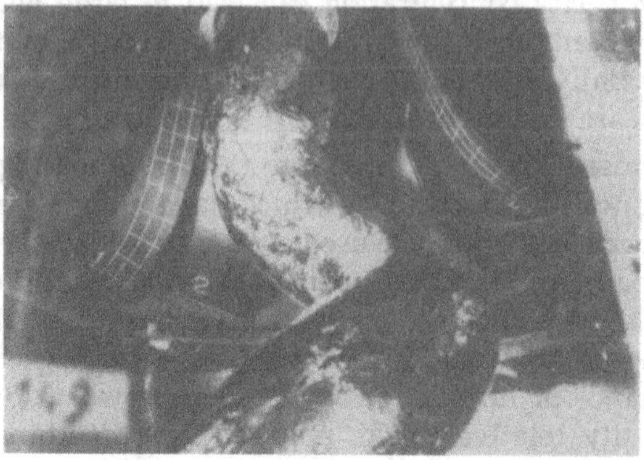

Fig. 3 Example of a draft tube vortex in a Francis turbine visualised by natural cavitatin(by courtesy of Sté Neyrpic).

Concluding this discussion of vortex flows I would like to mention vortex concentration by 3D turbulence in a rotating fluid as was demonstrated in the experiments by Hopfinger [32]. This vortex concentration may be relevant to tornado formation. The ratio of the vorticity in the vortex to the background vorticity is of the same order in the laboratory and in tornadoes. Intense vortices in fast rotating young stars might be at the origin of the observed bright spots. Due to the depression of the isotherms, the radiative flux is focused into the vortex which makes them appear bright [33].

REFERENCES:

1. Pedlosky, J.: Geophysical Fluid Dynamics, Springer Verlag, 1979.
2. Spiegel, E.A.: Stellar fluid dynamics, in: WHOI Course Lecture Notes (Ed. B. E. DeRemer), WHOI report n° 91-03, 1990.
3. Greenspan, H.P.: The Theory of Rotaing Fluids, Cambridge Univ. Press,1968.
4. Mory, M. Inertial waves, this volume
5. Read, P. L.: Ekman layers and Stewartson layers, this volume.
6. Maxworthy, T.: Wave motions in a rotating and/or stratified fluid, this volume.

8. Klein, P.: Transition to chaos in unstable baroclinic systems: a review, Fluid Dyn. Res., 5 (1990), 235.
9. Zahn, J.-P.: Shear flow instability and the transition to turbulence, in: WHOI Course Lecture Notes (Ed. B. E. DeRemer), WHOI report n° 91-03, 1990.
10. Read, P.L.: Long-lived eddies in the atmosphere of major planets, this volume.
11. Maxworthy, T., Redekopp, L. G. and Weidman, P.: On the production and interaction of planetary solitary waves: application to the Jovian atmosphere, Icarus, 33 (1978) , 388.
12. Sommeria, J., Meyers, S.D. and Swinney, H.L.: Laboratory simulation of Jupiter's great red spot, Nature, 331 (1988), 1.
13. Dowling, T. E. and Ingersoll, A.P.: Potential vorticity and layer thickness variations in the flow around Jupiter's Great Red Spot and White Oval BC, J. Atmos. Sci. 45 (1988), 1380.
14. Griffiths, R.W. and Hopfinger, E.J.: Coalescing of geostrophic vortices. J. Fluid Mech. 178 (1987), 73.
15. Verron, J., Hopfinger, E.J. and McWilliams, J. C.: Sensitivity to initial conditions in the merging of two-layer baroclinic vortices. Phys Fluids A, 2 (6) (1990), 886.
16. Kloosterziel, R.C. and van Heijst, G.J.F.: An experimental study of unstable barotropic vortices in a rotating fluid. J. Fluid Mech. 223 (1991), 1.
17. Carton,X. J., Flierl, G.R. and Polvani, L.M.: The generation of tripoles from unstable axisymmetric solated vortex structures. Europhys. Lett. 9 (1989), 339.
18. Nagib, H.M., Lavan, Z. and Fejer, A.A.: Stability of pipe flow with superposed solid body rotation, Phys Fluids, 14 (1971), 766.
19. Maslowe, S.A.: Instability of rigid rotating flows to non-axisymmetric disturbances, J.Fluid Mech., 64 (1974), 307.
20. Lessen, M., Singh,P.J. and Paillet, F.: The stability of a trailing line vortex, Part 1, inviscid theory, J.Fluid Mech., 63 (1974), 723.
21. Johnston, J. P., Halleen, R.M. and Lezius, J.D.: Effect of spanwise rotation on the structure of twodimensional fully developed turbulent channel
22. Maxworthy, T.: Convective and shear flow turbulence with rotation, this volume.
23. Bradshaw, P.: The analogy between streamline curvature and buoyancy in turbulent shear flow, J. Fluid Mech. 36 (1969), 177.
24. Matsson, O. J. E. and Alfredsson, H.I.: Curvature and rotation induced instability in channel flow, J. Fluid Mech., 210 (1990), 537.
25. Leibovich, S.: Vortex stability and breakdown: survey and extension, AIAA 22 (1984), 1192.
26. Moore, D. W.: Dynamics of vortex filaments, this volume.
27. Escudier, M.: Confined vortices in flow machinery. Ann. Rev. Fluid

Mech.,19 (1987), 27.

28. Brombach, H.: Vortex flow controllers in sanetary engineering, Trans. ASME J. Dyn. Sys. Meas. Control, 106 (1984), 129.
29. Greenspan G.H. and Ungarish, M.: 0n the enhancement of centrifugal separation, J. Fluid Mech. 178 (1985), 73.
30. Bergström, L.: Wakes behind the caulks in a disc stack centrifuge. Presented at Euromech 245, Cambridge (see report J.Fluid Mech. 211, p. 417), 1989.
31. Spall, R.E. Gatski, T.B. and Grosch, C.E.: A criterion for vortex breakdown, Phys. Fluids 30 (1987), 3434.
32. Hopfinger, E.J., Browand, F.K. and Gagne, Y.: Turbulence and waves in a rotating tank. J. Fluid Mech 125 (1982), 505.
33. Spiegel, E.A.: Lecture III. Photogasdynamics, in: WHOI Course Lecture Notes (Ed. B. E. DeRemer), WHOI report n° 91-03, 1990.

PART II

FUNDAMENTALS

Chapter II.1

PARAMETERS, SCALES AND GEOSTROPHIC BALANCE

E. J. Hopfinger
University J.F. and CNRS, Grenoble Cedex, France

1. INTRODUCTION

The presence of rotation introduces new parameters and constraints on fluid motions discussed in general terms before [1]. It is of interest to develop here in more detail the physical implications of these rotation effects. To those working with rotating fluids, the material presented in this paper is not new, except, perhaps, for some aspects of the section on turbulence scales. It is nevertheless useful to recall the fundamental processes. To others working in fluid mechanics but who are not so familiar with rotating fluids, the contents of this paper is a good introduction to what follows in later chapters in this volume.

The principal parameters and their physical meanings are introduced by way of the governing equations and the expression for the internal Rossby deformation radius follows from the balance between Coriolis and buoyancy forces. Geostrophic flow is a consequence of the balance beween Coriolis force and the horizontal pressure gradient. The vertical invariance of geostrophic flow is demonstrated by the Taylor column [2]. It will be shown that this flow condition and the accompaning hydrostatic approximation requires that H/(L Ro) <<1 and not the more restrictive condition H<<L. Of fundamental importance are also the problem of the simulation of β effects in laboratory systems and the related problems of rotational stiffness and Sverdrup flow.

The parameters and scales characterizing rotation effects on turbulence are discussed by analogy with stratified turbulence. The idea of an equivalent Ozmidov scale L_Ω, introduced by Mory and Hopfinger [3] has recently been extended by Gibson [4]. Mention is made of the relation of this scale with the turbulent Rossby number used previously [5].

2. PARAMETERS AND GEOSTROPHIC BALANCE

2.1 Parameters:

The physical significance of the different parameters related with rotation is best illustrated by writing down the governing equations. For a rotating fluid in the rotating frame of reference the momentum and continuity equations are:

$$\underset{1}{\frac{\partial \mathbf{u}}{\partial t}} + \underset{2}{\mathbf{u}.\nabla\mathbf{u}} + \underset{3}{2\Omega\mathbf{k} \times \mathbf{u}} = \underset{4}{-\frac{\nabla p}{\rho_0}} + \underset{5}{g\,\mathbf{k}\rho'/\rho_0} + \underset{6}{\nu\nabla^2\mathbf{u}} \qquad (1)$$

$$\nabla.\mathbf{u} = 0 \qquad (2)$$

The Coriolis force $-2\Omega\mathbf{k} \times \mathbf{u}$ is a deflecting force (in northern hemisphere it deflects to the right of the direction of motion), g the gravitational acceleration and $\rho = \rho_0(1+\rho'/\rho_0)$. The centrifugal acceleration $\Omega\mathbf{k} \times (\Omega\,\mathbf{k} \times \mathbf{r})$ is, as usual, included in the pressure and the Boussinesq approximation was made in equations (1) and (2).

We now introduce characteristic length and velocity scales and also a time scale T. It is necessary to distinguish between horizontal length L and velocity scale U and vertical length H and velocity scale W. Nondimensionalization by these scales of the variables in equation (1) gives the order of magnitude of the different terms, namely

1:3	2:3	3:3	4:3	5:3	6:3
$o(1/fT)$	$o(U/fL)$	$o(1)$	$o(1)$	$o(g'H/fUL)$	$o(\nu/fH^2)$

where $g' = g\,\Delta\rho/\rho_0$ and $f = 2\Omega$. The parameter

$U/fL = Ro$ is the Rossby number

$\nu/fH^2 = E$ is the Ekman number

$1/fT = \gamma$ is a scaled frequency.

The term $g'H/fUL = RoRi$, with $Ri = g'H/U^2$. The boundary between rotation dominated and stratification dominated flows is given by $RoRi = 1$. When $RoRi < 1$ the flow is rotation dominated and when $RoRi > 1$ it is stratification dominated. The condition $RoRi=1$ gives $L = \Lambda = g' H/f U$. From conservation of angular momentum, when a density front moves radially distance L, we obtain $U = f L$. Substitution of this value for U leads to the internal radius of deformation in the form:

$$\Lambda = (g'H)^{1/2}/f \tag{3}$$

The continuity equation gives $W \sim UH/L$, where the aspect ratio

H/L is the scale parameter.

When H/L is small, the vertical velocity scale is also small compared with horizontal velocities. This condition is at the origin of the hydrostatic approximation. The condition $H/L \ll 1$ is generally satisfied in geophysical flows but rarely in laboratory experiments. It is easily seen that in the laboratory, because of the small horizontal scale, f is necessarily large so that when H is small Ekman friction will dominate the flow. For this reason, in the laboratory we have often $H \sim L$. This does not violate the hydrostatic approximation when Ro is small, because from the vorticity equation we obtain that the vertical velocity scale is actually

$$W \sim U\, Ro\, H/L.$$

2.2 Geostrophic flow:

Geostrophic flow is a consequence of a balance between the Coriolis force and the horizontal pressure gradient, that is between the terms of o(1) in (1). All the other terms of order γ, Ro, E and RoRi are negligible. Equation (1) reduces to:

$$f\,\mathbf{k} \times \mathbf{u} = -\nabla p/\rho_0 \tag{4}$$

Writing this equation in (x,y,z) coordinates gives

$$u = -\frac{1}{f\rho_0}\frac{\partial p}{\partial y} \tag{4a}$$

$$v = \frac{1}{f\rho_0}\frac{\partial p}{\partial x} \tag{4b}$$

$$0 = \frac{1}{f\,\rho_0}\frac{\partial p}{\partial z} \tag{4c}$$

Equations (4a) and (4 b) give

$$\frac{\partial u}{\partial x} + \frac{\partial v}{\partial y} = 0 \quad \text{hence, by continuity (2), we have} \quad \frac{\partial w}{\partial z} = 0$$

Equation (4c) imposes $\frac{\partial u}{\partial z} = \frac{\partial v}{\partial z} = 0$.

Geostrophic flow is, therefore, invariant with z which is the Taylor-Proudman (TP) theorem

$$f(\mathbf{k}\cdot\nabla)\mathbf{u} = 0. \tag{5}$$

The experiment by G.I. Taylor [2] is shown schematically in Fig. 1

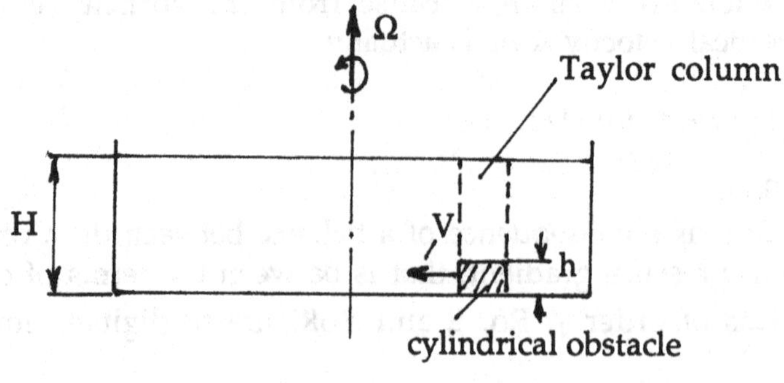

Fig. 1 Taylor's experiment demonstrating the Taylor-Proudman theorem.

Taylor's experiment showed qualitatively the validity of the TP theorem. Quantitative experiments were later conducted by Hide and Ibbetson [6] , where the influence of the small but finite values of Ro, E and γ was studied.

The TP theorem does not say anything about any limitation on fluid layer depth because it is a two-dimensional reasoning. A higher order development of the equation in terms of the small parameters would be required. Information is passed through the fluid layer depth by inertial waves in time scale t_W and this time must be much less than the characteristic time scale of the flow U/L. Neglecting viscous and stratification effects, one of the condition for a Taylor column to exist is then $H \ll L\ Ro^{-1}$.

3. BETA EFFECT AND ROTATIONAL STIFFNESS

3.1 Beta (β) effect:

On a sphere the Coriolis frequency varies with latitude and an eddy which moves in a longitudinal direction y, say north-south in the northern hemisphere will gain relative vorticty by conserving its potential vorticity. The change in Coriolis frequency in the linear approximation is

$$\Delta f = \frac{1}{R}\frac{\partial f}{\partial \theta}\ \Delta y = \beta\ \Delta y$$

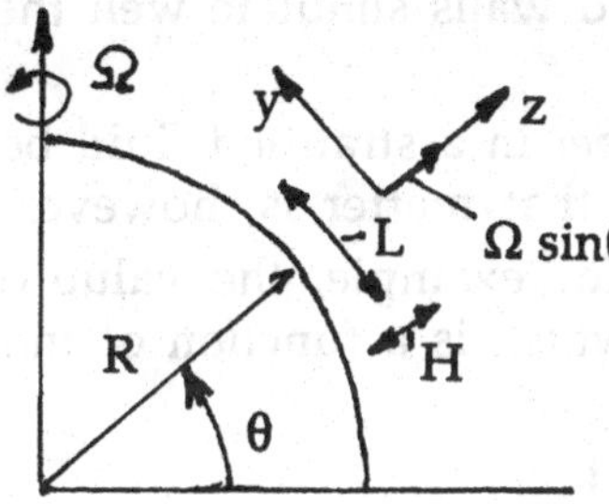

Fig. 2 Coordinates on a sphere

Since $f = 2\Omega \sin\theta$ we have $\beta = (2\Omega \cos\theta)/R$. The change in relative vorticity ζ when the fluid layer depth H is constant is then, by conservation of potential vorticity,

$$\Delta\zeta = -\Delta f = -\beta\ \Delta y. \tag{6}$$

In the laboratory, the β effect can be simulated in a container by a bottom and /or top cover inclined at an angle α (see Fig. 3). The expression for the angle α is, when α is small

$$\alpha = -\frac{\partial H}{\partial y}$$

hence

$$\Delta H = -\alpha \, \Delta y$$

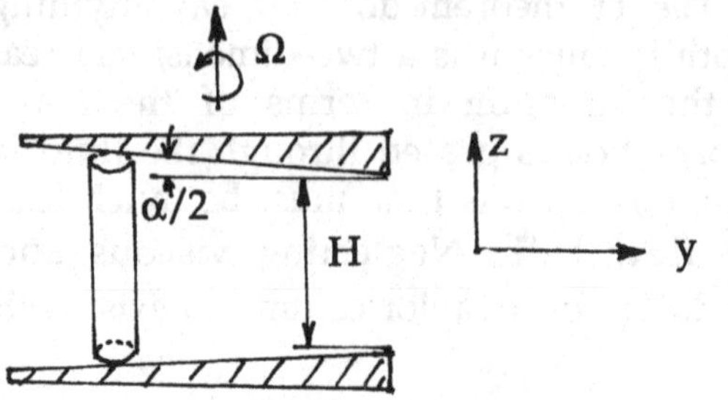

Fig.3 Wedge-shaped container with homogeneous fluid

From conservation of potential vorticity the change in relative vorticity is then when $\Delta H/H << 1$,

$$\Delta\zeta = \Delta H\,(\zeta + f)/H \tag{7}$$

giving

$$\beta = \alpha(\zeta + f)/H. \tag{8}$$

Since $\zeta/f \sim o(Ro)$, inclined end walls simulate well the β effect when Ro is small.

The expression for the β effect in a stratified fluid between wedge-shaped boundaries or on a planet for that matter is, however, no longer explicitly given. For a two layer fluid for example, the value of β depends on the interface shape or inclination which is a function of the relative velocities in the two layers.

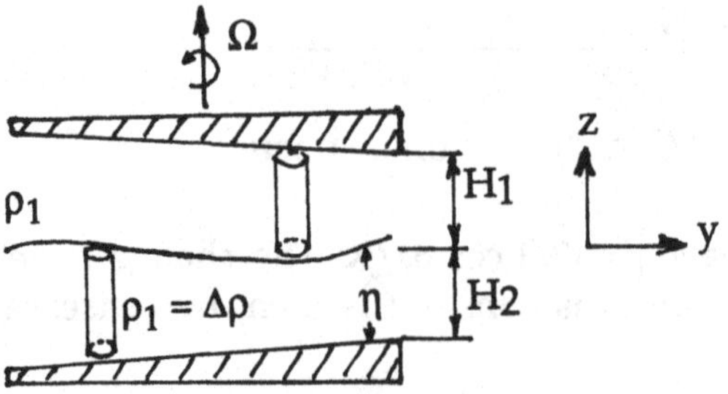

Fig. 4 Two-layer stratified β plane

3.2 Swerdrup Flow and Rotational Stiffness:

Slow vertical forcing of a geostrophic flow on a β plane generates a flow along the longitude, generally known as Sverdrup flow. Suppose we impose on the upper surface of the wedge (see Fig. 3) a gentle downward motion w_0, either by moving the lid, by fluid sources or by Ekman convergence. Since by virtue of the TP theorem the vertical velocity w_0 must be the same throughout the fluid column, the fluid column cannot contract and must, therefore, move to greater depth. For the negative w_0 imposed on the upper lid, the fluid column moves southward at the velocity -V (see Fig. 3) given by

$$V = w_0/\alpha$$

For the Sverdrup flow in a two-layer stratified fluid (Fig. 4) and equal depth layers, $H_1 = H_2 = H$, the corresponding relation is [8]:

$$\frac{1}{2}(V_1 + V_2) = w_0/\alpha \, .$$

We have seen in the above example that geostrophic flow is rotationally stiff that is to say no stretching or contraction of vortex lines is possible in this limit. It is interesting to know what is the pressure required to contract or stretch vortex lines in quasi geostrophic flow. Following the simple model of Rhines [7] let us suppose that the lower boundary is forced periodically with vertical velocity

$$w(0) = w_0 \sin(x/L) \sin(t/T)$$

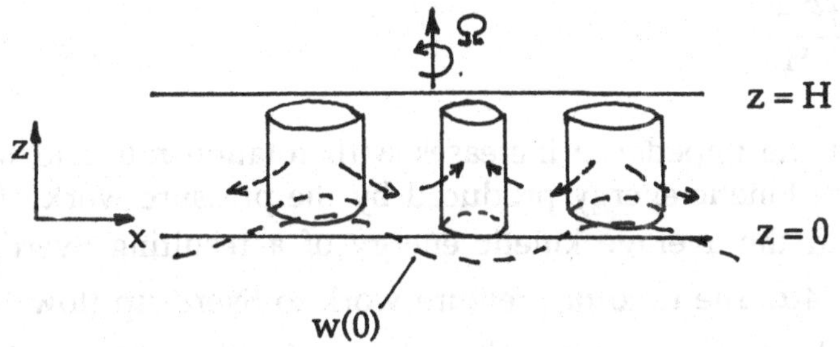

Fig. 5 Schematic of periodic forcing of a fluid layer

Integration from z= 0 to H of the vorticity equation for the vertical vorticity component ζ gives:

$$H\frac{\partial \zeta}{\partial t} = -f\, w_0 \sin(x/L)\sin(t/T).$$

Integrating this equation with respect to time gives:

$$\zeta H = f\, T w_0 \sin(x/L)\cos(t/T)$$

Expressing the vorticity component in terms of the pressure gradient by using the geostrophic balance

$$\zeta = \frac{1}{f\rho_0}\frac{\partial^2 p}{\partial x^2}$$

and integrating twice with respect to time leads to ($P = p/\rho_0$):

$$p/\rho_0 = \frac{f^2L^2T\, w_0}{H}\sin(x/L)\cos(t/T) \qquad (9)$$

The impedance of the flow to vertical stretching is then:

$$\frac{P}{w_0} = \frac{f^2L^2T}{H} \qquad (10)$$

It is seen that the impedance increases with rotation rate and horizontal scale. The average kinetic energy produced by the pressure work $\langle Pw(0)\rangle$ is $(fTw_0L)^2/2H$ and the average kinetic energy of a resulting Sverdrup flow would be $Hw_0^2/4\alpha$. The ratio of pressure work to Sverdrup flow energy is then $2(\beta LT)^2$ and when this ratio is less than unity the eddies will yield to vertical stretching because this requires less energy then displacing the eddy. Sverdrup flow gives first way to Rossby waves when $\beta LT<1$. For more intense and rapid forcing, nonlinearity and non-negligible vorticity are generated and linear waves give way to geostrophic turbulence. Replacing T by L/U the condition for the existence of turbulence is $\beta L^2/U<1$.

4. TURBULENCE SCALES

The scales characteristic of turbulence are the r.m.s. turbulent velocity u and a typical integral or eddy scale L_t. The other important parameter is the energy dissipation $\varepsilon = \nu<\zeta'^2>$, where ζ' is the vorticity fluctuation. The dissipation scale is then the Kolmogorov scale $L_K=(\nu^3/\varepsilon)^{1/4}$. This is the scale where lateral friction becomes important. By analogy with stratified flows it is useful to define a

$$\text{Coriolis scale } L_C=u/f$$

and a scale equivalent to the Ozmidov scale [3]

$$L_\Omega=(\varepsilon/f^3)^{1/2}$$

The ratio of L_C/L_t is the turbulent Rossby number ($P = p/\rho$)

$$Ro_t=u/fL_t$$

which is the equivalent of the turbulent Froude number in stratified fluids. When $L_t = L_\Omega$ or $Ro_t=1$, 3D turbulence starts to be affected by rotation and when $L_t= c_1L_\Omega$ or $Ro_t=c_2$ turbulence is quasi 2D on a long time scale L_t/u with superposed inertial waves on a short time scale f^{-1}. The value of c_1 is of order one, probably about 4 to 5 and c_2 is about 0.2 to 0.3 [5]. In the special case of $\varepsilon=u^3/L_t$, the reasonings in terms of L_Ω and Ro_t are equivalent. The evolution of the scales is shown schematically in Fig.6.

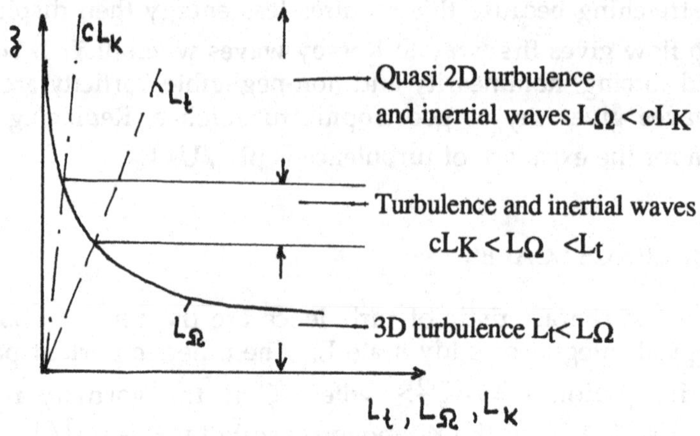

Fig. 6 A typical variation of the length scales L_t, L_Ω and L_K in turbulence produced by an oscillating grid for example.

A physical justification for the use of L_Ω was given by Mory and Hopfinger[3]. The essential idea is that when $L_t > L_\Omega$ energy radiation by inertial waves becomes more efficient than energy dissipation through the cascade process to small scales. Inertial waves are affected by viscosity when their wave length is less than $L_v = (\nu/f)^{1/2}$ which is recognized as the Ekman layer thickness.

When the turbulence is quasi 2D, enstrophy (vorticity squared) is cascading to small scales at the enstrophy transfer rate $\varpi \sim (u/L_t)^3$. Enstrophy is then dissipated at the rate $\varpi = \nu \langle (\nabla\zeta')^2 \rangle$. Vorticity gradients increase at small scales by an elongation of vorticity contours as sketched below. Because of this stretching iso-vorticity lines move closer together hence an increase in $\nabla\zeta'$. A dissipation scale equivalent to the Kolmogorov scale can be defined as $L_\zeta = (\nu^3/\varpi)^{1/6}$.

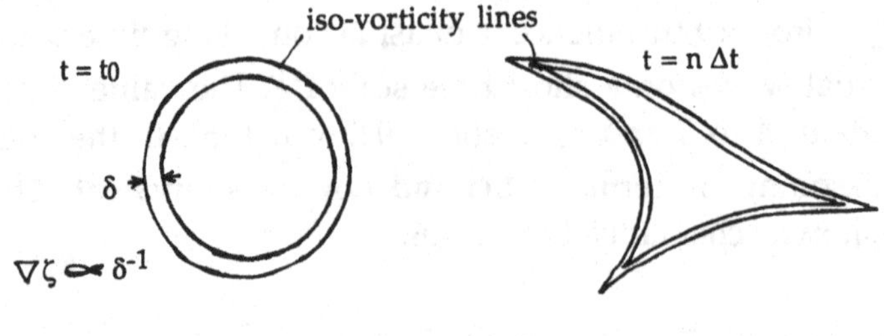

Fig.7 Intensification of vorticity in 2D turbulence

The energy cascade is an inverse cascade with a scale doubling in a time of about 30 initial turnover time scales. On an f-plane, the presence of end boundaries sets an upper bound to the turbulence scale due to Ekman dissipation. The limiting upper scale L_E being determined by the equality of the Ekman spin-down time τ and turnover time L_t/u, giving $L_t < L_E \approx u\tau$.

When the boundaries form a β-plane, the turbulent energy can be carried away by Rossby waves. The limiting scale L_β, called the Rhines scale, is given by the equality of the turnover time scale L_t/u and the Rossby wave period, giving $L_t < L_\beta = \pi(2u/\beta)^{1/2}$.

Acknowledgement:

The influence of P. Rhines through his writings and in particular through his lectures given in Cambridge in 1980 is found in these notes.

REFERENCES:

1. Hopfinger, E.J.: General concepts and examples of rotating fluids, this volume.
2. Taylor, G.I.: Experiments on the motion of solid bodies in rotating fluids, Proc. Roy. Soc. (London), A104 (1923), 213.
3. Mory, M. and Hopfinger, E.J.: Rotating turbulence evolving freely from an initial qusi 2D state, in: Lecture Notes in Physics, N 230, 16-18, Proc. of Macroscopic modelling of turbulent flows (Eds. U. Frisch, J.B. Keller, G. Papanicolaou and O. Pironneau), 1984.
4. Gibson, C.: Laboratory, numerical and oceanic studies of fossil turbulence in rotating and stratified flows, submitted J.G.R. 1990.
5. Hopfinger, E.J., Browand, F.K. and Gagne, Y.: Turbulence and waves in a rotating tank, J. Fluid Mech., 125 (1982), 505.
6. Hide, R . & Ibbetson, A.: An experimental study of taylor columns, Icarus, 5 (1966), 279.
7. Rhines, P.: The dynamics of unsteady currents. In "The Sea: Ideas and observations on progress in the study of the sea" (E.D. Goldberg Ed.) John Wiley & Sons, Inc 1977.

Chapter II.2

VORTICITY, POTENTIAL VORTICITY

M. Mory
University J.F. and CNRS, Grenoble Cedex, France

This lecture recalls the basic properties of vorticity and provides a summary of the main conservation laws associated with vorticity and its related quantities. It will be stated as clearly as possible how these different properties are used in industrial situations and in geophysical flows. The content is very classic; details of calculations may be found in chapter 2 of Pedlosky's book [1] and in chapter 1 of Greenspan's book [2]. The subject of vorticity dynamics is one which is attracting much attention at the moment; this presentation of vorticity dynamics is restricted to some basic properties that are used in this volume. More complex conservation laws, like for instance the conservation of helicity in a vortical flow (Moffatt [3]), are not discussed. The dynamics of point vortices (Aref *et al* [4]) is also not treated in this volume. An advanced account of the dynamics of vortex filament dynamics is given in this volume by Moore [5].

1. CONSERVATION LAWS ASSOCIATED WITH VORTICITY

11. Vorticity and circulation

Two definitions should first be introduced: the vorticity vector, defined as the curl of velocity:

$$\vec{\omega} = \nabla \times \vec{u} \tag{1}$$

and the circulation Γ around a closed line C contained in the fluid:

$$\Gamma = \int_C \vec{u} \, . \, d\vec{r} \quad . \tag{2}$$

The two quantities are related by the Stokes theorem: the circulation around any closed line C contained in the fluid is equal to the flux of vorticity across any surface S limited by this closed line:

$$\int_C \vec{u} \, . \, d\vec{r} = \int_S \vec{\omega} \, . \, \vec{n} \; ds \quad . \tag{3}$$

When a fluid element follows a closed trajectory C in a steady flow, the circulation along C is non-zero. However, a fluid element may have a curved trajectory without having any vorticity; non-zero vorticity means that the fluid element is rotating around itself. One classic example of a rotating flow in which fluid elements have no vorticity is the potential vortex in two dimensions:

$$u=0, \; v(r) = \Gamma / 2\pi r \; , \; w=0 \quad . \tag{4}$$

The velocity components (u,v,w) are written in the cylindrical coordinate system (O,r,θ,z). The circulation along any circle of radius r centered on the axis Oz is Γ, but the vorticity is zero everywhere, except on the axis Oz where vorticity and velocity are both infinite.

Equation (3) implies that vorticity exists somewhere when the circulation is non-zero. Vorticity is of major importance because the flow must conserve circulation when viscous and external forces are absent. The conservation laws associated with vorticity are important constraints applied on the flow.

12. Vortex line - vortex tube

A vortex line (or vortex filament) is a line in the fluid which is parallel to the vorticity vector at each point of the line. A vortex tube corresponds to the surface formed by the vortex filaments that pass through a closed curve C. The concepts of vortex filament and vortex tube have simple physical consequences on the spatial distribution of vorticity, since the vorticity field is non-divergent. Indeed, div $\vec{\omega}$ = div (curl $\vec{u}$) = 0 follows from from the very general properties of divergence and curl operators. The vorticity flux across any surface limited by the surface of a vortex tube is therefore constant:

$$\int_s \vec{\omega} \cdot \vec{n} \, ds = \text{const} \tag{5}$$

Eq. (5), which is a spatial conservation law, implies that a vortex tube cannot vanish inside the fluid. Either a vortex tube is attached to the boundaries of the fluid domain, or it is closed on itself (this is the classic case of a vortex ring). This property explains, on the one hand, the existence of the tip vortex attached on a wing of finite length. The wing experiences a lift force when the circulation around the profile is non-zero. The vorticity that is inside the vortex tube containing the boundary layer on the wing is pursued in a vortex tube trailing behind the wing. On the other hand, the vortex tube containing the boundary layer on a two-dimensional profile attached on both sides to walls is also attached to the wall.

13. The equation for vorticity

In order to focus more clearly on vorticity conservation laws, it is useful to derive the equation for the vorticity vector. We start with the Navier-Stokes equation with an external force $\vec{F}$

$$\frac{\partial \vec{u}}{\partial t} + \vec{u}.\nabla\vec{u} = -\frac{1}{\rho}\nabla p + \nu \, \Delta\vec{u} + \vec{F} \tag{6}$$

The flow field is also assumed to be non-divergent:

$$\nabla.\vec{u} = 0 \tag{7}$$

The equation for the vorticity is obtained by applying the curl operator to the Navier-Stokes equation. This gives (Pedlosky [1], p 35):

$$\frac{\partial \vec{\omega}}{\partial t} + \vec{u}.\nabla\vec{\omega} = \vec{\omega}.\nabla\vec{u} + \nu\,\Delta\vec{\omega} + \nabla \times \vec{F} + \frac{1}{\rho^2}\,\nabla\rho \times \nabla p \tag{8}$$

$$\nabla.\vec{\omega} = 0 \tag{9}$$

Geophysical flows concern fluids having density stratification, and therefore the "baroclinic vector"

$$\frac{1}{\rho^2}\,\nabla\rho \times \nabla p \tag{10}$$

is retained in (8). The discussion of this term is left to part 2, devoted to geophysical flows.

When the fluid is inviscid, when no external force is applied on the flow and when the baroclinic vector vanishes, it is worth discussing the remaining term on the right-hand side of (8) that allows changes in the vorticity as the fluid element moves. Two contributions need to be distinguished:

$$\vec{\omega}\,.\,\nabla\,\vec{u} = \begin{pmatrix} \omega_x\,\partial u/\partial x \\ \omega_y\,\partial v/\partial y \\ \omega_z\,\partial w/\partial z \end{pmatrix} + \begin{pmatrix} \omega_y\,\partial u/\partial y + \omega_z\,\partial u/\partial z \\ \omega_x\,\partial v/\partial x + \omega_z\,\partial v/\partial z \\ \omega_x\,\partial w/\partial x + \omega_y\,\partial w/\partial y \end{pmatrix} . \tag{11}$$

The terms in the first bracket help to increase or decrease the vorticity at the point considered due to the convergence or divergence of the velocity field at this point. However, there is no vorticity transfer from one component to another. This is the so-called "effect of stretching" of a vortex tube. On the other hand the terms in the second bracket serve to transfer vorticity from one component to another under the effect of an appropriate velocity gradient. This is the "effect of tilting" of the vorticity. Figure 1 summarizes the processes of vortex stretching and vortex tilting.

The particular case of two-dimensional motion is especially informative. For a two-dimensional flow in the plane Oxy, i.e. (u(x,y), v(x,y), w=0), the vorticity vector is along the Oz axis:

$$\vec{\omega} = \left(\frac{\partial v}{\partial x} - \frac{\partial u}{\partial y}\right)\vec{k}\,, \tag{12}$$

$\vec{k}$ being the unit vector along the Oz axis. For an inviscid fluid, the vorticity equation then reduces to

$$\frac{d\vec{\omega}}{dt} = \frac{\partial\vec{\omega}}{\partial t} + \vec{u} \,.\, \nabla \vec{\omega} = 0, \tag{13}$$

when no external force is applied to the fluid. The vorticity is therefore conserved as the particle moves.

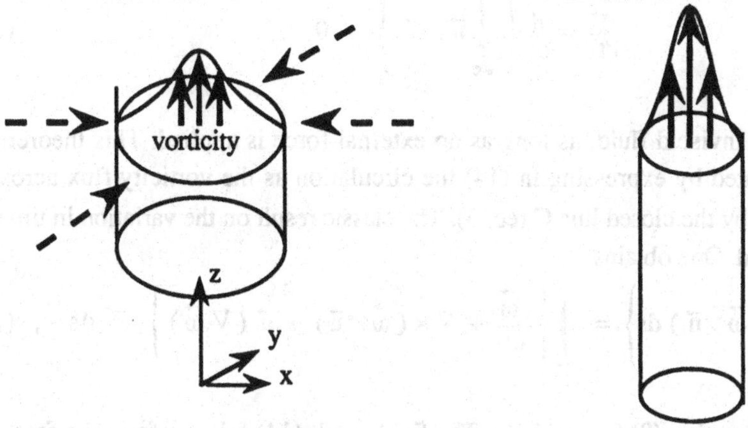

Figure 1a: ***vortex stretching*****: Convergence (or divergence) of the flow increases (or decreases) the vorticity without changing its direction along Oz. Solid arrows indicate the vorticity field in the vortex tube. Dashed arrows show the convergent flow in the xy plane.**

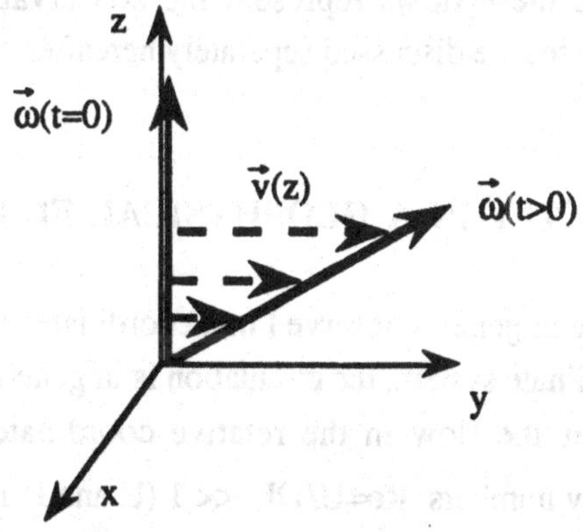

Figure 1b: ***vortex tilting*****: the vorticity is initially along Oz. The existence of a velocity gradient $\partial v/\partial z$ produces a vorticity component along Oy.**

14. The conservation of circulation

As was shown above, the vorticity is in general not conserved; vortex stretching may modify the magnitude of the vorticity vector, whereas vortex tilting may change the orientation of the vorticity vector. However, the circulation along a closed line is conserved as this closed line is followed in its motion. The conservation in time of the circulation, which is complementary to the spatial conservation of circulation (eq. 5), is known as Kelvin's theorem

$$\frac{d\Gamma}{dt} = \frac{d}{dt}\left\{ \int_C \vec{u} \,.\, d\vec{r} \right\} = 0 \quad . \tag{14}$$

It is valid for an inviscid fluid, as long as no external force is applied. This theorem is easily demonstrated by expressing in (14) the circulation as the vorticity flux across a surface bounded by the closed line C (eq. 3). The classic result on the variation in time of a flux is then used. One obtains

$$\frac{d\Gamma}{dt} = \frac{d}{dt}\left\{ \iint_s (\vec{\omega} \,.\, \vec{n}) \, ds \right\} = \iint_s \left\{ \frac{\partial \vec{\omega}}{\partial t} + \nabla \times (\vec{\omega} \times \vec{u}) + \vec{u} \, (\nabla . \vec{\omega}) \right\} . \, \vec{n} \; ds \quad , \tag{15}$$

combined with equation (8) for vorticity. The final result (14) follows from the fact that the velocity and vorticity fields are non-divergent.

The conservation of circulation is a very strong constraint imposed on the flow. However, the way this constraint applies in the flow will in general differ very much, depending on whether one addresses an industrial problem or a geophysical situation. The tools used to describe these flows represent the conservation of circulation in different ways. These two cases are discussed separately hereafter.

2. POTENTIAL VORTICITY IN A GEOPHYSICAL FLOW

Geophysical flows are in general observed in a coordinate system rotating around an axis. In the relative coordinate system, the circulation is in general not conserved. It is however useful to consider the flow in the relative coordinate system, when it is characterized by low Rossby numbers, $Ro = U/\Omega L \ll 1$ (U and L are the typical velocity and lengthscale of motion in a plane perpendicular to the rotation vector $\vec{\Omega}$ of the coordinate system). The flow is not far from being two-dimensional in this case, as it verifies the geostrophic equilibrium (see the paper by E. Hopfinger [6]). We noted in § 13 that vorticity is conserved for a two-dimensional flow. For a nearly two-dimensional flow, it is useful to introduce a new quantity, the potential vorticity, that is conserved in a way similar to the vorticity in a two-dimensional flow. The conservation of this quantity is much simpler to interpret than that of circulation.

21. Flow in shallow water

Geophysical flows are often confined to shallow water layers. This means that the depth D of the fluid layer is small compared to the lengthscale L of motion (D/L <<1). Choosing the reference coordinate system with the axis Oz indicating the depth, the Navier-Stokes equations are written as follows for an inviscid incompressible fluid:

$$\frac{\partial u}{\partial t} + u\frac{\partial u}{\partial x} + v\frac{\partial u}{\partial y} + w\frac{\partial u}{\partial z} = -\frac{1}{\rho}\frac{\partial p}{\partial x} \tag{16}$$

$$\frac{\partial v}{\partial t} + u\frac{\partial v}{\partial x} + v\frac{\partial v}{\partial y} + w\frac{\partial v}{\partial z} = -\frac{1}{\rho}\frac{\partial p}{\partial y} \tag{17}$$

$$\frac{\partial w}{\partial t} + u\frac{\partial w}{\partial x} + v\frac{\partial w}{\partial y} + w\frac{\partial w}{\partial z} = -\frac{1}{\rho}\frac{\partial p}{\partial z} \tag{18}$$

In a shallow water layer, the continuity equation

$$\frac{\partial u}{\partial x} + \frac{\partial v}{\partial y} + \frac{\partial w}{\partial z} = 0 \tag{19}$$

implies that the velocity scale W of the velocity component along Oz is much smaller than the velocity scale U of the velocity components perpendicular to that axis: $W = U\,O(D/L)$. After differentiating (16) and (17) with respect to z, (18) allows the pressure to be eliminated from the first two equations. Comparison of the various orders of magnitude then leads to

$$\frac{\partial u}{\partial z} = \frac{U}{L}\,O\!\left(\frac{D}{L}\right) \quad \text{and} \quad \frac{\partial v}{\partial z} = \frac{U}{L}\,O\!\left(\frac{D}{L}\right) . \tag{20}$$

This relationship states that the flow is two-dimensional to the first order in the small parameter D/L. The continuity equation implies that the velocity component w depends linearly on z to the D/L order.

The vorticity vector is then at leading order

$$\vec{\omega} = \left(\frac{\partial v}{\partial x} - \frac{\partial u}{\partial y} \right) \vec{k} , \tag{21}$$

all other terms being at most of order U/L O(D/L). The main component of vorticity is along the direction of depth of the layer.

22. Vorticity equation in the rotating coordinate system

In order to establish the vorticity equation for motion in the rotating coordinate system, it is easier to start again from the Navier-Stokes equation written in that coordinate system:

$$\frac{\partial \vec{u}}{\partial t} + \vec{u}.\nabla\vec{u} + 2\,\vec{\Omega} \times \vec{u} = -\frac{1}{\rho}\nabla p + \vec{F} + \nu\,\Delta\vec{u} + \Omega^2\,\vec{r}_\perp , \tag{22}$$

associated with incompressibility

$$\nabla . \vec{u} = 0 . \tag{23}$$

The centrifugal force $\Omega^2\,\vec{r}_\perp$ ($\vec{r}_\perp$ is the vector position of the point considered from the axis of rotation) can be written as a spatial gradient, and is later implicitly accounted for in the pressure. The equation for the relative vorticity is obtained by applying the curl operator to (22). This equation differs from equation (8) derived in the absolute coordinate system only through the effect of the Coriolis force

$$\frac{\partial \vec{\omega}}{\partial t} + \vec{u}.\nabla\vec{\omega} = (\vec{\omega} + 2\,\vec{\Omega}).\nabla\vec{u} + \nu\,\Delta\vec{\omega} + \nabla \times \vec{F} + \frac{1}{\rho^2}\nabla\rho \times \nabla p . \tag{24}$$

For a shallow water flow, variations of the vorticity component $\omega_z + 2\Omega_z$ to the first order of the small parameter D/L are only due to stretching, as is deduced from (11), when viscosity, external forces and effects of baroclinicity are negligible. The main equation for vorticity, which concerns the component ω_z of vorticity, becomes

$$\frac{\partial \omega_z}{\partial t} + \vec{u}.\nabla\omega_z = (\omega_z + 2\,\Omega_z)\,\frac{\partial w}{\partial z} . \tag{25}$$

23. Potential vorticity in homogeneous fluids

The shallow water motion equations (16 to 19) show that the velocity component w varies linearly with depth z. The boundary conditions on top and at the bottom of the layer then imply that

$$\frac{dh}{dt} = h \frac{\partial w}{\partial z} \ . \tag{26}$$

h is the depth at the point considered of the material fluid column whose motion is being followed. Eq. (25) is then modified into

$$\frac{d\omega_z}{dt} - (\omega_z + 2\,\Omega_z)\,\frac{1}{h}\frac{dh}{dt} = 0 \ . \tag{27}$$

This equation leads to a simple conservation relationship when Ω_z is constant, i.e. when the projection of the mean vorticity vector $\vec{\Omega}$ on the axis z does not vary with the point considered. One then obtains

$$\frac{d}{dt}\left\{ \frac{\omega_z + 2\Omega_z}{h} \right\} = 0 \ . \tag{28}$$

The quantity inside the brackets is conserved during motion. This quantity, which does not have the dimension of vorticity, is called the potential vorticity. Eq. (28) implies that the relative vorticity ω_z decreases due to stretching when the depth of the layer decreases and increases when the depth of the layer increases. This phenomenon is associated with the divergence or convergence of the flow that concentrates the vorticity by stretching. The relative vorticity of the flow varies when the flow does not follow isodepth lines. For this reason it will often be useful to compare the trajectories with isodepth lines.

24. The beta effect

We consider now the case when the projection Ω_z of the vorticity vector $\vec{\Omega}$ varies in the fluid layer. This case occurs in geophysical flows with large-scale motion when the curvature of the earth or of the planet can no longer be neglected. The so-called beta effect accounts for the variation in the vorticity of the rotating coordinate system that is experienced by the particles when their latitude on the rotating sphere changes during motion.

Figure 2 shows the typical coordinate system chosen to describe flow in a shallow water layer placed on a curved planet. The axis Oz, as defined in the framework of the shallow water equation, is perpendicular to the upper and lower boundaries of the fluid layer. The mean vorticity vector has therefore two components in the coordinate system

(O,x,y,z), $\Omega_y = \Omega \cos \phi$ and $\Omega_z = \Omega \sin \phi$, which vary when the latitude angle ϕ of the point changes during motion. Starting from the position y=0, where $2\Omega_z = 2\Omega \sin \phi_o$, the corresponding quantity when the particle has moved to the position y is

$$2\Omega_z(x,y) \approx 2\Omega \sin\phi_0 \left(1 + \mathrm{cotg}\phi_0 \frac{y}{R} \right) \quad . \tag{29}$$

The variation of the quantity $2\Omega_z$ around a given latitude ϕ_o is usually written in the form

$$2\Omega_z(x,y) = f_0 + \beta y \quad . \tag{30}$$

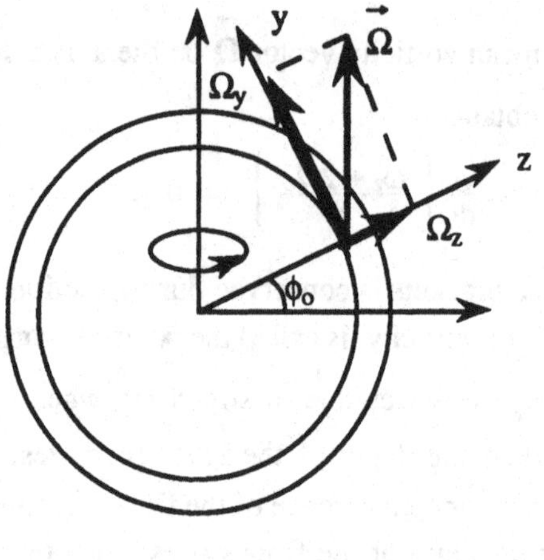

Figure 2: Coordinate system for a shallow water layer on a rotating sphere.

The equations of motion in a spherical coordinate system have to be expanded to the different orders of the three small parameters $R_o = U/Lf_o$, D/L and $\beta L/f_o$ in order to derive the shallow water motion equations in the coordinate system (O,x,y,z). This lengthy calculation is given in chapter 6 of Pedlosky's book [1]. The conservation of potential vorticity equivalent to (27) is finally

$$\frac{d}{dt}(\omega_z + 2\,\Omega_z) - (\omega_z + 2\,\Omega_z)\,\frac{1}{h}\frac{dh}{dt} = h\,\frac{d}{dt}\left\{\frac{\omega_z + f_0 + \beta y}{h}\right\} = 0 \quad . \tag{31}$$

The quantity

$$\Pi = \frac{\omega_z + f_0 + \beta y}{h} \tag{32}$$

is the potential vorticity. Eq. (31) differs from (27) by the term $d(2\Omega_z)/dt$. The origin of this term can be heuristically understood by returning to the Euler equation in the coordinate system (O,x,y,z), for the velocity vector $\vec{u}_\perp = u\,\vec{i} + v\,\vec{j}$ perpendicular to Oz

$$\frac{d\vec{u}_\perp}{dt} + 2\Omega_z\,\vec{k}\times\vec{u}_\perp + 2\Omega_y\,w\,\vec{i} = -\frac{1}{\rho}\nabla_\perp p \ . \tag{33}$$

In the shallow water approximation, the third term on the left hand side of (33), which is of the order of O(ΩUD/L), is neglected. Applying the curl to (33) gives

$$\frac{d\omega_z}{dt} - (\omega_z + 2\,\Omega_z)\frac{\partial w}{\partial z} + v\,\frac{\partial(2\Omega_z)}{\partial y} = 0 \ , \tag{34}$$

which is exactly (31), when the variation of Ω_z is represented (eq. 30).

The conservation of potential vorticity implies, for instance, that a particle increases its relative vorticity when moving southward in a layer of constant depth. For small changes of the layer depth, i.e. $h = H + \zeta$ with $\zeta << H$,

$$\Pi = \frac{\omega_z + f_0 + \beta y - f_0\,\zeta/H}{H} \ . \tag{35}$$

Changes in latitude and changes in the depth of the layer have similar effects on flow, as is shown by Maxworthy [7]. The beta effect is responsible for the existence of Rossby waves, whereas topographic changes inducing changes in the layer depth produce topographic waves. Experiments reported in the literature reproduce the beta effect by introducing a slope in the tank that gives a linear variation of the layer depth.

The scale at which the beta effect becomes significant is in general large. This means that mean vorticity changes βL must be of the order of magnitude of the relative vorticity U/L. Thus,

$$L \approx \sqrt{U/\beta}. \tag{36}$$

On the earth at mid-latitude, $f_o = 10^{-4}\ s^{-1}$ and $\beta = 10^{-11}\ m^{-1}\ s^{-1}$. In the ocean, L = O(100 km) is deduced for $U = 5\ cm.s^{-1}$. In the atmosphere, L = O(1000 km) for $U = 10\ m.s^{-1}$.

25. Potential vorticity in the presence of stratification

The principles developed above are modified in the presence of stratification. Actually, the elegant theorem established by Ertel (see Pedlosky [1], §2.5) is a more general proof of the conservation of potential vorticity. We will not reproduce this demonstration here, being more inclined to discuss the different forms of potential vorticity in the various cases.

Two cases need be considered, depending on whether stratification is continuous or whether different layers of homogeneous density can be identified in the fluid. In the first case, stratification becomes a leading constraint on flow when the horizontal scale is not too large,

$$L \leq R_i = \sqrt{-\frac{1}{\rho}\frac{\partial \rho}{\partial z} g}\ \frac{D}{2\,\Omega_z} = \frac{N}{2\,\Omega_z}\, D\,. \tag{37}$$

N is the Brünt-Väisälä frequency, which enters into the scaling of internal waves. In the second case the corresponding condition is

$$L \leq R_i = \sqrt{\frac{\Delta\rho\, g\, D}{\rho\, 4\,\Omega_z^2}}\ . \tag{38}$$

$\Delta\rho$ is the density jump across an interface and D is the typical depth of a layer (assuming that all layers are of similar depth). R_i is called the internal deformation radius. When the horizontal lengthscale of the motion is larger that the internal radius of deformation, stratification becomes unimportant; the flow is two-dimensional as in shallow water theory. The potential vorticity (32) is then conserved.

When the lengthscale of motion is smaller than the internal deformation radius, the w component of velocity is no longer simply related to the depth h of the layer. At least when the fluid can be separated into several layers of homogeneous density, the potential vorticity as expressed by (32) is conserved in each layer, when the depth h designates the depth of the layer considered. The case of continuous stratification is more complicated. The equation for the vorticity component parallel to Oz is now,

$$\left\{\frac{\partial}{\partial t} + u\frac{\partial}{\partial x} + v\frac{\partial}{\partial y} + w\frac{\partial}{\partial z}\right\}(\omega_z+2\Omega_z) = (\omega_z+2\Omega_z)\frac{\partial w}{\partial z} + \frac{1}{\rho^2}\left\{\frac{\partial p}{\partial y}\frac{\partial \rho}{\partial x} - \frac{\partial p}{\partial x}\frac{\partial \rho}{\partial y}\right\} \tag{39}$$

The effect of the baroclinic term on the evolution of the vorticity component parallel to Oz is in general negligible when the Rossby number is small and the layer shallow (Pedlosky [1], § 2.6 & § 2.8). This term is neglected in what follows.

In incompressible flow, the derivative of w with respect to z is deduced from the conservation of particle density. We thus deduce

$$\left\{\frac{\partial}{\partial t}+u\frac{\partial}{\partial x}+v\frac{\partial}{\partial y}+w\frac{\partial}{\partial z}\right\}\frac{\partial \rho}{\partial z} = -\frac{\partial u}{\partial z}\frac{\partial \rho}{\partial x}-\frac{\partial v}{\partial z}\frac{\partial \rho}{\partial y}-\frac{\partial w}{\partial z}\frac{\partial \rho}{\partial z} \quad . \tag{40}$$

Again, horizontal density gradients are negligible as compared to the vertical density gradient. Equations (39) and (40) reduce to

$$\left\{\frac{\partial}{\partial t}+u\frac{\partial}{\partial x}+v\frac{\partial}{\partial y}+w\frac{\partial}{\partial z}\right\}(\omega_z+2\Omega_z) = (\omega_z+2\Omega_z)\frac{\partial w}{\partial z} \quad , \tag{41}$$

$$\left\{\frac{\partial}{\partial t}+u\frac{\partial}{\partial x}+v\frac{\partial}{\partial y}+w\frac{\partial}{\partial z}\right\}\frac{\partial \rho}{\partial z} = -\frac{\partial w}{\partial z}\frac{\partial \rho}{\partial z} \quad . \tag{42}$$

Multiplying (41) and (42) by $\partial\rho/\partial z$ and $\omega_z+2\Omega_z$, respectively, and adding the two equations, we obtain

$$\left\{\frac{\partial}{\partial t}+u\frac{\partial}{\partial x}+v\frac{\partial}{\partial y}+w\frac{\partial}{\partial z}\right\}\left\{(\omega_z+2\Omega_z)\frac{\partial \rho}{\partial z}\right\} = 0 \quad . \tag{43}$$

In incompressible fluids, conservation of the quantity in brackets in (43) is usually presented as the conservation of potential vorticity defined as follows

$$\Pi = \frac{(\omega_z+2\Omega_z)\dfrac{\partial \rho}{\partial z}}{\rho} \quad . \tag{44}$$

The theorem of conservation of potential energy is generalized for compressible fluids by replacing the density gradient by the potential temperature gradient.

26. Potential vorticity in the framework of geostrophy

Potential vorticity takes a simpler form when the flow is geostrophic (i.e. the Rossby number is small). The variation of density resulting from fluid motion is distinguished from the mean density stratification,

$$\rho\,(x,y,z) = \bar{\rho}\,(z)+\tilde{\rho}\,(x,y,z) \quad . \tag{45}$$

Assuming that the density variations $\tilde{\rho}$ are small compared to the density $\bar{\rho}$ in the absence of motion, the equations of motion for the geostrophic component of the flow are

$$-2\Omega\,\rho_0\,v = -\frac{\partial\tilde{p}}{\partial x} \tag{46}$$

$$2\Omega\,\rho_0\,u = -\frac{\partial\tilde{p}}{\partial y} \tag{47}$$

$$0 = -\frac{\partial\tilde{p}}{\partial z} - \tilde{\rho}\,g\,. \tag{48}$$

Using the stream function

$$\psi = \tilde{p}\,/\,2\Omega\rho_0 \tag{49}$$

the potential vorticity (44) is approximated with the first order in the small Rossby number parameter as follows

$$\Pi \approx \left\{2\Omega_z + \frac{\partial^2\psi}{\partial x^2} + \frac{\partial^2\psi}{\partial x^2}\right\}\frac{1}{\rho_0}\frac{\partial\bar{\rho}}{\partial z} - \frac{(2\Omega_z)^2}{g}\frac{\partial^2\psi}{\partial z^2}\,. \tag{50}$$

The usual form of potential vorticity is derived by dividing (50) by $1/\rho_0\ \partial\bar{\rho}/\partial z$. The internal deformation radius R_i (eq. 37) then appears explicitly. The final form of potential vorticity is therefore

$$\Pi = f_0 + \beta\,y + \frac{\partial^2\psi}{\partial x^2} + \frac{\partial^2\psi}{\partial x^2} + \frac{D^2}{R_i^2}\frac{\partial^2\psi}{\partial z^2}\,, \tag{51}$$

whereas conservation of potential vorticity is written

$$\left\{\frac{\partial}{\partial t} - \frac{\partial\psi}{\partial y}\frac{\partial}{\partial x} + \frac{\partial\psi}{\partial x}\frac{\partial}{\partial y}\right\}\left\{f_0 + \beta y + \frac{\partial^2\psi}{\partial x^2} + \frac{\partial^2\psi}{\partial x^2} + \frac{D^2}{R_i^2}\frac{\partial^2\psi}{\partial z^2}\right\} = 0\,. \tag{52}$$

The major interest of this single equation is that it allows complete determination of the stream function of flow in a plane perpendicular to the z axis.

Geophysical flows are studied in a relative coordinate system rotating around an axis. The flow in this coordinate system is in general concerned with Rossby numbers that are small compared to one or O(1). We have focused here on flows with small Rossby numbers, which are in geostrophic equilibrium. Flow with a Rossby number of the order of O(1) occurs in certain important conditions, for example near a density front or near the coast, where the depth of the layer becomes zero. The conservation of potential vorticity as defined by (32) then implies that the relative vorticity should become of the order of magnitude of the mean vorticity. The flow is therefore no longer in geostrophic equilibrium. Nevertheless, effects of rotation, referred to as "ageotrophic effects", may be very strong. Such effects are not discussed in the present paper.

3. CONSERVATION OF CIRCULATION IN AN INDUSTRIAL FLOW

Observing a flow in industrial situations in a rotating coordinate system does not usually provide a simpler insight, precisely because the flow often does not rotate in all parts of the domain considered. In particular, it is often the case that some boundary is not rotating; the flow in the boundary layer is not suitably modeled in a rotating coordinate system.

In this section we discuss the case of axisymmetric flow rotating around an axis Oz. This is a very good illustration of the conservation of circulation. Such flows are common in turbomachines, vortex chambers, combustors, and, more generally, in vortex flows. The 20-year old report "Review of confined vortex flows" by Lewellen [8] is certainly still one of the most complete references on this subject.

3.1. Incompressible, axisymmetric, laminar flow

Consider the axisymmetric flow around an axis Oz. In the cylindrical coordinate system (O,r,θ,z), the velocity components are the radial, azimuthal and axial components, respectively u,v and w. All components depend only on the r and z coordinates. The Navier-Stokes equations of motion in the cylindrical coordinate system are then simplified

$$\frac{\partial u}{\partial t} + u\frac{\partial u}{\partial r} + w\frac{\partial u}{\partial z} - \frac{v^2}{r} = -\frac{1}{\rho}\frac{\partial p}{\partial r} + \nu\left[\frac{\partial}{\partial r}\left\{\frac{1}{r}\frac{\partial (ru)}{\partial r}\right\} + \frac{\partial^2 u}{\partial z^2}\right] \tag{53}$$

$$\frac{\partial v}{\partial t} + \frac{u}{r}\frac{\partial (rv)}{\partial r} + w\frac{\partial v}{\partial z} = \nu\left[\frac{\partial}{\partial r}\left\{\frac{1}{r}\frac{\partial (rv)}{\partial r}\right\} + \frac{\partial^2 v}{\partial z^2}\right] \tag{54}$$

$$\frac{\partial w}{\partial t} + u\frac{\partial w}{\partial r} + w\frac{\partial w}{\partial z} = -\frac{1}{\rho}\frac{\partial p}{\partial z} + \nu\left[\frac{1}{r}\frac{\partial}{\partial r}\left\{r\frac{\partial w}{\partial r}\right\} + \frac{\partial^2 w}{\partial z^2}\right] . \tag{55}$$

The equation of continuity

$$\frac{1}{r}\frac{\partial(ru)}{\partial r} + \frac{\partial w}{\partial z} = 0 \tag{56}$$

leads to the introduction of a stream function which describes the flow in a plane crossing the axis Oz:

$$u = -\frac{1}{r}\frac{\partial \psi}{\partial z} \quad \text{and} \quad w = \frac{1}{r}\frac{\partial \psi}{\partial r} . \tag{57}$$

It is also common to introduce the circulation C(r,z), which is $1/2\pi$ the circulation along a circle of radius r centered on the axis

$$C(r,z) = r\, v(r,z) . \tag{58}$$

Differentiating (53) with respect to z and (55) with respect to r, pressure is eliminated by subtracting the two resulting equations. A set of two differential equations is then obtained from (53-58), the solution of which determines flow as a function of circulation C and stream function ψ. One obtains (Lewellen [8])

$$\left\{\frac{\partial}{\partial t} - \psi_z\frac{\partial}{\partial y} + \psi_y\frac{\partial}{\partial z}\right\}\left\{\psi_{yy} + \frac{1}{2y}\psi_{zz}\right\} + \frac{C\,C_z}{2\,y^2} = \nu\left\{(2y\,\psi_{yy}+\psi_{zz})_{yy} + \left(\psi_{yy} + \frac{1}{2y}\psi_{zz}\right)_{zz}\right\} \tag{59}$$

$$\left\{\frac{\partial}{\partial t} - \psi_z\frac{\partial}{\partial y} + \psi_y\frac{\partial}{\partial z}\right\} C = \nu\,\{\, 2y\, C_{yy} + C_{zz} \,\} . \tag{60}$$

In (59) and (60) the radial coordinate r has been replaced by $y = r^2/2$. When the fluid is inviscid, equation (60) states the conservation of circulation along a circle of radius r centered on the axis Oz. This is a particular form of Kelvin's theorem. The quantity $\psi_{yy}+\psi_{zz}/2y$ is $-\omega_\theta/\sqrt{2y}$, where ω_θ is the vorticity component along the $\vec{e_\theta}$ vector of the cylindrical coordinate system. Equation (59) is therefore an equation for vorticity. The term $CC_z/2y^2$ corresponds both to the tilting of the vorticity component parallel to Oz by the gradient of the azimuthal velocity component v along direction z and to the tilting of the vorticity component parallel to Or by the gradient of the azimuthal velocity component v along direction r. Figure 3 gives a sketch of these vorticity tilting processes.

Equations (59) and (60) have proved to be very useful. They serve to determine rotating flow in many situations. As shown by (60), circulation depends on the shape of streamlines in meridian planes passing though the axis Oz and is subject to viscous diffusion. When flow is steady and inviscid, circulation is constant along any streamline

ψ = constant. Equations (59) and (60) provide a set of equations to study axisymmetric vortex waves (Benjamin, [9]) and axisymmetric flows in vortex chambers. When non-axisymmetric vortex flows are considered, a much more sophisticated approach has to be used. Interpreting the relevant equations of motion in terms of vorticity dynamics is however not so simple. The case of vortex filaments having non-axisymmetric deformation is discussed in this volume by Moore [5].

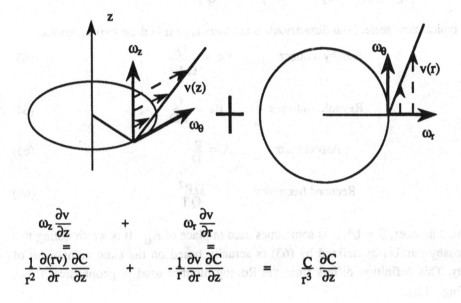

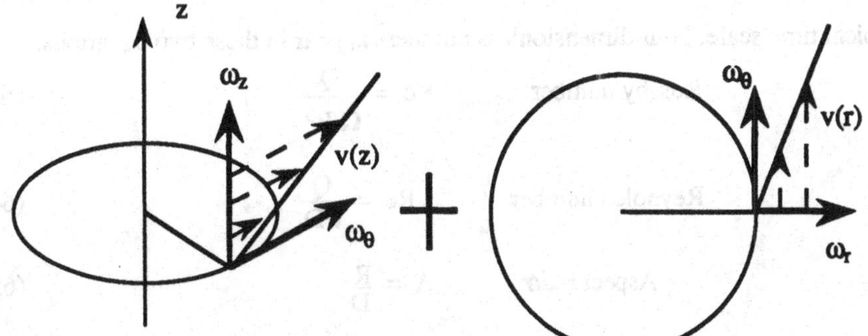

$$\omega_z \frac{\partial v}{\partial z} \quad + \quad \omega_r \frac{\partial v}{\partial r}$$

$$= \qquad\qquad =$$

$$\frac{1}{r^2}\frac{\partial (rv)}{\partial r}\frac{\partial C}{\partial z} \quad + \quad -\frac{1}{r}\frac{\partial v}{\partial r}\frac{\partial C}{\partial z} \quad = \quad \frac{C}{r^3}\frac{\partial C}{\partial z}$$

Figure 3: The tilting of vorticity components in an axisymmetric flow with velocity gradients.

32. Dimensionless parameters

Introducing the characteristic lengthscales D and R along the axial and radial coordinates, respectively, the flow rate Q across a disc of radius R centered on the axis (i.e. $\psi = O(Q)$) and the characteristic rotation rate Ω (i.e. $C = O(\Omega R^2)$), equations (59) and (60) are written in dimensionless form

$$\left\{\left[\frac{DR^2}{QT}\right]\frac{\partial}{\partial t}-\psi_z\frac{\partial}{\partial y}+\psi_y\frac{\partial}{\partial z}\right\}\left\{\psi_{yy}+\frac{1}{2y}\Lambda^2\psi_{zz}\right\}+\left[\frac{\Omega R^3}{Q}\right]^2\frac{C C_z}{2y^2}=$$
$$\left[\frac{\nu D}{Q}\right]\left\{\left(2y\,\psi_{yy}+\Lambda^2\psi_{zz}\right)_{yy}+\Lambda^2\left(\psi_{yy}+\frac{1}{2y}\Lambda^2\psi_{zz}\right)_{zz}\right\} \tag{61}$$

$$\left\{\left[\frac{DR^2}{QT}\right]\frac{\partial}{\partial t}-\psi_z\frac{\partial}{\partial y}+\psi_y\frac{\partial}{\partial z}\right\}C=\left[\frac{\nu D}{Q}\right]\left\{2y\,C_{yy}+\Lambda^2 C_{zz}\right\}\ . \tag{62}$$

T is a typical time scale. Four dimensionless numbers appear in these two equations:

Rossby number $$Ro=\frac{Q}{\Omega R^3}\ , \tag{63}$$

Reynolds number $$Re=\frac{Q}{\nu D}\ , \tag{64}$$

Aspect ratio $$\Lambda=\frac{R}{D}\ , \tag{65}$$

Reduced frequency $$\frac{DR^2}{QT}\ . \tag{66}$$

The swirl number, $S = 1/R_{o,}$ is sometimes used in place of R_o. It is worth noting that the Rossby number as defined by (63) is actually based on the axial component of velocity. This definition differs from the Rossby number used in geophysical flows (Hopfinger [6]).

Several consequences may be drawn, depending on the relative magnitude of the different numbers. We consider the common case with a large Reynolds number and an aspect ratio of the order of unity. The Rossby number influences the evolution of circulation C through its indirect effect on the stream function. Circulation is advected by secondary flow in meridian planes and diffuses. The difficult question that is usually to be solved is the determination of the secondary flow. In most industrial cases, the Rossby number is of the order of 1 or small compared to unity. Equation (61) then shows that circulation will apply a constraint on the secondary flow because the term involving C is not negligible. When the Rossby number is large, (61) shows that the stream function does not depend, in principle, on circulation. However, the existence of walls and boundary layers imposes strong constraints of rotation on the secondary flow outside boundary layers, even when the Rossby number is large. Boundary layers, when they are dominated by rotation effects, can transport large flow masses which they pump from and reinject into the flow domain. They therefore interact extensively with the outside rotating flow (Rott & Lewellen [10]). An example of such flow is the flow produced by a rotating disc, which is considered in this volume by Mory and Spohn [11]. The importance of boundary layers in rotating flows is also emphasized by Read [12] in this volume.

REFERENCES

1. Pedlosky J.: Geophysical Fluid Dynamics, Springer Verlag, 1977.
2. Greenspan H.P.: The Theory of Rotating Fluids, Cambridge University Press, 1968.
3. Moffatt H.K.: The degree of knottedness of tangled vortex lines, J. Fluid Mech., 35 (1969), 117-129.
4. Aref H., Kadtke J.B., Zawadski I. & L.J. Campbell: Point vortex dynamics: recent results and open problems, Fluid Dyn. Res., 3 (1988), 63-74.
5. Moore D.W.: Dynamics of vortex filaments, this volume.
6. Hopfinger E.J.: Parameters, scales and geostrophic balance. This volume
7. Maxworthy T.: Wave motions in a rotating and/or stratified fluid, this volume.
8. Lewellen W.S.: Review of confined vortex flows, Space Propulsion Laboratory, M.I.T., Cambridge, Mass., 1971.
9. Benjamin T.B.: Theory of vortex breakdown, J. Fluid Mech., 14 (1962), 593-629.
10. Rott N. & Lewellen W.S.: Boundary layers and heir interactions in rotating flows, Prog. Aero. Sci., 7 (1966), 111-144.
11. Mory, M. & Spohn, A.: Vortex flow generated by a rotating disc. This volume.
12. Read P.L.: Dynamics and Instabilities of Ekman and Stewartson layers. This volume.

PART III

INSTABILITIES

Chapter III.1

DYNAMICS AND INSTABILITIES OF EKMAN AND STEWARTSON LAYERS

P.L. Read
University of Oxford, Oxford, UK

ABSTRACT

We review the basic dynamical properties and structures of the steady, laminar forms of the two principal rotationally-dominated boundary layers encountered in a homogeneous rotating fluid; the Ekman and Stewartson layers. The modifying influence of density stratification (due to thermal contrasts) is considered, and the main modes of instability exhibited by these boundary layers in the laboratory and in nature are discussed with regard to both observations and theory.

1. INTRODUCTION

Previous discussion [1] has highlighted the importance of geostrophic balance in rapidly rotating flow, in which the horizontal pressure gradient is in local balance with the Coriolis acceleration acting on the horizontal motion. It is important to note, however, that this balance is of lower order than the original equations of motion. Geostrophic balance cannot therefore be expected to hold everywhere. In particular, geostrophic flow is unlikely to satisfy the most general forms of boundary conditions. Viscosity is likely to play an important role as the next most significant force in considering how a geostrophic flow can be rendered compatible with boundary conditions such as the no-slip condition at a rigid surface.

Since the viscous term in the momentum equation depends upon higher spatial derivatives than the pressure gradient and Coriolis forces, we anticipate that viscosity will become important in narrow regions adjacent to physical boundaries - in a *boundary layer*. In this chapter, we consider the effect of introducing viscous effects into a flow in which Coriolis accelerations are dominant. This will require an encounter with some aspects of boundary layer theory - a mathematically complex subject in many contexts, but in the case of rotating flow somewhat more straightforward, at least for quasi-horizontal and certain types of vertical boundaries. In the following discussions, we consider two types of 'classical' boundary layer which have become well known in the literature of rotating fluids. The first type comprise horizontal boundary layers in which viscous and Coriolis accelerations are comparable in magnitude - an important class of boundary layers associated with the name of the Norwegian oceanographer V. W. Ekman, who first discussed a special case of this problem in the context of the wind-driven boundary layer at the surface of the oceans[2],[3]. In the second case, we consider a class of boundary/shear layers which are essentially vertical, and act to match the boundary conditions appropriate to a vertical boundary to a geostrophic interior, taking into account constraints posed by factors such as mass conservation. This problem was first analysed by the British mathematician K. Stewartson[4] in the 1950s, from whom this class of boundary layers take their name.

The following chapter is divided into five broad sections. Sections 2 and 3 discuss in turn the structure and essential dynamics of the steady, laminar forms of Ekman and Stewartson layers in a homogenous, rotating fluid. Section 4 briefly considers the effects of stratification on the above. The intrinsic stability of these boundary layers is considered in Section 5, in which the principal modes of instability observed in laboratory experiments are discussed together with the relevant theory, and we consider their possible manifestations in nature.

2. DYNAMICS OF THE EKMAN LAYER

2.1 *Scale Analysis of the laminar Ekman layer*

We suppose that the flow is characterised by length L, a velocity U and a time T. We can then introduce dimensionless coordinates by means of the transformation:

$$\mathbf{x} = L\mathbf{x}^*, t = Tt^*, \ \mathbf{u} = U\mathbf{u}^*, \boldsymbol{\Omega} = \Omega\mathbf{k}, \ p = \rho\Omega LUp^*. \qquad (1)$$

We may then define the Rossby number Ro by

$$Ro = U/L\Omega \tag{2}$$

and the Ekman number E by

$$E = \nu/L^2\Omega \tag{3}$$

The equations of motion then become

$$(\Omega T)^{-1}\partial\mathbf{u}^*/\partial t^* + Ro(\mathbf{u}^*.\nabla)\mathbf{u}^* + 2\mathbf{k}\times\mathbf{u}^* = -\nabla p^* + E\nabla^2\mathbf{u}^* \tag{4}$$

and

$$\nabla.\mathbf{u}^* = 0 \tag{5}$$

We consider cases for which Ro << 1 and E << 1, corresponding to weak inertial and viscous effects relative to the Coriolis and pressure gradient forces.

2.2 *The laminar Ekman layer*

We suppose there to be a geostrophic flow

$$\mathbf{u_I} = [u_I(x,y),\ v_I(x,y),\ 0], \tag{6}$$

above a rigid horizontal plane boundary at z = 0. For Ro and E small, the geostrophic flow field is a good approximation to the total flow, but cannot satisfy the no-slip boundary conditions on the plane:-

$$u = v = w = 0, \quad \text{on } z = 0. \tag{7}$$

We consider the case where the viscous term in (4) is the next most significant (ignoring the nonlinear inertial term for simplicity), so that we have (dropping superscripts * hereafter)

$$2\mathbf{k}\times\mathbf{u} = -\nabla p + E\nabla^2\mathbf{u} \tag{8}$$

which is, in component form

$$-2v = -\partial p/\partial x + E\nabla^2 u \tag{9a}$$

$$2u = -\partial p/\partial y + E\nabla^2 v \tag{9b}$$

$$0 = -\partial p/\partial z + E\nabla^2 w \tag{9c}$$

with the continuity equation

$$\partial u/\partial x + \partial v/\partial y + \partial w/\partial z = 0 \tag{10}$$

We see from (9a) and (9b) that u and v are O(1) and $\partial/\partial x$ and $\partial/\partial y$ to be O(1) by definition, so p must be O(1). Since E is small, the viscous terms can be comparable with Coriolis accelerations and pressure gradients only if the second (z-)derivatives of (u,v,w) are large. This will be the case if a boundary layer is formed with a vertical length scale δ which is much smaller than L. Then $\partial/\partial z$ is $O(\delta^{-1})$ and the condition that viscous and Coriolis terms be in balance requires that

$$E(L/\delta)^2 \sim 1, \tag{11}$$

so that

$$\delta \sim E^{1/2}L \tag{12}$$

i.e. viscous effects are significant only in a thin region of dimensional thickness $E^{1/2}L$. From the continuity equation (10) $w \sim E^{1/2}$ and so the viscous term in (9c) must be $O(E^{1/2})$. Thus, $\partial p/\partial z \sim 0$ and so the pressure in the boundary layer will take the geostrophic value in the geostrophic interior to a good approximation [O(E)]. The governing equations thus reduce to

$$-2v = -\partial p_I/\partial x + E\nabla^2 u \tag{13a}$$

$$2u = -\partial p_I/\partial y + E\nabla^2 v \tag{13b}$$

together with the continuity equation (10) as before. Since p_I is the geostrophic pressure, we may use

$$2(u_I, -v_I) = -(\partial p_I/\partial x, \partial p_I/\partial y) \tag{14}$$

so that (13a,b) become

$$-2(v - v_I) = E\partial^2 u/\partial z^2 \tag{15a}$$

$$2(u - u_I) = E\partial^2 v/\partial z^2 \tag{15b}$$

with the boundary conditions

$$u = v = 0 \quad \text{at } z = 0 \tag{16a}$$

$$u \rightarrow u_I,\ v \rightarrow v_I, \ \text{as } z \rightarrow \infty \tag{16b}$$

The latter is the critical matching condition, which states that the boundary layer solution must merge smoothly with the geostrophic flow outside the boundary layer. We could obtain the solution to (15)-(16) by eliminating u or v to obtain a fourth order differential equation, but the system can be kept at second order by working in terms of the complex variable Z defined by

$$Z = u + iv \tag{17}$$

Taking (15a) plus i times (15b) leads to

$$d^2Z/dz^2 \ -2i(Z - Z_I)/E = 0 \tag{18}$$

where

$$Z_I = u_I + i\, v_I \tag{19}$$

with the boundary conditions

$$Z = 0 \quad \text{at } z = 0$$

$$Z \rightarrow Z_I \ \text{as } z \rightarrow \infty \tag{20}$$

This has the solution

$$Z = Z_I[1 - \exp\{(2i/E)^{1/2}z\}] \tag{21}$$

or, taking real and imaginary parts

$$u = -v_I e^{-\xi}\sin \xi + u_I[1 - e^{-\xi}\cos \xi], \tag{22a}$$

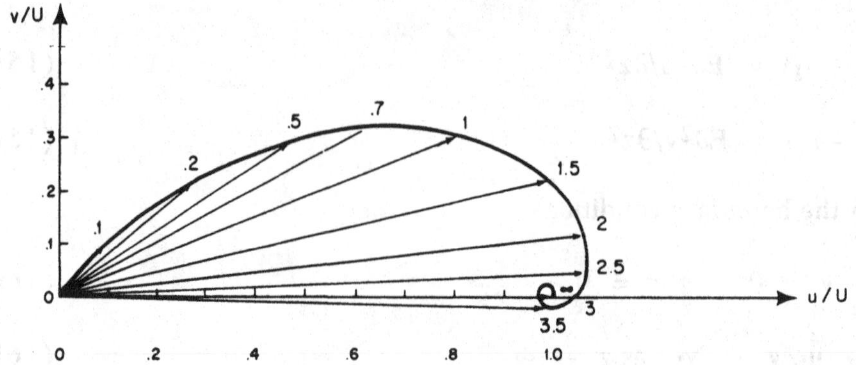

Figure 1: Velocity vector within the Ekman layer. The locus of the tip of the velocity vector traces the Ekman spiral, showing the value of z/δ corresponding to each vector.

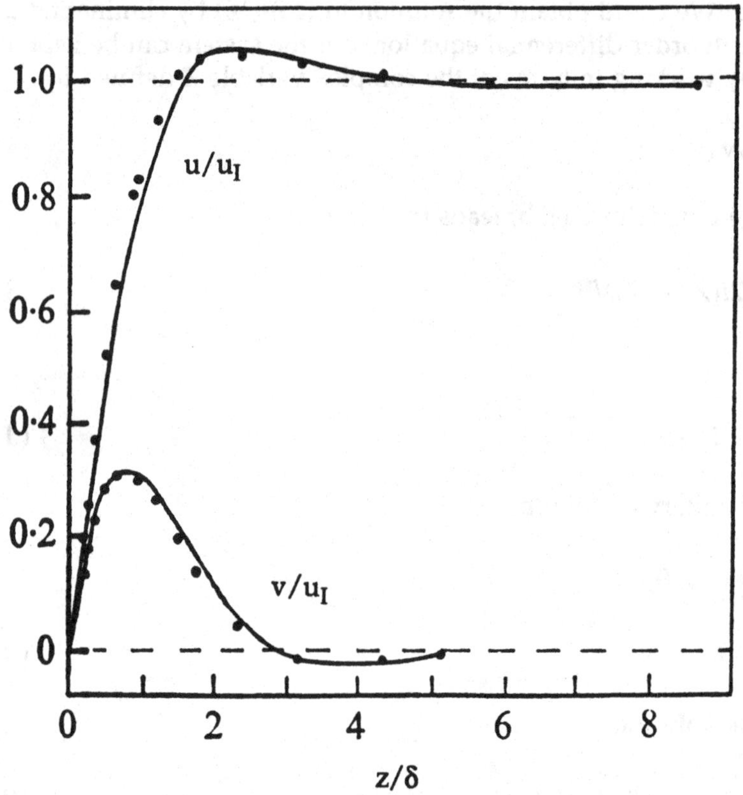

Figure 2: Theoretical and experimental velocity distributions in an Ekman layer (data from [5]).

$$v = u_I e^{-\xi} \sin \xi + v_I [1 - e^{-\xi} \cos \xi], \tag{22b}$$

where

$$\xi = (2/E)^{1/2}z \tag{23}$$

which is the complete solution for the laminar Ekman layer. Note the decaying oscillatory nature of u and v as z becomes large, forming the well known Ekman spiral (illustrated below for $v_I = 0$).

Flow near the boundary is at 45° to the interior geostrophic velocity, becoming parallel to u_I at ξ (= z/δ) = π. Laboratory measurements, though difficult to make in typical Ekman layers (with thicknesses of a few mm), are generally in good agreement with this solution (see Fig. 2 and [5]).

2.3 *Ekman transports and vertical motion*

The above analysis has shown that viscous effects result in a component of flow confined to a narrow layer adjacent to a boundary, which is not in general parallel to the interior geostrophic velocity. The Ekman layer will thus result in a volume transport of fluid related to, and in addition to the geostrophic transport. We can calculate this transport by integrating (22a,b) in ξ from 0 to ∞ to obtain the *Ekman transports*

$$(U_E, V_E) = \int_0^\infty (u - u_I, v - v_I)\, dz$$

$$= (E/2)^{1/2}\,(-[u_I + v_I],\ [u_I - v_I]) \tag{24}$$

Note that these horizontal transports depend only upon the geostrophic velocities outside the boundary layer and the Ekman number. It is also straightforward to show that (U_E, V_E) is directed at 45° to the geostrophic flow.

The vertical velocity component w is of considerable importance in relation to the influence exerted by the Ekman layer on the geostrophic interior flow. We can obtain w from the continuity equation (10) using expressions (22a,b) for u and v to obtain

$$\partial w/\partial z = (\partial v_I/\partial x - \partial u_I/\partial y)e^{-\xi}\sin\xi\ -$$

$$(\partial u_I/\partial x + \partial v_I/\partial y)[1 - e^{-\xi}\cos \xi] \qquad (25)$$

Note, however, that the continuity equation for the geostrophic flow is

$$\partial u_I/\partial x + \partial v_I/\partial y = 0 \qquad (26)$$

Thus, since the boundary condition at the surface is $w = 0$ on $z = 0$, we may integrate (25) to obtain

$$w(x,y,z) = (\partial v_I/\partial x - \partial u_I/\partial y) \int_0^z e^{-\xi}\sin \xi \, d\xi \qquad (27)$$

For $\xi \to \infty$, we have at the top of the Ekman layer

$$w_I = 1/2\, E^{1/2} (\partial v_I/\partial x - \partial u_I/\partial y) = 1/2\, E^{1/2} \zeta_I \qquad (28)$$

where ζ_I is the vertical component of vorticity in the geostrophic flow. Note that the dynamics of the Ekman layer effectively places a constraint on the geostrophic flow above (or below) it. Equation (28) is frequently referred to as the 'Ekman compatibility condition'[6],[7],[8], and expresses the control the Ekman layer may exert upon the interior flow through the 'pumping' action associated with the vorticity of the interior flow. We go on to consider a problem in which this effect determines the behaviour of the flow.

2.4 Ekman layer formation and spin-up

The above discussion has considered steady flows, but it is important also to consider the way in which these steady flows are set up from initial conditions. We consider a fluid inside a container initially at rest, and switch on rotation at $t = 0$. Within the Ekman layer, the time-dependent form of the equations is

$$\partial u/\partial t - 2(v - v_I) = \partial^2 u/\partial \xi^2 \qquad (29a)$$

$$\partial v/\partial t + 2(u - u_I) = \partial^2 v/\partial \xi^2 \qquad (29b)$$

These can be solved[9] (e.g. using Laplace Transforms), but since each term as written is O(1), the timescale over which the time-dependence decays to a steady state is also O(1). Since the timescale on which we have scaled is $O(\Omega^{-1})$, we find that the steady Ekman layer is formed within a couple of revolutions of the container. This is not hard to understand, since the Ekman layer depth is the length scale over which viscous diffusion will penetrate during one rotation period.

Once formed, the Ekman layer can begin to exert an influence on the interior flow by means of the circulation induced by Ekman suction. If we consider the time-dependent form of the equations describing the interior flow (4), assuming Ro ~ 0 (not strictly valid for spin-up from rest, since then Ro ~ 1, but this assumption makes the mathematics simpler! - cf [10]), we obtain

$$\partial \mathbf{u}/\partial t + 2\mathbf{k}\times\mathbf{u} = -\nabla p^* + E\nabla^2\mathbf{u}^* \quad (30)$$

The corresponding O(1) equation for the vertical vorticity is obtained by taking $\mathbf{k}.\nabla\times$ (30) to obtain

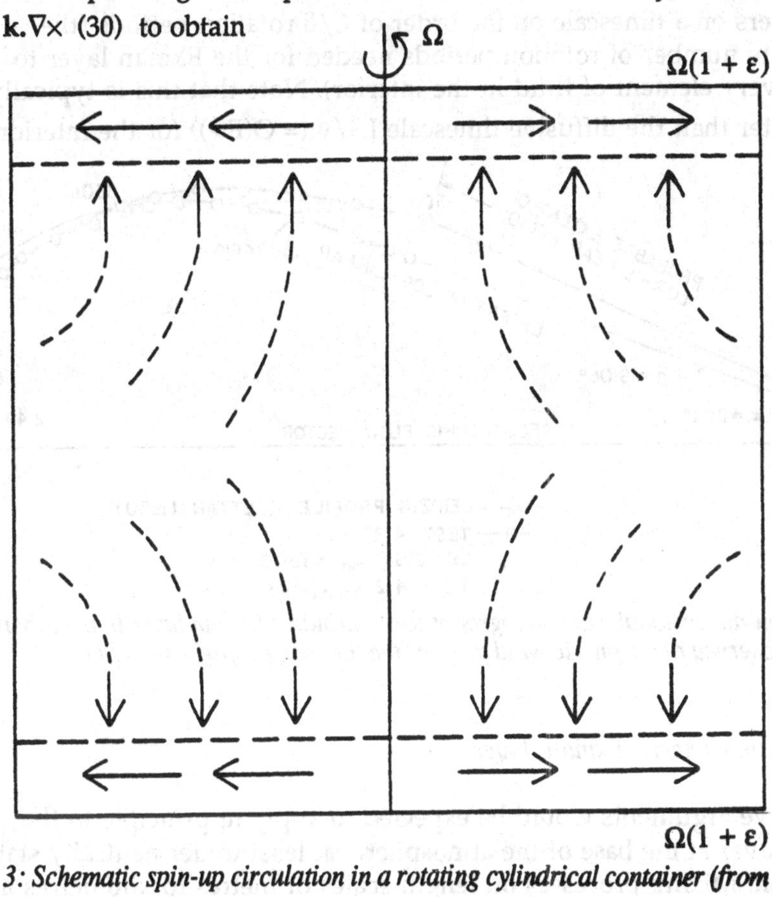

Figure 3: Schematic spin-up circulation in a rotating cylindrical container (from [10]).

$$\partial \zeta_I/\partial t - 2\,\partial w/\partial z = 0 \quad (31)$$

This can be integrated in the vertical, using the Ekman compatibility condition (28) to obtain

$$\partial\bar{\zeta}_I/\partial t + E^{1/2}\,\bar{\zeta}_I = 0 \qquad (32)$$

(since w driven by the Ekman suction is of opposite sign at the top and bottom boundaries). The interior vorticity will thus adjust on a timescale set by Ekman suction, i.e. on a timescale $O(E^{-1/2})$. In physical terms (see Fig. 3 above), as the container is spun-up from Ω to $\Omega(1+\varepsilon)$, the Ekman layers adjust within one or two revolutions. Fluid is spun rapidly outwards within the Ekman layers and friction acts to increase its angular velocity. A small vertical velocity is induced in the interior as the Ekman layer 'sucks in' fluid from the interior to replace fluid spun outwards. This suction stretches vortex lines which increase the vorticity of the interior, bringing it into co-rotation with the Ekman layers on a timescale on the order of L/δ rotation periods (the approximate number of rotation periods needed for the Ekman layer to 'process' every element of fluid in the interior). Note that this is typically much shorter than the diffusive timescale L^2/ν (= $O(E^{-1})$) for the interior region.

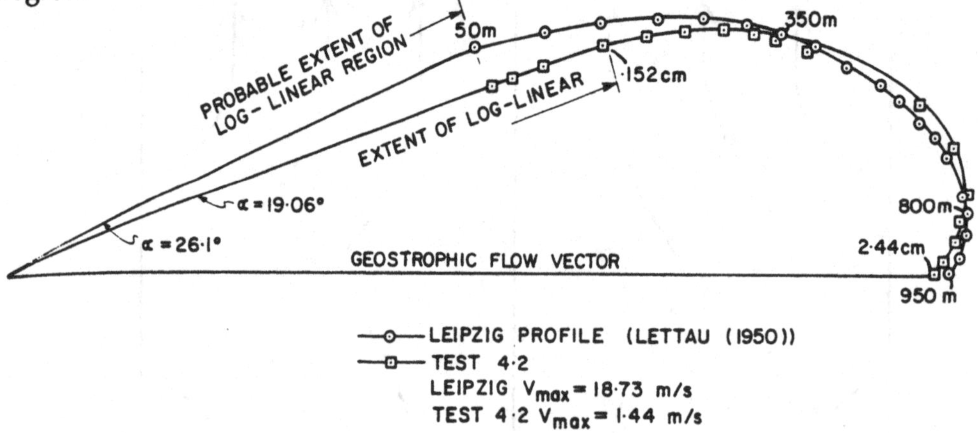

Figure 4: Non-dimensional wind hodographs for a turbulent Ekman layer in the laboratory and for an observed atmospheric wind profile (the 'Leipzig profile' - see [11]).

2.5 The 'Atmospheric Ekman layer'

The above arguments would be expected to apply in principle to the boundary layer at the base of the atmosphere, at least under neutrally stable stratification. Mixing processes on length scales of metres to kilometres are governed by small-scale turbulent eddies rather than the molecular viscosity. The effective diffusion coefficient for momentum is therefore roughly representative of an 'eddy viscosity' $K \sim 10\ m^2\ s^{-2}$. With $2\Omega = 7\times10^{-5}\ s^{-1}$ this leads to an Ekman depth δ of around 1km (in the oceans the equivalent boundary depth is ~ 50-100m). The resulting wind profile observed in the lowest km of the atmosphere (see Fig. 4) roughly follows the Ekman spiral, though not very accurately - especially near the ground (e.g. see [11]). This is mainly because K is not constant with height, but more closely follows a logarithmic profile in the lowest 'friction layer' characteristic of a self-similar turbulent layer. Nevertheless, as a qualitative description the Ekman layer solution is a useful one. Some further details of measurements in turbulent Ekman layers produced in the laboratory can be found in [11] and [12].

2.6 Ekman layers on a free surface

The discussion so far has considered the friction layer in the vicinity of a rigid, non-slip boundary appropriate for rigid containers in the laboratory, and for the boundary layers at the bottom of the oceans or atmosphere. At the top of the oceans, however, the surface is free, and the appropriate boundary conditions are continuity of pressure and of frictional stress across the boundary [6],[7].

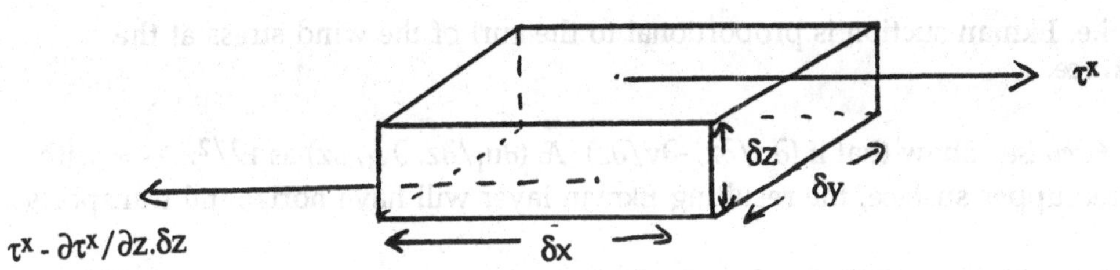

Figure 5: Schematic elemental layer of fluid across which a surface stress is applied.

We consider a fluid to be made up of thin 'plates'. The upper 'plate' exerts a force/unit area τ on the sample 'plate'. The 'plate' in turn exerts a force/unit area on the 'plate' below of ($\tau - \partial\tau/\partial z$), which exerts a reaction force on the sample 'plate'. The net force/unit area is then ($\partial\tau/\partial z.dz$) and the net force/unit mass is

$$[\partial\tau/\partial z.\delta x.\delta y.\delta z]/[\rho\ \delta x.\delta y.\delta z] = 1/\rho\ \partial\tau/\partial z \qquad (33)$$

If we assume the boundary layer is thin as before, so that z derivatives dominate, the momentum equations then require extra terms of the form $(1/r\ \partial\tau^x/\partial z, 1/r\ \partial\tau^y/\partial z)$.

Consider $u = u_I + u_E$, $v = v_I + v_E$ as before, so that in dimensionless form for a steady state

$$(2u_E, -2v_E) = (\partial\tau^y/\partial z, \partial\tau^x/\partial z)\, E^{1/2} \quad (34)$$

Integrating over the layer from $E^{1/2}z = 0 \rightarrow \infty$ to obtain the transports

$$(U_E, V_E) = E^{1/2}/2\, (\tau^y_{bottom} - \tau^y_{top}, \tau^x_{top} - \tau^x_{bottom}) \quad (35)$$

In the case of the oceans, $\tau^x_{bottom} = 0$ and τ^x_{top} is the stress exerted on the oceans by the atmosphere - a vitally important process driving the ocean circulation. Note that in this case the Ekman transport is *perpendicular* to the applied stress - in the northern hemisphere ocean transport is at an angle of $\pi/2$ to the *right* of the wind stress. We can also integrate the continuity equation to show that

$$w_I = E^{1/2}\, (\partial\tau^y/\partial x - \partial\tau^x/\partial y) \quad (36)$$

i.e. Ekman suction is proportional to the curl of the wind stress at the surface.

Exercise: Show that if $(\partial u/\partial z, -\partial v/\partial z)$ Æ $(\partial u_I/\partial z, \partial v_I/\partial z)$ as $E^{1/2}z \rightarrow \infty$ with a free upper surface, the resulting Ekman layer will have horizontal transports

$$(U_E, V_E) = E/2\, (\partial u_I/\partial z, -\partial v_I/\partial z) \quad (37)$$

and vertical suction velocity

$$w_I = E/2\, \partial\zeta_I/\partial z \quad (38)$$

(cf [13]).

2.7 *Ekman layers on inclined surfaces*

For many purposes we may need to consider the boundary layer adjacent to a surface which is inclined with respect to the horizontal (e.g. on a surface of revolution). To do this, we consider [7] a plane defined by $z^P = 0$, where z^P is the coordinate measured along the direction of a unit vector $\mathbf{k}^P$ perpendicular to the plane which is inclined to the horizontal at an angle α.

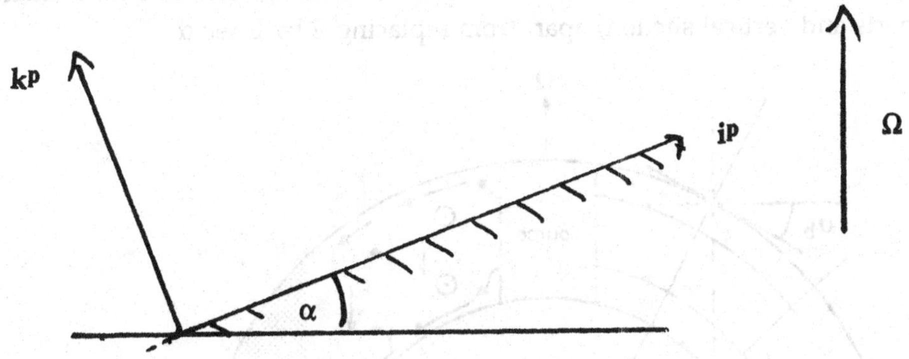

Figure 6: Orientation sketch for Ekman layer adjacent to a boundary inclined at angle α.

We may then write

$$\Omega = \Omega\,(\mathbf{i}^P \sin\alpha,\ \mathbf{j}^P \cos\alpha) \tag{39}$$

and the Ekman layer equations become

$$-2v^P \cos\alpha = -\partial p/\partial x^P + E\partial^2 u^P/\partial z^{P2} \tag{40a}$$

$$2u^P \cos\alpha - 2w^P \sin\alpha = -\partial p/\partial y^P + E\partial^2 v^P/\partial z^{P2} \tag{40b}$$

$$2v^P \sin\alpha = -\partial p/\partial z^P + E\partial^2 w^P/\partial z^{P2} \tag{40c}$$

together with the continuity equation

$$\partial u^P/\partial x^P + \partial v^P/\partial y^P + \partial w^P/\partial z^P = 0 \tag{41}$$

and boundary conditions

$$u^P = v^P = w^P = 0 \qquad \text{at } z^P = 0 \tag{42a}$$

$$u^p \rightarrow u_I^p,\ v^p \rightarrow v_I^p,\ w^p \rightarrow w_I^p \text{ as } E^{1/2}z^p \rightarrow \infty \tag{42b}$$

The continuity equation can be integrated from a point inside the boundary layer to a point in the geostrophic interior to show that $w^p - w_I^p = O(E^{1/2})$. Hence it is possible to show[7] that the theory for the sloping Ekman layer reduces entirely to the form in Sections 2.3-2.5 (including formulae for Ekman transports and vertical suction) apart from replacing E by E sec α.

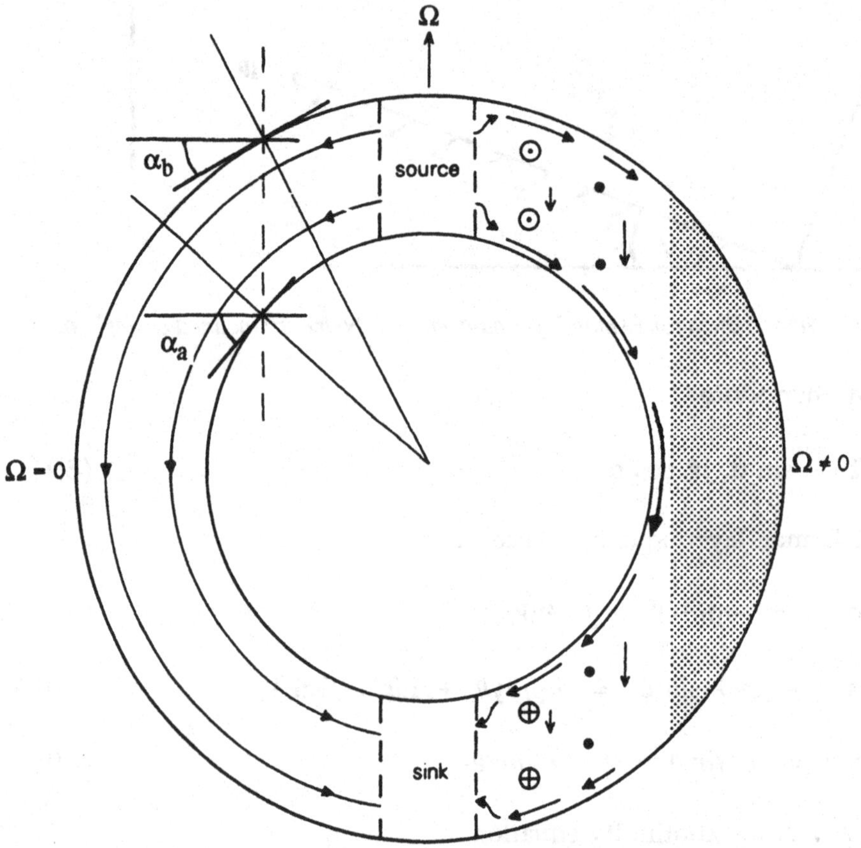

Figure 7: Axisymmetric source-sink flow in a spherical shell of fluid (after [14]). Left side shows streamlines for the case Ω = 0, when fluid moves directly from source to sink at the northern and southern poles respectively. Right side shows the rapidly rotating case, in which flow takes place mainly via Ekman boundary layers on the inner and outer spherical boundaries. Arrows indicate direction of flow in the meridional plane. Also shown is the direction of azimuthal flow, which is in geostrophic balance and vanishes on the equator.

Exercise: Show that an Ekman layer at a boundary inclined at an angle α to the horizontal leads to Ekman transports of the form

$$(U_E, V_E) = 1/2\,(E \sec\alpha)^{1/2}\,(-[u_I + v_I],\,[u_I - v_I]) \qquad (43)$$

and vertical velocity

$$w_I = u_I \tan\alpha + 1/2\,(E \sec\alpha)^{1/2}\,\zeta_I \qquad (44)$$

An illustration of the use of these ideas can be seen in the spherical source-sink problem considered by Hide [14] (see Fig. 7). Fluid is contained in a hollow shell between two concentric spheres of radii a and b respectively, and pumped between the 'north' and 'south' poles, while the whole system is rotated about the two poles. In the absence of rotation, flow takes place uniformly from one pole to the other at all latitudes. When $W \neq 0$, Coriolis accelerations inhibit direct poleward flow, which then becomes confined to Ekman layers at the boundaries of the two spheres. However, since $|\alpha_b| < |\alpha_a|$, the boundary layer on r = b must transfer less mass than at r = a, leading to a compensating axial flow in the interior between the spheres. Azimuthal flow will occur in the interior in geostrophic balance, but vanishes in the stippled region beyond r = b in accordance with the Proudman-Taylor theorem and Ekman suction formulae. All the mass transport must then take place in a *vertical* boundary layer near the equator, where the Ekman theory breaks down. Full analysis of such vertical boundary layers is mathematically very complicated[8],[10], and beyond the scope of the present course (though see Section 3 below on Stewartson layers).

3. DYNAMICS OF THE STEWARTSON LAYER

In Section 2, we have considered the characteristic properties of quasi-horizontal boundary layers in which viscous forces become comparable with Coriolis accelerations. Such boundary layers were found under some circumstances to be 'active' - in the sense that they directly modify the flow in the geostrophic interior via 'Ekman suction'; the component of vertical motion $O(E^{1/2})$ and proportional to the vertical component of relative vorticity in the interior. Although such boundary layers can be generalised to surfaces inclined to the horizontal, we have seen that the Ekman theory analysis breaks down altogether when the bounding surface is essentially vertical.

In this case(e.g. see [7],[8],[10]), viscous shear layers form with rather different, and more complicated, properties than Ekman layers. Their formation is necessary (a) to allow the geostrophic interior to match to the boundary conditions imposed at a vertical boundary, and (b) to return the $O(E^{1/2})$ vertical mass flux which may be generated in the interior by Ekman suction. The analysis applies not only to sidewall boundary layers (e.g. in cylindrical containers), but also to detached shear layers which form in association with sharp changes in tangential velocity in the fluid interior. The latter problem, in which a shear flow is driven in a cylindrical container by

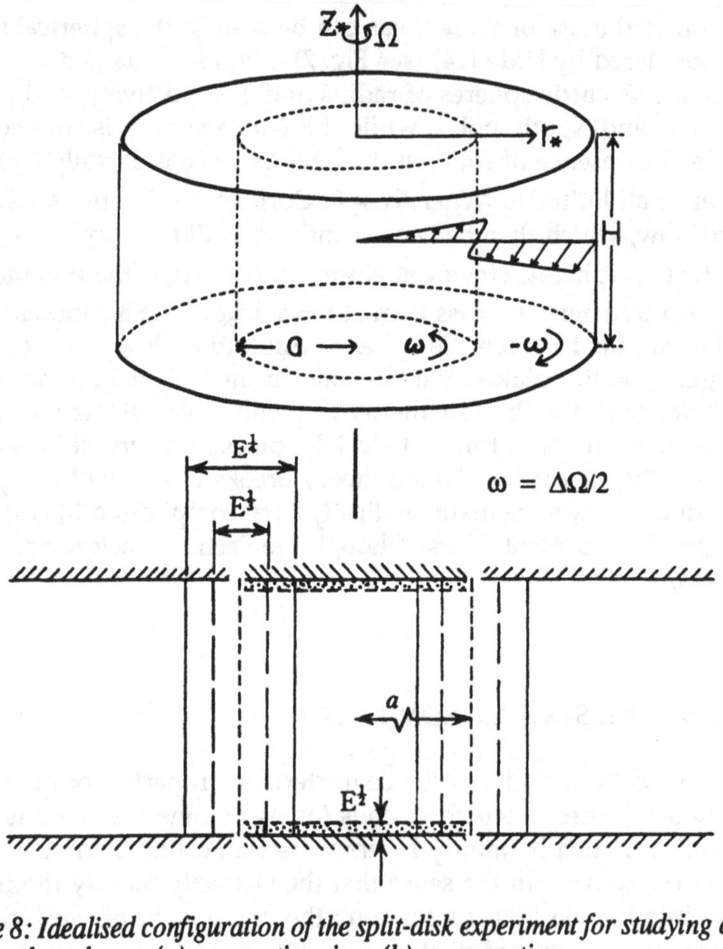

Figure 8: Idealised configuration of the split-disk experiment for studying detached Stewartson shear layers (a) perspective view, (b) cross-section.

differential rotation of concentric disks on the horizontal boundaries (see below) was first studied by Stewartson [4], who pioneered the type of analysis needed to understand their dynamics. For this reason, this class of vertical shear layers are usually known as 'Stewartson layers'.

3.1 Scale analysis for the laminar Stewartson layer

We take as a starting point the steady equations of motion, scaled as in Section 2.1 and linearised by setting Ro = 0, to obtain (8) and (10). It is convenient to reduce these equations to a single equation in a scalar variable (albeit of higher order). We take **k**.(8) to obtain

$$\partial p/\partial z = E\nabla^2(\mathbf{k}.\mathbf{u}) \quad (45)$$

Take **k**.($\nabla\times$(8)) and use (10) to obtain a vorticity equation

$$-2\partial/\partial z\ [\mathbf{k}.\mathbf{u}] = E\nabla^2(\mathbf{k}.(\nabla\times\mathbf{u})) \quad (46)$$

We combine (45) and (46) to obtain

$$\partial^2 p/\partial z^2 = E^2\nabla^4(\mathbf{k}.(\nabla\times\mathbf{u})) \quad (47)$$

From the divergence of (8) we obtain

$$2\mathbf{k}.(\nabla\times\mathbf{u}) = \nabla^2 p \quad (48)$$

which can be combined with (47) to give

$$4\partial^2 p/\partial z^2 + E^2\nabla^6 p = 0 \quad (49)$$

which effectively represents a balance between Coriolis accelerations and viscous forces. We go on to consider the two principal boundary layers discussed by Stewartson by taking various limits of (49) and consider their respective characteristics.

3.2 The $E^{1/3}$ Layer

We anticipate that any viscous-dominated vertical shear layer will be characterised by a horizontal length scale $O(E^\theta)L$ (assumed << L). We reduce our cylindrical system to a Cartesian system by defining a stretched cross-layer coordinate x by

$$xE^\theta = r - a \quad (50)$$

(where a is a reference radius) so that $\partial/\partial r = E^{-\theta}\partial/\partial x$ and streamwise and vertical coordinates (y,z) respectively, where $y = a\theta$. Assuming $E << 1$, the dominant terms in (49) become

$$E^{2-6\theta}\,\partial^6 p/\partial x^6 + 4\partial^2 p/\partial z^2 = 0 \tag{51}$$

Balance between these terms implies $\theta = 1/3$, hence we deduce that a vertical shear layer exists with dimensionless thickness $O(E^{1/3})$. As mentioned above, at least one of the vertical shear layers is required to accommodate a vertical mass flux $O(E^{1/2})$ to return the flux through the geostrophic interior generated via Ekman suction. We suppose the vertical velocity in this layer to be $O(E^{a})$ such that $E^{a+1/3} = O(E^{1/2})$, implying $a = 1/6$. Taking the rest of the flow to be axisymmetric ($\partial/\partial y = 0$), the continuity equation becomes

$$E^{-1/3}\,\partial u/\partial x + \partial w/\partial z = 0 \tag{52}$$

A balance of terms requires $u = O(E^{1/2})$. We can therefore write the three components of the momentum equation as

$$-2v = -E^{-1/3}\,\partial p/\partial x + E^{1/3}\,\partial^2 u/\partial x^2 \tag{53a}$$

$$2u = E^{1/3}\,\partial^2 v/\partial x^2 \tag{53b}$$

$$0 = -\partial p/\partial z + E^{1/3}\,\partial^2 w/\partial x^2 \tag{53c}$$

Thus, a balance of terms in (53b,c) requires $v = O(E^{1/6})$ and $p = O(E^{1/2})$. This implies that geostrophic balance must hold within the $E^{1/3}$ layer for the tangential flow v *only*, but (53c) implies a significant departure from satisfying the Taylor-Proudman theorem. If we represent each variable in the shear layer as an infinite series in powers of $E^{1/6}$, e.g.

$$s = s_0 + s_1 E^{1/6} + s_2 E^{1/3} + \ldots\ldots.. \tag{54}$$

the momentum and continuity equations reduce at leading order to

$$-2v_1 = -\partial p_3/\partial x \tag{55a}$$

$$2u_3 = \partial^2 v_1/\partial x^2 \tag{55b}$$

$$0 = -\partial p_3/\partial z + \partial^2 w_1/\partial x^2 \tag{55c}$$

$$\partial u_3/\partial x + \partial w_1/\partial z = 0 \tag{55d}$$

Equations (55d) and (55c) also imply significant vertical structure within the $E^{1/3}$ layer. Note, however, that since $v = O(E^{1/6})$, we cannot match the shear-layer tangential velocity with the interior v, which is O(1) (to be compatible with the Ekman suction condition that vertical mass flux $= O(E^{1/2})$). The $E^{1/3}$ layer is therefore incomplete as a description of the shear-layer structure.

3.3 *The $E^{1/4}$ Layer*

We consider a further vertical shear layer in which the tangential velocity component is defined to be O(1), so as to match with the interior. We define a new cross-layer coordinate λ such that

$$\lambda E^{\beta} = r - a, \text{ where } \beta \neq 1/3 \tag{56}$$

and $\partial/\partial r = E^{-\beta}\partial/\partial\lambda$. The dominant terms in the pressure equation (51) now give

$$E^{2-6\beta}\, \partial^6 p/\partial\lambda^6 + 4\partial^2 p/\partial z^2 = 0 \tag{57}$$

Two possibilities must be considered, depending upon whether β is greater than or less than 1/3. If $\beta > 1/3$ then (57) reduces to

$$\partial^6 p/\partial\lambda^6 = 0$$

This possibility, however, cannot satisfy the correct boundary conditions on v via geostrophic balance while at the same time ensuring that $\lim_{\lambda\to\infty} p = 0$

Hence β must be less than 1/3, and (57) reduces to

$$\partial^2 p/\partial z^2 = 0 \tag{58}$$

The components of the momentum equation then become

$$-2v = -E^{-\beta}\,\partial p/\partial\lambda + E^{1-2\beta}\,\partial^2 u/\partial\lambda^2 \qquad (59a)$$

$$2u = E^{1-2\beta}\,\partial^2 v/\partial\lambda^2 \qquad (59b)$$

$$0 = -\partial p/\partial z + E^{1-2\beta}\,\partial^2 w/\partial\lambda^2 \qquad (59c)$$

with the continuity equation

$$E^{-\beta}\,\partial u/\partial\lambda + \partial w/\partial z = 0 \qquad (60)$$

Since we require $v = O(1)$, balance in (59b) requires $u = O(E^{1-2\beta})$, and from (60) $w = O(E^{1-3\beta})$. If $\beta < 1/3$, the first term on the rhs of Equation (59a) $E^{-\beta}\,\partial p/\partial\lambda$ must dominate over the second term. Hence balance with the O(1) term 2v requires $p = O(E^{\beta})$ and (59c) reduces to

$$\partial p/\partial z = 0 \qquad (61)$$

implying that this layer, in contrast to the $E^{1/3}$ layer, satisfies a form of the Taylor-Proudman theorem. If we also impose the condition that the vertical mass flux in this layer is again $O(E^{1/2})$, we obtain from (60) that $E^{1-3\beta}\,E^{\beta} = E^{1/2}$ so that $\beta = 1/4$. Thus, this further shear-layer has a dimensionless thickness $O(E^{1/4})$, but with much greater vertical uniformity than the $E^{1/3}$ layer. If we represent each variable in the shear layer as an infinite series in powers of $E^{1/4}$, the momentum and continuity equations reduce at leading order to

$$-2v_0 = -\partial p_1/\partial\lambda \qquad (62a)$$

$$2u_2 = \partial^2 v_0/\partial\lambda^2 \qquad (62b)$$

$$0 = -\partial p_1/\partial z \qquad (62c)$$

$$\partial u_2/\partial\lambda + \partial w_1/\partial z = 0 \qquad (62d)$$

Thus, progressing inwards from an interior with essentially no vertical structure, we first encounter the $E^{1/4}$ layer in which u and v are independent of z and w depends linearly on z, leading into the innermost $E^{1/3}$ layer with considerable vertical structure in all fields.

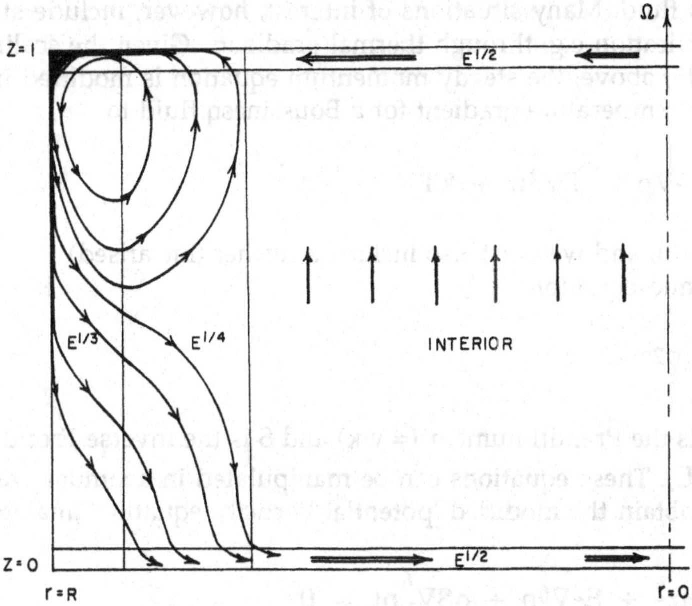

Figure 9: Schematic circulation showing the typical nested structures of the $E^{1/4}$ and $E^{1/3}$ boundary layers matching into thin ($E^{1/2}$) Ekman layers on the horizontal boundaries (after [8]).

Exercise: Solve (59a-c) and (60) for the boundary conditions comprising the Ekman suction condition on z = 0,1

$$w_1 = \pm 1/2\ \partial v_0/\partial\lambda \quad \text{at } z = 0,1$$

and

$$v_0 = 0 \qquad \text{at } \lambda = 0$$

$$v_0 \to v_I \qquad \text{as } \lambda \to \infty$$

hence show that the *volume-averaged* vertical mass flux in the $E^{1/4}$ layer is identically zero. This implies that the $E^{1/3}$ layer is necessary to accommodate the $O(E^{1/2})$ mass flux returned from the interior.

4. THE MODIFYING ROLE OF STABLE STRATIFICATION

All of the discussion so far of Ekman and Stewartson layers has been for a homogenous fluid. Many situations of interest, however, include stable density stratification e.g. through thermal gradients. Given the scaling used in (1)-(2) and (45) above, the steady momentum equation is modified in the presence of a temperature gradient for a Boussinesq fluid to

$$2\mathbf{k}\times\mathbf{u} = -\nabla p + E\nabla^2\mathbf{u} + \mathbf{k}T \quad (63)$$

(e.g. [8], [10]) and we must also include a further (linearised) thermodynamic equation

$$\sigma S w = E\nabla^2 T \quad (64)$$

where σ is the Prandtl number ($= \nu/\kappa$) and S is the inverse Froude number ($= g\alpha\, \Delta T/\Omega^2 L$). These equations can be manipulated in a similar way to Section 2 to obtain the modified 'potential vorticity equation' analogous to (49)

$$\nabla^2[4\partial^2 p/\partial z^2 + E^2\nabla^6 p + \sigma S\nabla_h^6 p] = 0 \quad (65)$$

(a) (b) (c)

(where ∇_h is the horizontal gradient operator) *[Exercise: Derive Eq. (65) from Eqs (1-2) and (63-64)]* which is the principal device for assessing the critical stratification at which homogenous Ekman or Stewartson theory must be modified. Thus, for the Ekman layer in which $\partial/\partial z = O(E^{-1/2})$, terms (a) and (b) are $O(E^{-1})$, so that (c) is negligible unless $\sigma S = O(E^{-1})$. Since this is typically very large, Ekman theory is usually applicable even in a stratified flow.

For the $E^{1/3}$ layer in which $\partial/\partial x = O(E^{-1/3})$, terms (a) and (b) in (65) are O(1), so that (c) is negligible unless $\sigma S = O(E^{2/3})$. This criterion is the first to be met as σS increases from 0, mainly because the $E^{1/3}$ layer has strong vertical motion readily modified by stratification. When the $E^{1/3}$ layer is modified, a new boundary layer scale δ_S is applicable. At the critical stratification, (c) ~ (b) (i.e. buoyancy forces are balanced by viscosity) so that $E^2/\delta_S^6 \sim \sigma S/\delta_S^2$, hence

$$\delta_S = E^{1/2}/(\sigma S)^{1/4}L = (\nu\kappa/g\alpha\Delta T L^3)^{1/4}L = Ra^{-1/4}L \quad (66)$$

where Ra is the Rayleigh number - NB a boundary layer *independent of rotation*. Other examples of rotation-dominated boundary layers modified by stratification are discussed in more detail e.g. in [8],[10].

5. BOUNDARY LAYER INSTABILITIES

5.1 Ekman layer Instabilities: Laboratory Experiments

Discussion so far has focussed upon the steady laminar Ekman layer as a major component of the circulation in many rotating fluid systems. In practice, however, many flows are unsteady. We consider here possible sources of unsteadiness which may originate in the Ekman layer itself. The stability of Ekman layers in a homogenous fluid was originally studied by Faller & Kaylor [15], who set up a radial mass flow in a cylindrical container (2m in radius) by withdrawing fluid from the centre of the container and introducing fluid at the rim. Flow in a 5mm thick Ekman layer at the bottom of a fluid layer (depth 20cm) was visualized by dropping crystals of potassium permanganate into the tank and observing from above the dye streaks against a light background (see Fig. 10 below).

The flow was found to comprise a steady laminar inflow (with streaks close to the bottom of the tank oriented at 45° to the predominantly azimuthal flow in the geostrophic interior. As the mass flow rate was increased, a series of roll-like disturbances was found to occur (designated 'Class A' by Faller & Kaylor [15]), oriented around 15° to the left of the interior flow and fairly widely spaced (with a wavelength ~ 20-30δ). At even higher flow rates, a second set of rolls was found to set in, oriented at ~ 15° to the *right* and with a shorter wavelength (~ 10δ). The latter were designated 'Class B' disturbances by [15]. At yet higher flow rates, the flow in the boundary layer become highly disordered and turbulent. The onset of these disturbances could be ordered with reference to a dimensionless measure of the flow speed based on a Reynolds number R_E, using the depth of the Ekman layer as the principal length scale. i.e.

$$R_E = E^{1/2} U_I L/\nu \qquad (67)$$

The 'Class A' disturbances occur for $R_E > 55$, but 'Class B' rolls dominate for $R_E > 120$. Turbulent flows (initially in the form of isolated 'patches') occur for

R_E much larger than 150 or so. The main properties of these instabilities are summarised in Table 1 (from [5]).

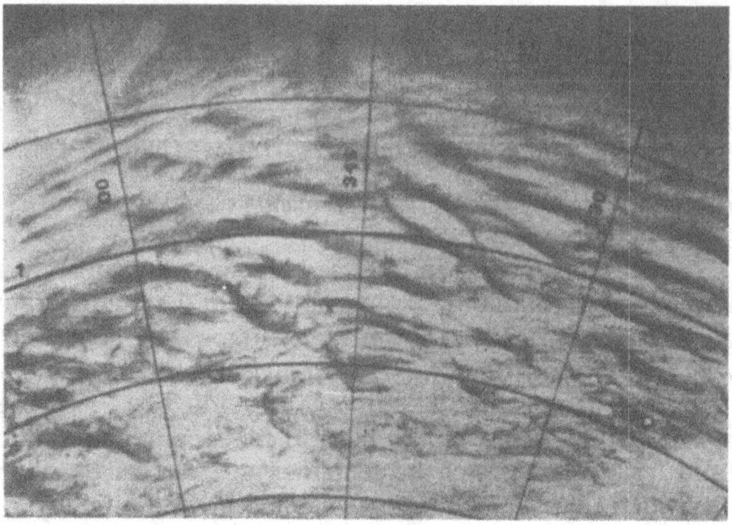

Figure 10: An example of the simultaneous occurrence of the Class A and Class B modes of instability of the laminar Ekman boundary layer, observed in the laboratory by [15]. Rolls visualised using potassium permanganate dye released from crystals dropped into a rotating tank of fluid.

5.2 Ekman Layer Instabilities: Theory

We define the Reynolds number in terms of other dimensionless parameters as $R_E = Ro\, E^{-1/2}$, and follow Greenspan [9] by rescaling the equations of motion and continuity using

$$\mathbf{x} = E^{1/2}L\mathbf{x}^*,\ t = E^{1/2}L/U\ t^*,\ \ \mathbf{u} = U\mathbf{u}^* \qquad (68)$$

to obtain

$$R_E(\partial\mathbf{u}^*/\partial t^* + \mathbf{u}^*.\nabla\mathbf{u}^*) + 2\mathbf{k}\times\mathbf{u}^* = -\nabla p^* + \nabla^2\mathbf{u}^* \qquad (69a)$$

$$\nabla.\mathbf{u}^* = 0 \qquad (69b)$$

Table 1: Summary of Ekman layer instabilities [5]

Type of instability	*Quantity*	Theory *Faller & Kaylor [15]*	Theory *Lilly [23]*	Experiment *Faller & Kaylor [15]*	Experiment *Tatro et al. [24]*	Experiment *Caldwell et al. [5]*
Class A	Critical Reynolds no.	55	55	< 70	56.3 + 116.8 Ro	56.7
	Wavelength Ekman depths)	24	21	22-33	27.8 ± 2.0	
	Inclination of rolls η	-15°	-20°	$+5^o$ to -20°	0 to -8°	
	Velocity/ geostrophic	0.50	0.57	-	0.16	
Class B	Critical Reynolds no.	118	110	125 ± 5	124.5 + 7.32 Ro	
	Wavelength Ekman depths)	11	11.9	10.9	11.8	
	Inclination of rolls η	$+10^o$-12°	$+8^o$	$+14.5^o \pm 2.0^o$	$+14.8^o \pm 0.8^o$	
	Velocity/ geostrophic	0.33	0.094	0.023	0.034	

Consider a steady laminar flow $\mathbf{u}_\ell$ (x,y,z) which satisfies appropriate boundary conditions. We apply a small perturbation [$\mathbf{u}'$(x,y,z,t), p'(x,y,z,t)] and seek growing solutions. We linearize (69a,b) about $\mathbf{u}_\ell$ to obtain

$$R_E(\partial \mathbf{u}'/\partial t + \mathbf{u}_\ell . \nabla \mathbf{u}' + \mathbf{u}' . \nabla \mathbf{u}_\ell) + 2\mathbf{k}\times\mathbf{u}' = -\nabla p' + \nabla^2 \mathbf{u}' \qquad (70a)$$

$$\nabla . \mathbf{u}' = 0 \qquad (70b)$$

Since the experiments [15],[5] suggest that 2D roll-like disturbances are the favoured form of instability, we consider wave-like perturbations relative to a coordinate system referred to the interior flow $\mathbf{u}_I$, such that

$$\mathbf{u}_I = (\cos\eta\, \mathbf{i}^q - \sin\eta\, \mathbf{j}^q) \qquad (71)$$

hence

$$\mathbf{u_I} = [\cos\eta - e^{-z}\cos(\eta + z)]\,\mathbf{i^q} - [\sin\eta - e^{-z}\sin(\eta + z)]\,\mathbf{j^q}$$

$$= u_\ell\,\mathbf{i^q} + v_\ell\,\mathbf{j^q} \tag{72}$$

The perturbations are in the form of 2D waves, independent of x^q. Thus we can define a streamfunction χ by

$$v' = \partial\chi/\partial y, \quad w' = -\partial\chi/\partial z \tag{73}$$

and seek solutions of the form

$$u'(y^q,z,t) = U(z)\,e^{ik(y^q - ct)} \tag{74a}$$

$$\chi'(y^q,z,t) = X(z)\,e^{ik(y^q - ct)} \tag{74b}$$

Upon substitution into the x^p components of the linearised momentum and vorticity equations we obtain a pair of coupled ODEs

$$d^2U/dz^2 - k^2U + 2\,dX/dz = ikR_E[(v_\ell - c)U - X\,du_\ell/dz] \tag{75a}$$

$$(d^2/dz^2 - k^2)^2 X = ikR_E[(v_\ell - c)(d^2/dz^2 - k^2)X - X\,d^2v_\ell/dz^2] + 2\,dU/dz \tag{75b}$$

Appropriate boundary conditions are

$$U = 0,\ X = dX/dz = 0 \quad \text{at } z = 0 \tag{76a}$$

$$(U, X, dX/dz) \rightarrow 0 \qquad \text{as } z \rightarrow \infty \tag{76b}$$

The instability problem then reduces to one of seeking conditions under which $Im(c) > 0$. Again laboratory experience suggests that instabilities set in when viscous effects are small relative to inertial effects, i.e. as $R_E \rightarrow \infty$. We therefore consider the inviscid forms of (75a,b). In this case, (75b) becomes

$$(v_\ell - c)(d^2/dz^2 - k^2)X - X\,d^2v_\ell/dz^2 = 0 \tag{77}$$

The study of this form of normal mode equation has a long history dating back to Lord Rayleigh (e.g. see [16]). Following Rayleigh, we can find a necessary (though not sufficient) condition for instability, by taking (77) and its complex conjugate and multiplying respectively by $X^+/(v_\ell - c^+)$ and $X/(v_\ell - c)$. We then integrate over the entire domain in z to obtain

$$Im(c) \int_0^\infty XX^+/[(v_\ell - c^+)(v_\ell - c)]\, d^2v_\ell/dz^2\, dz = 0 \tag{78}$$

If $Im(c) > 0$, then the integrand cannot be positive everywhere, hence d^2v_ℓ/dz^2 must change sign (passing through zero) somewhere - i.e. instability requires an *inflection point* in the flow. This occurs naturally within the laminar Ekman layer at a height of $z = n\pi$ (n integer), so many sources of instability may be present.

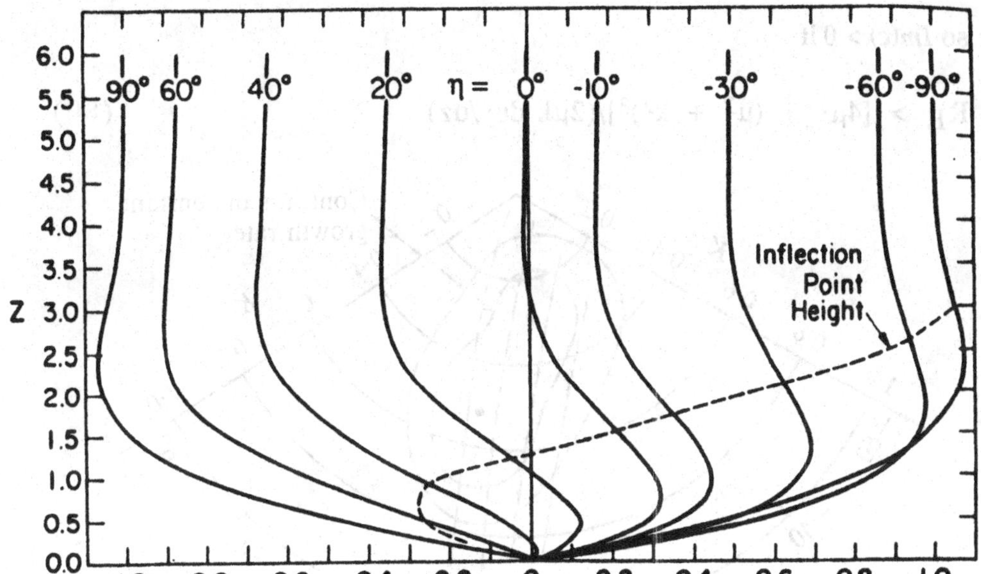

Figure 11: Theoretical Ekman layer velocity profiles in planes at various angles to the free stream flow (after [18]).

Detailed investigations of the nature of these inviscid instabilities indicate that they are consistent with observations of the Class B waves found for $R_E >$ 125, though inviscid theory predicts too large a wavenumber. Viscosity must be included properly in the selection of k. The full viscous form of (75b) without the 2 dU/dz term is identical to the Orr-Sommerfeld (O-S) equation, well known in classical shear-flow instability theory (e.g. see [16], [17]).

Numerical work confirms that Class B waves are essentially governed by the O-S equation.

Class A instabilities seem to be of a different type, involving an interaction between the Coriolis acceleration and (ageostrophic) shear in the initial flow (and are hence to be distinguished from Tollmien-Schlichting waves, as noted recently by Faller [25]). A form of the instability can be demonstrated in a simple model flow (suggested by J. T. Stuart; see [9]) in which $v_{\mathfrak{c}} = 0$ and $du_{\mathfrak{c}}/dz$ = constant. If we substitute these conditions into (75b), taking $X = e^{-\mu z}$, to obtain

$$-ick = 1/R_E \, [-(\mu^2 + k^2) \pm \{(2\mu k R_E \, du_{\mathfrak{c}}/dz - 4\mu^2)/(\mu^2 + k^2)\}^{1/2}] \tag{79}$$

so $Im(c) > 0$ if

$$R_E > [4\mu^2 + (\mu^2 + k^2)^3]/(2\mu k \, du_{\mathfrak{c}}/dz) \tag{80}$$

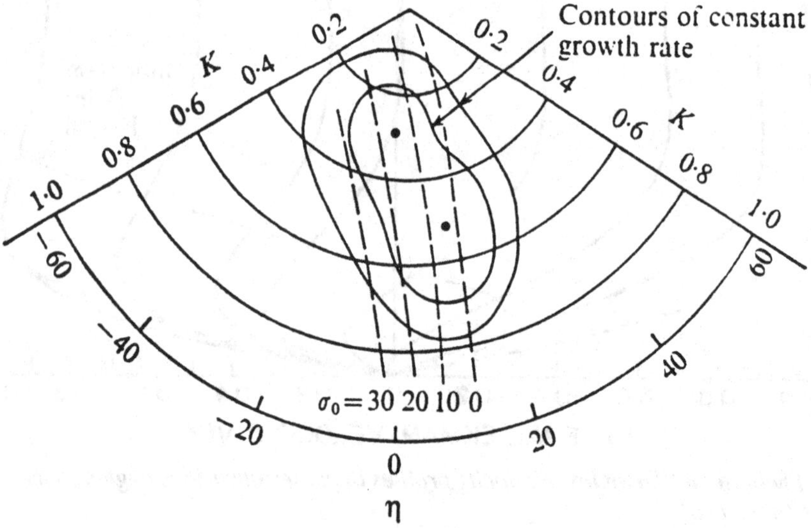

Figure 12: Contours of growth rate and frequencies predicted by linear theory [23] for Ekman layer instabilities with a Reynolds number of 150.

However, note that $|Im(c)| \to 0$ as $R_E \to \infty$. Detailed numerical solutions of (75b) confirm the existence of two classes of unstable mode, one of which is centred near the first inflection point in the Ekman profile (Class A) and the other lower down, where the shear of the zonal flow is large (Class B).

5.3 Applications to the Planetary Boundary Layer

As was discussed in Section 2.5 above, the atmospheric boundary layer has a number of features in common with the laminar Ekman layer. Roll-like disturbances are sometimes observed within the lowest 1km altitude, resulting in the formation of parallel linear arrays of cloud patterns aligned with the large-scale wind (referred to as 'cloud streets' - see [18] for a review). The previous two sections have indicated at least two possible mechanisms for the formation of rolls in an Ekman boundary layer, and it is of interest to examine whether these processes could be responsible for formation of 'cloud streets'.

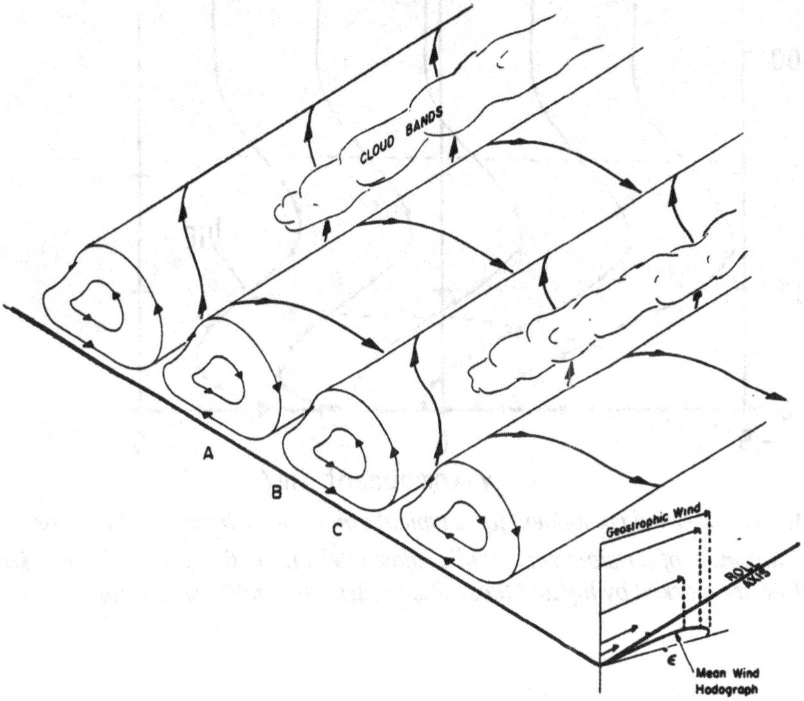

Figure 13: Schematic sketch of the atmospheric boundary layer containing large roll eddies, showing a typical secondary flow (modified Ekman layer) - after [18].

Ekman-like inflection points are sometimes found to occur in observations of atmospheric boundary layer structure under statically neutral conditions, indicating one source of instability similar to the Ekman layer. As reviewed by [19], however, other possible sources of instability are often observed at the same time. Cloud streets are often observed under convectively unstable conditions, so rolls could be formed and organised by virtue of the effect of shear in a mean wind on Benard-like thermal convection. In addition, inflection points in the wind profile may occur for reasons other than the Ekman-type boundary layer structure, such as in association with the advection of heat in the lower atmosphere, which causes the wind direction to turn with height. Thus, Ekman layer instabilities may provide but one example of a process relevant to the range of roll-like disturbances in the atmosphere.

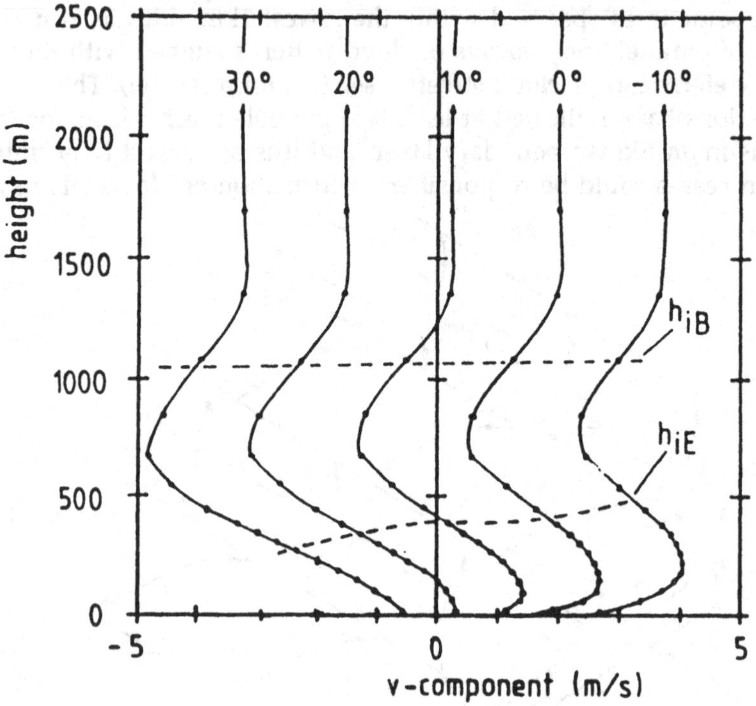

Figure 14: Mean cross-wind component for a typical atmospheric boundary layer for different orientations η of possible vortex rolls (after [19]). Inflection points due to Ekman layer shear flow are marked by h_{iE} and those due to thermal wind shear by h_{iB}.

5.4 *Stewartson Layer instability: Experiments*

Various workers have studied a realisation of the split-disk experiment in the laboratory, which largely verify the theory of Section 3 in terms of the shear-layer structure. Similar analyses also apply to other situations (a) in which two distinct regions of geostrophic flow are separated by fronts or other discontinuities where viscous effects become important e.g. in the boundaries of a Taylor column, or (b) adjacent to the sidewalls of cylindrical or spheroidal containers. A further result of these studies is the observation that shear-layers of this type may become unstable to various kinds of disturbances. We consider here some of the principal mechanisms responsible for such an instability.

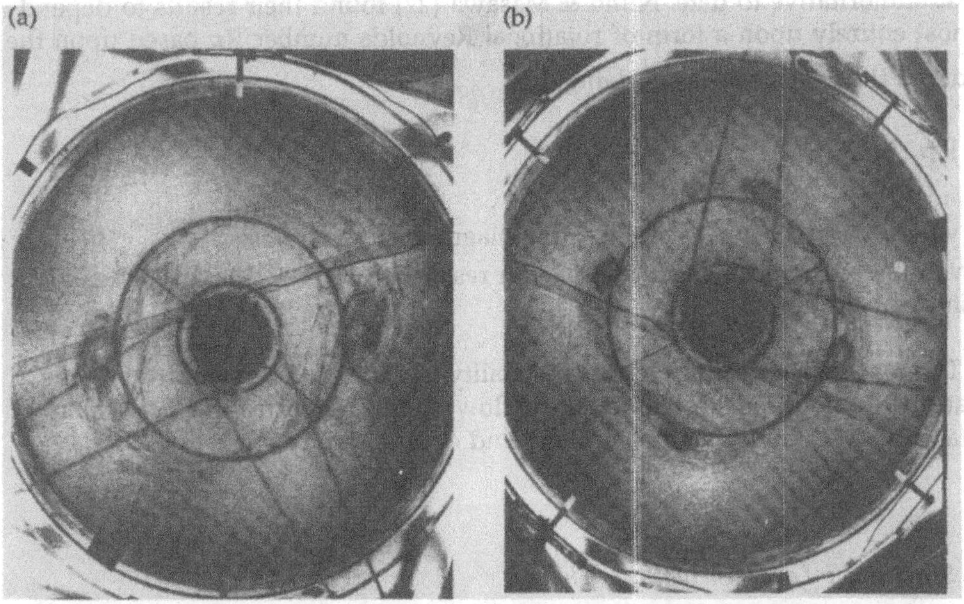

Figure 15: Typical eddy fields in the split-disk experiment [22], showing fully-developed vortices arising from the instability of a detached Stewartson shear layer at different values of the rotational Reynolds number R_E. (a) $R_E = 50.0$, (b) $R_E = 34.0$.

The first kind of instability is observed e.g. when a rotating cylindrical container, containing fluid initially in solid body co-rotation, is slowed down or stopped impulsively. A series of spiral rolls or vortices is found to develop in a sidewall boundary layer with axes aligned roughly horizontally in the azimuthal direction - somewhat like a stack of 'doughnuts' (of the US variety!). These are known as 'Taylor-Görtler' vortices, after their discoverers, and arise from a centrifugal instability known since the time of Rayleigh. The essential criterion for instability is for specific absolute angular momentum (= $\Omega(\Omega r + v)$) to *decrease* with r (e.g. see [20]).

Another type of instability is observed in the split-disk experiment [21],[22], in which trains of waves or vortices develop with axes parallel to the rotation axis. Some examples are illustrated in Fig. 15. Most studies to date have investigated the regular non-axisymmetric flow regimes, in which the wave pattern is m-fold symmetrical about the rotation axis (where m is the azimuthal wavenumber) and steady (apart from a slow drift around the apparatus). The onset of the instability and the wavenumber observed is found to depend on the basic rotation rate Ω and the differential rotation rate $\Delta\Omega$ of the disks, as well as on the geometric properties of the container (aspect ratios, curvature etc.). Some disagreement as to the dominant parameter governing the instability, but the recent work by Niino & Misawa [22] seems to be the most authoritative to date. Niino & Masawa [22] found their results to depend almost entirely upon a form of rotational Reynolds number R_E based upon the width of the $E^{1/4}$ shear-layer, defined by

$$R_E \; = \; V(E/4)^{1/4}L/\nu \tag{81}$$

where $V = r\Delta\Omega/2$. A typical regime diagram showing observed wavenumber (non-dimensionalised with respect to $(E/4)^{1/4}L$) plotted against R is shown below.

The mechanism leading to this instability is widely understood to be the rotationally-modified horizontal shear-flow or quasi-geostrophic *barotropic* instability, well known in meteorology and oceanography [6].

5.6 Stewartson Layer Instability: Theory

The essence of the barotropic instability mechanism can be understood with reference to the quasi-geostrophic vorticity equation[6]. We form the vorticity equation and isolate the vertical component, then average in the vertical by performing $1/H \int_0^H dz$ and include the effect of Ekman suction. Advection is taken due to the geostrophic component only, so that with ψ as the geostrophic streamfunction for the horizontal flow and the same non-dimensionalisation as before in the $E^{1/4}$ layer we have

$$\partial/\partial t\,(\nabla^2\psi) \; + \; \partial\psi/\partial y\,.\partial/\partial x(\nabla^2\psi) \; - \; \partial\psi/\partial x.\partial/\partial y(\nabla^2\psi) \; + \; \nabla^2\psi$$

$$+ \; 1/2\;\partial V_B/\partial x \; = \; \nabla^4\psi \tag{82}$$

Note that in this case the Ekman friction term $\nabla^2\psi$ is of the same order as the internal viscous diffusion $\nabla^4\psi$. This equation can be linearised about the Stewartson $E^{1/4}$ layer solution for V(x) e.g. of [9]

$$V(x) = 1 - e^{x}, \quad x < 0 \tag{83a}$$

$$V(x) = e^{-x} - 1, \quad x > 0 \tag{83b}$$

which we recall is in geostrophic balance.

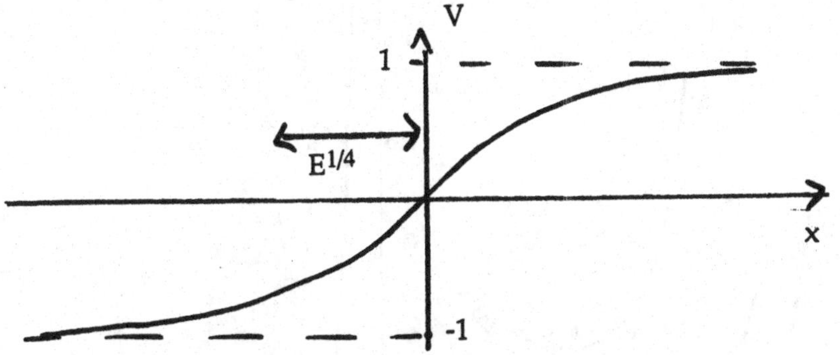

Figure 16: Idealised azimuthal flow profile in Greenspan's [9] solution for the steady $E^{1/4}$ layer.

The resulting form of (82) is

$$\partial/\partial t\,(\nabla^2\psi') + V(x).\partial/\partial y(\nabla^2\psi') - d^2V/dx^2.\partial\psi'/\partial y + \nabla^2\psi' = \nabla^4\psi' \tag{84}$$

We seek normal mode perturbations of the form

$$\psi' = \phi(x).\exp\{ik(y - R_E ct\} \tag{85}$$

so that (84) reduces to

$$[ikR_E(V - c) + 1](d^2\phi/dx^2 - k^2\phi) - ikR_E d^2V/dx^2.\phi = (d^2/dx^2 - k^2)^2\phi \tag{86}$$

This is another form closely related to the Orr-Sommerfeld equation, modified to include the Ekman friction term. We can see that, like the Class A disturbances in the Ekman layer instability problem, the instability pivots about the presence of an inflection point in the basic flow where d^2V/dx^2 passes through zero (cf (83)). This can be solved numerically, given appropriate boundary conditions (e.g. $\phi, d\phi/dx \rightarrow 0$ for $|x| \rightarrow \infty$). Niino & Masawa [22] extended the analysis for finite cylindrical curvature (with mean radius non-dimensionalised by the $E^{1/4}$ scale and represented by γ), and their results for neutral stability boundaries are shown below.

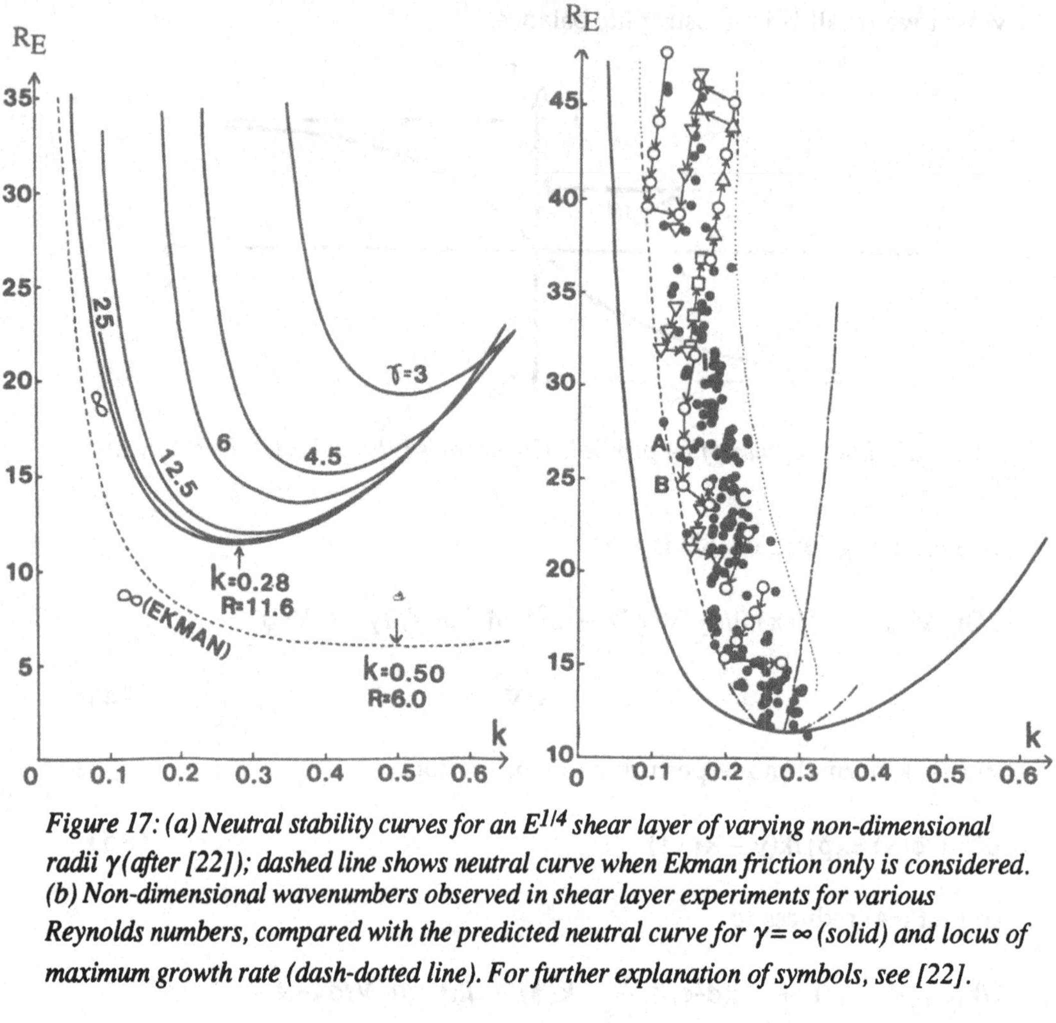

Figure 17: (a) Neutral stability curves for an $E^{1/4}$ shear layer of varying non-dimensional radii γ (after [22]); dashed line shows neutral curve when Ekman friction only is considered. (b) Non-dimensional wavenumbers observed in shear layer experiments for various Reynolds numbers, compared with the predicted neutral curve for $\gamma = \infty$ (solid) and locus of maximum growth rate (dash-dotted line). For further explanation of symbols, see [22].

These results are largely confirmed in experiments, at least for the case where $d\Omega(r)/dr < 0$. Hide & Titman [21] also investigated the case where $d\Omega(r)/dr > 0$ and found qualitatively different behaviour. Non-axisymmetries developed for $R_E >$ some critical value, but the form of the waves was quite different. The reason for this difference remains largely unresolved to date, though may be related to centrifugal effects.

REFERENCES

1. Hopfinger, E. J.: General concepts and examples of rotating fluids, this volume, 1992.

2. Ekman, V. W.: On the influence of the earth's rotation on ocean currents, *Ark. Math. Astr. Fys.*, **2** (1905), 1-52.

3. Walker, J. M.: Farthest north, dead water and the Ekman spiral. Part 1: An audacious adventure, *Weather*, **46** (1991), 103-107; Farthest north, dead water and the Ekman spiral. Part 2: Invisible waves and a new direction in current theory, *Weather*, **46** (1991), 158-164.

4. Stewartson, K.: On almost rigid rotations, *J. Fluid Mech.*, **3** (1957), 17-26.

5. Caldwell, D. R. & Van Atta, C. W.: Characteristics of Ekman boundary layer instabilities, *J. Fluid Mech.*, **44** (1970), 79-95.

6. Pedlosky, J.: *Geophysical Fluid Dynamics*, Springer-Verlag 1987, ppvii+624.

7. Moore, D. W.: Homogenous fluids in rotation A:viscous effects, in: *Rotating Fluids in Geophysics (ed. P. H. Roberts & A. Soward)*, Academic Press 1978, 29-66.

8. Fein, J. S. (ed.): *Boundary Layers in Homogenous and Stratified-Rotating Fluids*, University Presses of Florida 1978, ppxiii+128.

9. Greenspan, H. P.: *The Theory of Rotating Fluids*, Cambridge University Press 1968, ppxi + 327.

10. Friedlander, S.: *An Introduction to the Mathematical Theory of Geophysical Fluid Dynamics*, North-Holland 1980, ppviii+272.

11. Howroyd, G. C. & Slawson, P. R.: The characteristics of a laboratory produced turbulent Ekman layer', *Boundary Layer Met.*, **8** (1975), 201-219.

12. Caldwell, D. R., van Atta, C. W. & Helland, K. N.: A laboratory study of the turbulent Ekman layer, *Geophys. Fluid Dyn.*, **3** (1972), 125-160.

13. Hide, R.: The viscous boundary layer at the free surface of a rotating baroclinic fluid, *Tellus*, **16** (1964), 523-529.

14. Hide, R.: Experiments with rotating fluids, *Quart. J. R. Met. Soc.*, **103** (1977), 1-28.

15. Faller, A. J. & Kaylor, R. (1967) 'Instability of the Ekman spiral with applications to the Planetary Boundary Layers', *Phys. Fluids Suppl.* (1967), S212-219.

16. Lindzen, R. S.: Instability of plane parallel shear flow (toward a mechanistic picture of how it works), *PAGEOPH*, **126** (1988), 103-121.

17. Bayly, B. J., Orszag, S. A. & Herbert, T.: Instability mechanisms in shear flow transition, *Ann. Rev. Fluid Mech.*, **20** (1988), 359-391.

18. Brown, R. A.: Longitudinal instabilities and secondary flows in the Planetary Boundary Layer: A review, *Rev. Geophys. Space Phys.*, **18** (1980), 683-697.

19. Müller, D., Etling, D., Kottmeier, Ch. & Roth, R.: On the occurrence of cloud streets over northern Germany, *Quart. J. R. Met. Soc.*, **111** (1985), 761-772.

20. Chandrasekhar, S.: *Hydrodynamic and Hydromagnetic Instability*, Clarendon Press, Oxford 1961.

21. Hide, R. & Titman, C. W.: Detached shear layers in a rotating fluid, *J. Fluid Mech.*, **29** (1967), 39-60.

22. Niino, H. & Misawa, N.: An experimental and theoretical study of barotropic instability, *J. Atmos. Sci.*, **41** (1984), 1992-2011.

23. Lilly, D. K.: *J. Atmos. Sci.*, **23** (1966), 481-494.

24. Tatro, P. R. & Mollo-Christensen, E. L.: *J. Fluid Mech.*, **28** (1967), 531-544.

25. Faller, A. J.: Instability and transition of disturbed flow over a rotating disk, *J. Fluid Mech.*, **230** (1991), 245-269.

Chapter III.2

BAROTROPIC AND BAROCLINIC INSTABILITIES

P.F. Linden
University of Cambridge, Cambridge, UK

1. INTRODUCTION

The stability of a given flow field is a central question in the consideration of the realisability and persistence of a flow. This question is most simply addressed by considering the linear stability of the flow. A small perturbation is made to the given flow field and the tendency for this perturbation to grow or decay is determined. Since the perturbation is small the equations of motion are linearised about this basic state, and these linearised equations often yield mode-like solutions with time-dependence of the form $e^{i\sigma t}$, where t is time and σ the frequency of the mode. If $Im(\sigma) < 0$, the mode grows exponentially and the flow is said to be linearly unstable.

As the mode grows, finite amplitude effects become increasingly important and eventually may dominate the initial linearised behaviour. These effects may lead either to a equilibration of the instability with a stable finite amplitude motion, or the flow may continue to be unstable and to break up, possibly into a turbulent flow.

Unstable flows may evolve radically, changing both their structure and the dynamical balances from those that existed in the original flow. For example, our weather is dominated by high and low pressure systems which are manifestations of instabilities of the flow driven by the pole-equator temperature variation in the atmosphere. An understanding of the conditions for, and the nature of flow instabilities is, therefore, crucial to the study of flow systems. Rotating flows have their own special dynamics which make these considerations especially interesting.

The study of flow instabilities has a long and distinguished history and was begun in the nineteenth century by Helmholtz, Kelvin, Rayleigh and Reynolds, and continues to be an active area of research today. Many notable papers and books have been written on the subject, and a recent lucid discussion is given by Drazin & Reid [1]. Here I shall concentrate on some of the instabilities that arise in rotating flows, with the intention of exemplifying the basic physical processes relevant to geophysical flows.

In many geophysical and industrial applications viscous forces are small (except perhaps in some localised regions of the flow) and instability results when the flow may be altered so that the energy in the basic flow is reduced by the motion produced by a perturbation. There are two energy sources in a fluid, the kinetic energy of the motion and the potential energy associated with an external force field (eg gravity). In a rotating fluid, instabilities which gain their energy from the kinetic energy of the motion are called *barotropic* and those which derive their energy from the gravitational potential energy are called *baroclinic*. In order to elucidate these two basic mechanisms, I shall first discuss flow instability under the approximations of quasi-geostrophic (QG) theory.

2. LINEAR INSTABILITY OF QG FLOWS

Consider a Boussinesq fluid with buoyancy frequency $N(z)$ rotating with angular velocity $\frac{1}{2}f$ about the vertical z-axis. The equations of motion for an inviscid fluid are

$$u_t + \mathbf{u}.\nabla u - fv = -p_x \quad , \tag{1}$$

$$\nu_t + \mathbf{u}.\nabla v + fu = -p_y \quad , \tag{2}$$

$$w_t + \mathbf{u}.\nabla w - \sigma = -p_z \quad , \tag{3}$$

$$\sigma_t + \mathbf{u}.\nabla\sigma + N^2 w = 0 \quad , \tag{4}$$

$$u_x + v_y + w_z = 0 \; . \tag{5}$$

Referred to Cartesian coordinates (x, y, z) the velocity is (u, v, w), p is the pressure divided by the density and $\sigma \equiv -g\rho'/\rho$ is the perturbation buoyancy. For slow (timescale T, with $1/fT \ll 1$), small amplitude motions (with Rossby number $U/fL \ll 1$) the horizontal velocities are in geostrophic balance

$$u = -\frac{1}{f}p_y \; , \qquad v = \frac{1}{f}p_x \; . \tag{6}$$

If the horizontal scales of the flow are long and the motion is hydrostatic, (3) becomes

$$p_z = \sigma \quad , \tag{7}$$

and from (6) we obtain the 'thermal wind' balance

$$u_z = -\frac{1}{f}\sigma_y \quad v_z = \frac{1}{f}\sigma_x \ , \tag{8}$$

which expresses the relation between the *vertical* shear and the *horizontal* gradients of density

On a $\beta-$ plane where $f = f_o + \beta y$, (6) and (7) allow us to define a stream function $\psi(x, y, z)$ for the geostrophic motion by $\psi = p/f_o$ with

$$u = -\psi_y \ , \ v = \psi_x \ , \ \sigma = f_o\psi_z \ . \tag{9}$$

Then $\frac{\partial(2)}{\partial x} - \frac{\partial(1)}{\partial y}$ gives

$$\frac{\partial\zeta}{\partial t} + \mathbf{u}.\nabla\zeta + \beta v = -f_o w_z \ , \tag{10}$$

where $\zeta = v_x - u_y$ is the vertical component of relative vorticity. Use of (4) and (9) shows that, to leading order

$$\left(\frac{\partial}{\partial t} - \psi_y\frac{\partial}{\partial y} + \psi_x\frac{\partial}{\partial y}\right) Q = 0 \quad , \tag{11}$$

where

$$Q \equiv \left[\psi_{xx} + \psi_{yy} + \left(\frac{f_o^2}{N^2}\psi_z\right)_z + f_o + \beta y\right] \ , \tag{12}$$

is the quasigeostrophic potential vorticity, and (11) expresses its conservation following the geostrophic motion (the QGPVE : quasigeostrophic potential vorticity equation).

3. INSTABILITY OF ZONAL FLOWS

The simplest context in which to discuss the stability of these flows is to consider a time-independent flow $(U(y,z), 0, 0)$ in the x-direction. According to (8) there will be a slope of isopycnal surfaces in the y-direction, which in geophysical terms corresponds to a north-south gradient of density at a given height. The streamfunction for this basic flow is $\overline{\psi} = -Uy$, and we perturb this by a disturbance streamfunction $\psi'(x, y, z, t)$ so that

$$\psi = \overline{\psi} + \psi' \ . \tag{13}$$

Substitute (13) into (11) and, on dropping primes, we get after linearization

$$\left(\frac{\partial}{\partial t} + U\frac{\partial}{\partial x}\right)\left(\psi_{xx} + \psi_{yy} + \left(\frac{f_o^2}{N^2}\psi_z\right)_z\right) + \overline{Q}_y\psi_x = 0 \ , \tag{14}$$

where

$$\overline{Q} = \beta y + \overline{\psi}_{yy} + \left(\frac{f_o^2}{N^2}\overline{\psi}_z\right)_z \tag{15}$$

is the potential vorticity (PV) of the basic flow.

We further restrict our considerations to flow in a channel $0 \leq y \leq L$ of uniform depth $0 \leq z \leq -H$. The no-flux boundary condition on the four walls of the channel are

$$\psi = 0 \text{ on } y = 0, L \ , \tag{16}$$

$$\left(\frac{\partial}{\partial t} + U\frac{\partial}{\partial x}\right)\psi_z - U_z\psi_x = 0 \text{ on } z = 0, -H \ , \tag{17}$$

with the latter obtained by putting $w = 0$ in (3) and (4), eliminating σ and linearizing about the basic flow.

The boundary value problem (14)-(17) admits solutions of the form

$$\psi = Re\left\{\phi(y,z)e^{ik(x-ct)}\right\} \ , \tag{18}$$

and substitution gives

$$(U-c)\left[\frac{\partial^2}{\partial y^2} - k^2 + \frac{\partial}{\partial z}\left(\frac{f_o^2}{N^2}\frac{\partial}{\partial z}\right)\right]\phi + \overline{Q}_y\phi = 0 \ , \tag{19}$$

$$\phi = 0 \text{ on } y = 0, L \ ,$$

$$(U-c)\phi_z - v_z\phi = 0 \ , \text{ on } z = 0, -H \ . \tag{20}$$

Since we are looking for unstable modes we write $c = c_r + ic_i$. Instability occurs when $c_i > 0$ and since for these modes $U - c \neq 0$, we can divide by $U - c$, multiply (19) by ϕ^* (the conjugate of ϕ), integrate over the flow domain apply the boundary conditions (20). This procedure gives

$$\int_{-H}^{0}\int_{0}^{L}\frac{|\phi|^2\overline{Q}_y}{(U-c)}dydz + \int_0^L\left[\frac{f_o^2}{N^2}\frac{U_z|\phi|^2}{(U-c)}\right]_H^0 dy \ ,$$

$$= -\int_0^L\int_{-H}^0\left\{|\phi_y|^2 + \frac{f_o^2}{N^2}|\phi_z|^2k^2|\phi|^2\right\}dydz \ . \tag{21}$$

The right hand side of (21) is real and so equating imaginary parts we have

$$c_i\left\{\int_0^L\int_{-H}^0\frac{|\phi|^2\overline{Q}_y}{|U-c|^2}dydz + f_o^2\int_0^L\left[\frac{|\phi|^2U_z}{N^2|U-c|}\right]_{-H}^0 dy\right\} = 0 \ . \tag{22}$$

Since $c_i \neq 0$, the necessary condition for instability is that the term inside the brackets in (22) vanishes.

4. BAROTROPIC INSTABILITY

In an unstratified fluid ($N = 0$), in which the only source of energy is the kinetic energy of the basic flow, then (8) shows that all quantities are independent of height and (22) reduces to

$$\int_0^L \frac{|\phi|^2 \overline{Q}_y}{|U-c|^2} dy = 0 \ , \tag{23}$$

with

$$\overline{Q}_y = \beta - v_{yy} \ . \tag{24}$$

Equation (23) shows that a necessary condition for instability is that *the gradient of PV changes sign at least once in the flow.* When $\beta = 0$ (i.e. on a f-plane) rotation has no effect and (23) reduces to Rayleigh's criterion. We also see immediately from (24) that if β is large enough, $\overline{Q}_y$ may remain positive and so the variation of Coriolis parameter with latitude is a stabilizing influence as was recognised by Kuo [2].

The real part of (21), again with $N = 0$, implies that

$$\int_{-H}^0 \int_0^L \frac{|\phi|^2 \overline{Q}_y (U - c_r)}{|U-c|^2} dydz = -\int_0^L \int_{-H}^0 \{|\phi_y|^2 + k^2|\phi|^2\}\, dydz \ . \tag{25}$$

Let U^* be a point in the flow domain at which $U = U^* = c_r$ the phase speed of the disturbance. Then (25) implies that

$$\int_{-H}^0 \int_0^L \frac{|\phi|^2 \overline{Q}_y (U - U^*)}{|U-c|^2} dydz < 0 \ , \tag{26}$$

and so we have the further condition that

$$(U - U^*)\overline{Q}_y < 0 \ , \tag{27}$$

everywhere in the flow. Condition (27) is the equivalent of Fjortoft's theorem for a rotating flow.

We can interpret these two conditions (23) and (27) in the following way. For barotropic instability to occur we require:

(i) two y-intervals with opposing gradients of PV. This allows regions in which oppositely directed Rossby waves can propagate;

(ii) a frame of reference moving with speed U^* in which $U(y)$ opposes Rossby wave phase propagation in both regions.

5. BAROCLINIC INSTABILITY

Consider now the case where N is constant, $\beta = 0$ and the mean flow is a function of height only i.e. $U = U(z)$. Then from (15) we see that $\overline{Q}_y = 0$, and the necessary condition for instability (22) becomes

$$\int_0^L \left[\frac{|\phi|^2 U_z}{N^2|U-c|}\right]_{z=-H}^{0} dy = 0 \quad . \tag{28}$$

In this simplified case when the interior PV of the flow is uniform, the condition for instability is given by the properties on the top and bottom boundaries. This case is Eady's model [3].

A more revealing form of the solution is obtained by solving (19) directly in the case $\overline{Q}_y = 0$. The solution for ϕ is

$$\phi = \sin\frac{n\pi y}{L}\left\{A\cosh\left[\kappa\left(z/H+\frac{1}{2}\right)\right] + B\sinh\left[\kappa\left(z/H+\frac{1}{2}\right)\right]\right\} \quad , \tag{29}$$

where

$$\kappa = \frac{NH}{f}\left(\kappa^2 + \frac{n^2\pi^2}{l^2}\right)^{1/2} \quad , \quad n = 0, 1, 2\ldots \quad . \tag{30}$$

Application of the boundary conditions (20) on $z = 0$ and $z = -H$ gives the dispersion relation

$$c = \frac{U_0}{2} \pm \frac{U_0}{\kappa}\left[\left(\frac{\kappa}{2} - \tanh\frac{\kappa}{2}\right)\left(\frac{\kappa}{2} - \cosh\frac{\kappa}{2}\right)\right]^{1/2} \quad , \tag{31}$$

where $U_0 = U(0)$.

Instability occurs from large κ and sets in when $\kappa = 2\coth\kappa/2$ (i.e. $\kappa \simeq 2.4$). For wide channels (30) shows that instability occurs when the wavelength $\lambda = 2\pi/k$ satisfies

$$\lambda = 2.6\frac{NH}{f} = 2.6\Re \quad , \tag{32}$$

where $\Re$ is the internal Rossby deformation radius. Thus only motions larger than the scale of the deformation radius are unstable, and the fastest growing mode has a scale of about $4\Re$. The physical reason for this may be seen in figure 1 With a flow $U(z)$, there is a y-slope of the isopycnal surfaces given by the thermal wind balance (8), $v_z = -\frac{1}{f}\sigma_y$. These sloping density surfaces imply a source of potential energy which may be released if fluid motion occurs so that dense fluid is lowered and light fluid is raised. Figure 1 shows that for energy release to occur, the motion must occur within the 'wedge of instability' formed by the angle the isopycnals make with the horizontal. This range of trajectories requires that for a displacement $(0, \zeta, \eta)$

$$\frac{\eta}{\zeta} < -\frac{\sigma_y}{\sigma_z} \quad .$$

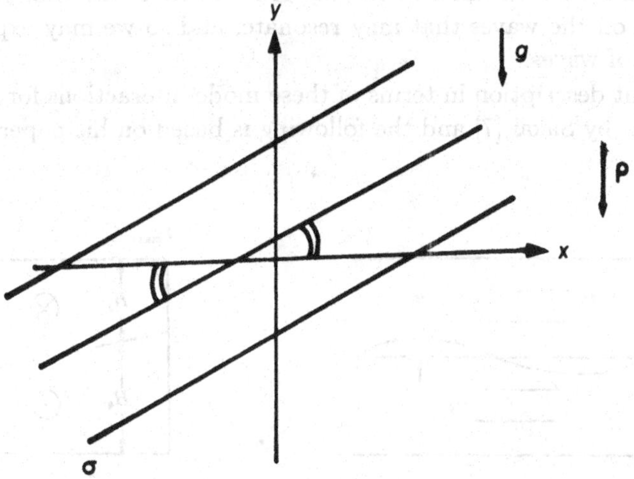

Fig. 1 Sloping isopycnals associated with the flow $U8z$). Parcel exchanges in the 'wedge of instability' marked by)) release potential energy from the stratification.

In most situations of geophysical interest both barotropic and baroclinic energy sources contribute to the instability of the basic flow. Killworth [4] presented a general study of QG linear instability of a parallel shear flow and found that the most important parameter governing the flow is the rotational Froude number F which is the square of the ratio of the horizontal lengthscale of the shear to the internal deformation radius. Two other parameters were also examined: viz, the vertical scale of the stratification compared with the scale of the shear, and the magnitude of the β-effect. In general, these latter two parameters are of secondary importance compared with the Froude number. For a given vertical shear, motion within the wedge of instability occurs for all scales greater than a given value and this range of scales is determined by F.

6. A GENERALISED APPROACH

In many circumstances of geophysical interest the restriction to quasi-geostrophic flow is too severe. Recently rotating flow instability mechanisms have been discussed in terms of resonant disturbance waves in a manner similar to that developed in the plasma physics literature.

The Rayleigh-Fjortoft theorem for QG flows has been generalised by Ripa [5] to the shallow water equations, and interpreted by Hayashi & Young [6]. Instability occurs when a wave having positive disturbance momentum resonates with a wave with negative disturbance momentum. The resulting unstable mode has zero disturbance momentum and so may grow without external forcing. There is no restriction on the waves that may resonate, and so we may expect to find new combinations of waves.

An elegant description in terms of these mode interactions for a two-layer flow has been given by Sakai [7] and the following is based on his paper.

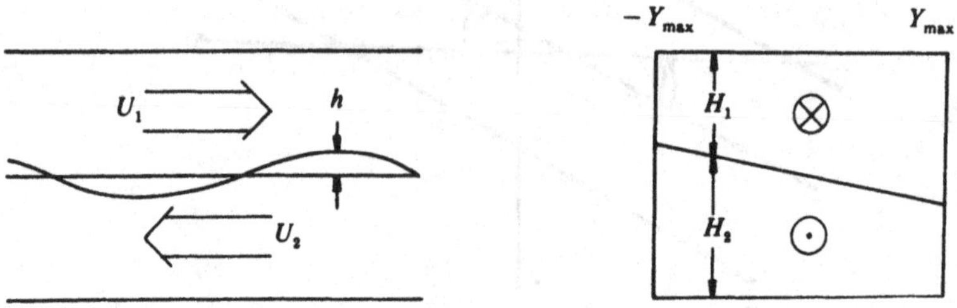

Fig. 2 Schematic diagrams of the model.

Consider a two-layer flow in a channel on an f-plane with uniform but different horizontal velocities U_1, U_2 as shown in figure 2. The linear perturbation equations for each layer ($i = 1, 2$) are

$$D_i u_i - f v_i = -p_{ix} \quad , \tag{33}$$
$$D_i u_i + f u_i = -p_{iy} \quad , \tag{34}$$
$$D_i h = \pm(H_i u_{ix} + (H_i v_i)_y) \quad , \tag{35}$$
$$g' h = p_i - p_2 \quad , \tag{36}$$

where $D_i = \partial_t + U_i \partial_x$ and $H_i = H_0 \mp sy$ is the depth of each layer. H_0 is the average depth and the slope s of the interface is $f(U_1 - U_2)/g'$.

From (33) - (36) we obtain PV conservation

$$D_i q_i + v_i Q_{iy} = 0 \quad , \tag{37}$$

where $Q_i = f/H_i$ is the background PV and

$$q_i = (v_{ix} - u_{iy} \pm Q_i h)/H \quad , \tag{38}$$

is the disturbance PV. If η_i is the particle displacement in the y-direction

$$D_i \eta_i = v_i \quad , \tag{39}$$

then (37) may be integrated to give

$$q_i = -\eta_i Q_{iy} \quad . \tag{40}$$

The disturbance momentum M may be defined as follows

$$M = M_R + M_g \quad , \tag{41}$$

where

$$\begin{aligned} M_R &= -\tfrac{1}{2} < H_1^2 \eta_1^2 Q_{1y} > -\tfrac{1}{2} < H_2^2 \eta_2^2 q_{2y} > \quad , \\ M_g &= < hu_1 > + < hu_2 > \quad , \end{aligned} \tag{42}$$

where <> denotes a spatial average. Hayashi & Young [6] show that the disturbance energy E is related to M by the result

$$E = cM \quad , \tag{43}$$

where c is the phase speed of the wave observed from a particular frame of reference.

Assuming a solution of the form

$$(u_i\ , v_i\ , p_i)^T = \hat{\mathbf{z}}_i(y) e^{i(kx-\omega t)} \quad , \tag{44}$$

(33) - (36) may be rewritten in the form

$$\hat{D}_1 \hat{\mathbf{z}}_1 - A_1 \hat{\mathbf{z}}_1 = \hat{D}_1 B \hat{\mathbf{z}}_2 \quad , \tag{45}$$

$$\hat{D}_2 \hat{\mathbf{z}}_2 - A_2 \hat{\mathbf{z}}_2 = \hat{D}_2 B \hat{\mathbf{z}}_1 \quad , \tag{46}$$

where

$$A_j = \begin{pmatrix} 0 & if & k \\ -if & 0 & -i\partial y \\ kg'H_j & -ig'\partial_y(H_j.) & 0 \end{pmatrix} \quad B = \begin{pmatrix} 0 & 0 & 0 \\ 0 & 0 & 0 \\ 0 & 0 & 1 \end{pmatrix} \quad , \tag{47}$$

where $\hat{D}_i = (\omega \pm kU_0)$ is an intrinsic frequency and $\partial_y(H_j.)$ operates on vj as $\partial_y(H_j v_j)$. The eigenvalue problem can be written in terms of an orthogonal basis. However, Sakai [7] has shown that it is possible to form two subproblems, one for each layer and to express the solution in terms of wavecoordinates for each layer separately. In this manner the modes within a given layer are orthogonal and only interact with those in the other layer, and the technique provides an elegant way of diagonalising the matrix.

For example, setting $p_2 = 0$ in (33) - (36) or $B = 0$ in (47), for the upper layer (45) becomes

$$\hat{D}_1 \hat{\mathbf{z}}_1 - A_1 \hat{\mathbf{z}}_1 = 0 \quad , \tag{48}$$

where $\hat{D}_1 = \omega - kU_0$. Each eigenvector $\mathbf{e}_{in}$ of (48) corresponds to a topographic Rossby wave or a gravity wave. The adjoint equation is

$$\hat{D}_1^* \hat{\mathbf{z}}_1 - A_1^+ \hat{\mathbf{z}}_1 = 0 \quad ,$$

$$A_1^+ \equiv \begin{pmatrix} 0 & if & kg'H_1 \\ -if & 0 & -ig'H_1\partial_y \\ k & -i\partial_y & 0 \end{pmatrix} \quad . \tag{49}$$

Then if $\mathbf{e}_{in} = (u_i \,, v_{in} \,, p_{in})^T$ is an eigenvector of (48)

$$\mathbf{e}_{in}^+ = (H_1 u_{1n} \,, H_1 v_{1n} \,, \frac{1}{g'} p_{1n})^T \quad , \tag{50}$$

is the corresponding eigenvector of (49) and $\mathbf{e}_{1n}^{*+} . \mathbf{e}_{1m} = d_{1n}\delta_{nm}$. In the basis $\mathbf{e}_{1n}$ the upper layer wave coordinate may be written as

$$\hat{\mathbf{z}}_1 = \sum \frac{1}{d_{1n}} X_n \mathbf{e}_{1n} \quad , \tag{51}$$

and similarly for the lower layer with a basis $\mathbf{e}_{2n}$ and coordinates Y_n.

In this coordinate system, each wave component is independent of the other wave components in the same layer, but interacts with the other layer through the pressure. Then (45) can be written as

$$\hat{D}_1 X_n - \tilde{\omega}_{in} X_n = \hat{D}_1 \sum_m \epsilon_{nm} Y_m \quad , \tag{52}$$

$$\hat{D}_2 Y_m - \tilde{\omega}_{2m} Y_m = \hat{D}_2 \sum_n \epsilon_{nm}^* X_n \quad , \tag{53}$$

$$\epsilon_{nm} = \frac{1}{g' d_2 d_{1n}} \int p_{1n}^* p_{2m} d_y \quad , \tag{54}$$

where $\tilde{\omega}_{1n}$ and $\tilde{\omega}_{2n}$ are the intrinsic frequencies of free modes for the two single-layer subproblems.

For the basic interaction between two components these equations reduce to

$$\hat{D}_1 X_n - \tilde{\omega}_{1n} X_n = \hat{D}_1 \sum_m \epsilon_{nm} Y_m \quad ,$$

$$\hat{D}_2 Y - \tilde{\omega}_2 X = \epsilon^* \hat{D}_2 X \quad .$$

These equations may be written as

$$\left(\omega - \frac{\tilde{\omega}_1 + \tilde{\omega}_2}{2(1 - \epsilon\epsilon^*)}\right)^2 - \mu = 0 \quad , \tag{55}$$

where

$$\mu = \left(kU_0 + \frac{\tilde{\omega}_1 - \tilde{\omega}_2}{2(1-\epsilon\epsilon^*)}\right)^2 + \frac{\epsilon\epsilon^*}{1-\epsilon\epsilon^*}\tilde{\omega}_1\tilde{\omega}_2 \quad . \tag{56}$$

Ripa's theorem gives $\tilde{\omega}_1$,$\tilde{\omega}_2$ real. If $\tilde{\omega}_1\tilde{\omega}_2 > 0$, $\mu > 0$ and (55) shows that ω is real. If $\tilde{\omega}_1\tilde{\omega}_2 < 0$, $\mu < 0$ when

$$kU_0 = -(\tilde{\omega}_1 - \tilde{\omega}_2)/(1-\epsilon\epsilon^*) \ ,$$

and for small ϵ, instability occurs when

$$\hat{\omega}_1 + kU_0 \simeq \omega_2 - kU_0 \quad .$$

Thus the flow is unstable if there is a pair of wave components such that

(i) they propagate in opposite directions relative to the basic flow ($\tilde{\omega}_1\tilde{\omega}_2 < 0$)
(ii) they have almost the same Doppler shifted frequency ($\tilde{\omega}_1 + kU_0 \simeq \tilde{\omega}_2 - kU_0$)
(iii) they can interact with each other ($\epsilon \neq 0$)

An example of this situation is shown schematically in figure 3.

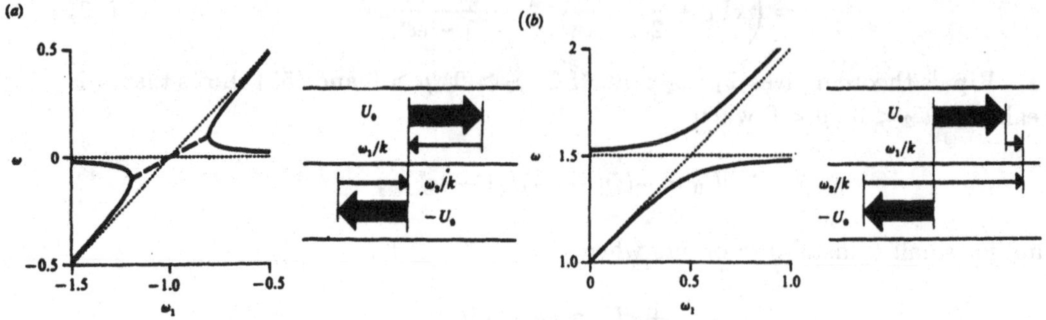

Fig. 3 Frequency of the wave modes when two wave components have almost the same frequencies (solid lines in left-hand panels) for (a) $\tilde{\omega}_1\tilde{\omega}_2 < 0$ and (b) $\tilde{\omega}_1\tilde{\omega}_2 > 0$. Dotted line shows Doppler frequencies $\tilde{\omega}_1 + kU_0$ and $\tilde{\omega}_2 - kU_0$ and the broken line shows the unstable mode. Right panels show schematically the relation between U_0, $\tilde{\omega}_2.kU_0 = 1$, $\omega_2 = 1$ for (a), $\tilde{\omega} = 2.5$ for (b) and only $\tilde{\omega}_1$ is varied.

7. BAROCLINIC INSTABILITY AND ROSSBY WAVES

We return now to consider the geostrophic limit where the variation in depth $\Delta H = sY_{\mathbf{max}}$ of the two layers is small. In this limit A is independent of y, and we seek a solution in the upper layer of the form

$$\hat{\mathbf{z}}_1 = X_n(u' \, , v' \, , p')^T e^{il_n y} \quad . \tag{57}$$

Substitution of (57) into the one-layer problem (48), with the QG approximation, gives an equation for a Rossby wave

$$\hat{D}_1 X_n - \tilde{\omega}_{1n} X_n = 0 \quad , \tag{58}$$

where

$$\tilde{\omega}_{1n} = \frac{-2kU_o}{2\Re^2|\mathbf{k}|^2 + 1} \quad , \quad \mathbf{k} = (k \, , l_n) \quad , \tag{59}$$

and the corresponding eigenvector is

$$\mathbf{e}_{1n} = \left(\frac{il_n}{f} \, , -\frac{ik}{f} \, , 1\right)^T e^{il_n y} \quad . \tag{60}$$

The Rossby wave components in the lower layer are identical, except that

$$\tilde{\omega}_{2n} = -\tilde{\omega}_{1n} \quad . \tag{61}$$

gives an equation for a Rossby wave

$$\hat{D}_1 X_n - \tilde{\omega}_{1n} X_n = 0 \quad , \tag{58}$$

where

$$\tilde{\omega}_{1n} = \frac{-2kU_o}{2\Re^2|\mathbf{k}|^2+1} \quad , \quad \mathbf{k} = (k\ , l_n) \quad , \tag{59}$$

and the corresponding eigenvector is

$$\mathbf{e}_{1n} = \left(\frac{il_n}{f}\ , -\frac{ik}{f}\ , 1\right)^T e^{il_n y} \quad . \tag{60}$$

The Rossby wave components in the lower layer are identical, except that

$$\tilde{\omega}_{2n} = -\tilde{\omega}_{1n} \quad . \tag{61}$$

The interactions between the Rossby waves are expressed by substituting (59) and (61) into (52)

$$\hat{D}_1 X_n - \tilde{\omega}_{1n} X_n = \epsilon_n \hat{D}_1 Y_n \quad , \tag{62}$$

$$\hat{D}_2 Y_n + \tilde{\omega}_{1n} Y_n = \epsilon_n \hat{D}_2 X_n \quad , \tag{63}$$

$$\epsilon_n = \frac{1}{2\Re^2|\mathbf{k}|^2+1} \quad . \tag{64}$$

Modes only interact when they have the same spatial structure and so X_n, Y_n may be eliminated from (62) and (63) to obtain the frequency of the wave mode

$$\omega = \pm kU_0 \left(1 - \frac{2}{\Re^2|\mathbf{k}|^2+1}\right)^{1/2} \quad . \tag{65}$$

Instability occurs for longwaves with $|\mathbf{k}|^2 < \frac{1}{\Re^2}$ in accordance with the standard result (Phillips, [8]; Pedlosky, [9]).

REFERENCES

1. Drazin, P.G. & Reid, W.H. 1991 Hydrodynamic stability *C.U.P.*
2. Kuo H.L. 1949. Dynamic instability of two-dimensional non-divergent flow in a barotropic atmosphere. *J. Meteorology* **6**, 105-122.
3. Eady, E.T. 1949. Long waves and cyclone waves. *Tellus* **1**, 33-52.
4. Killworth, P.D. 1980. Barotropic and baroclinic instability in rotating stratified fluids. *Dynamics of Atmospheres and Oceans* **4**, 143-184.

5. Ripa, P. 1983. General stability conditions for zonal flows in a one-layer model on the beta-plane of the sphere. *J. Fluid Mech.* **126**, 436-489.
6. Hayashi, Y. & Young, W.R. 1987. Stable and unstable shear modes on rotating parallelflows in shallow water. *J. Fluid Mech.* **184**, 477-504.
7. Sakai, S. 1989. Rossby-Kelvin instability: a new type of ageostrophic instability caused by a resonance between Rossby waves and gravity waves. *J. Fluid Mech.* **202**, 149-176.
8. Phillips, N.A. 1954. Energy transformations and meridional circulations associated with simple baroclinic waves in a two-level, quasigeostrophic model. *Tellus* **6**, 273-286.
9. Pedlosky, J., 1979. Geophysical Fluid Dynamics *(Springer-Verlag)*.

Chapter III.3

DYNAMICS OF FRONTS AND FRONTAL INSTABILITY

P.F. Linden
University of Cambridge, Cambridge, UK

1. INTRODUCTION

Fronts and eddies in the ocean are intimately linked. Eddies are produced at fronts often by instability processes, while the deformation fields associated with eddy motions can lead to frontogenesis. I shall discuss this interplay between fronts and eddies by examining the instabilities of fronts and by describing the eddies produced by these instabilities at finite amplitude.

Fronts in the ocean and atmosphere are difficult to define but involve large spatial gradients of temperature and/or salinity. While it is hard to give precise values of the magnitudes of these gradients, it seems pretty clear when a front is present. An excellent summary of recent oceanic observations can be found in Federov [1], and from these and other examples it is clear that fronts are usually produced at the boundary between two different water masses. They can be large, semi-permanent features such as the Gulf Stream front, or the subtropical convergence with typical horizontal scales of tens of kilometers , or they can be more transient features down to the scale of a few hundred meters.

In the atmosphere fronts are mostly associated with high and low pressure systems, and are results of the intensification of temperature gradients by these cyclonic and anticyclonic flows. Mesoscale features such as the sea-breeze and thunderstorm outflows can also produce well-defined fronts. In contrast to atmospheric fronts, in the ocean compensating temperature and salinity fields can lead to fronts with virtually no density signature. Under these circumstances, double-diffusive processes can be expected to be active, leading to the development of intrusive layers. These processes are discussed in Ruddick and Turner [2] and will not be described further here.

Since fronts do, in general, represent boundaries between water or air masses, it is important to determine rates of cross-frontal transfer. These rates often control exchanges of mass, heat and momentum and so determine the mixing rates. And, since spatial gradients are high in fronts, transfers are also correspondingly high. A major mechanism for cross-frontal mixing is the migration of eddies produced at a front. These eddies are nonlinear and advect fluid within closed streamlines and transport it away from the frontal zone (see figure 1).

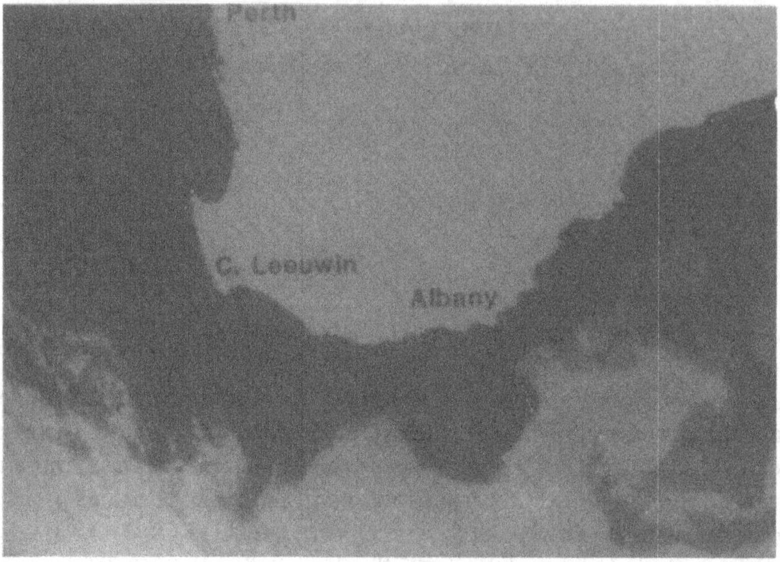

Fig. 1 Geometrically corrected image from 30 September 1982 showing a dipole on the outer edge of the Leeuwin Current. From Griffiths & Pearce [19].

2. FRONTS IN A TWO-LAYER FLUID

A simple way of producing a front is to allow a region of buoyant fluid to adjust under gravity in a rotating system. Initially we consider the buoyant fluid to be constrained against a vertical wall $y = 0$ in a strip $0 \leq y \leq L_0$, $-h_o \leq z \leq 0$ as shown in figure 2. We suppose that the lower layer is infinitely deep, and that its density exceeds that of the buoyant fluid by an amount $\Delta\rho$. Initially the fluid is at rest.

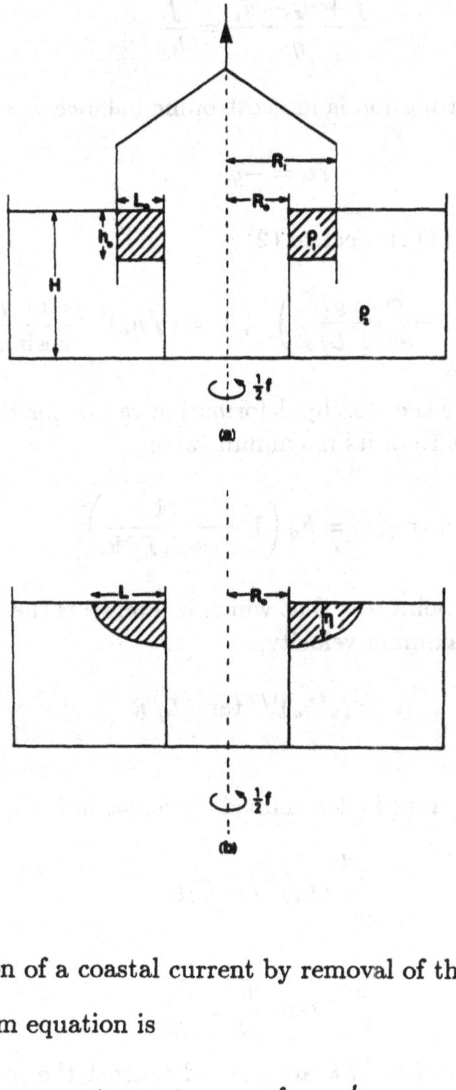

Fig. 2 The production of a coastal current by removal of the outer wall.

The x-momentum equation is

$$u_t + uu_x + vu_y - fv = g'\eta_x \quad , \tag{1}$$

where the horizontal velocity is (u, v), η is the depth of the upper layer, f is the Coriolis parameter and $g' = \frac{g\Delta\rho}{\rho}$ is the reduced gravity. We anticipate that the motion is independent of x and hence, at the wall where $v = 0$, (1) implies that $u_t = 0$. Since $u = 0$ initially, the boundary conditions at the wall are

$$u = v = 0 \quad \text{at } y = 0 \quad . \tag{2}$$

The motion is governed by conservation of potential vorticity

$$\frac{f + v_x - u_y}{\eta} = \frac{f}{h_o} \quad , \tag{3}$$

and the final along-front motion is in geostrophic balance

$$fu = -g'\eta_y \quad . \tag{4}$$

The solution of (3) and (4) subject to (2) is

$$\eta = h_o \left(1 - \frac{\cosh y/\Re}{\cosh L/\Re}\right) \quad , \; u = (g'h_o)^{1/2} \frac{\sinh y/\Re}{\cosh L/\Re} \tag{5}$$

where $\Re = (g'h_o)^{1/2}/f$ is the Rossby deformation radius for the two-layer case.

The interface slopes from its maximum value

$$h = \eta(o) = h_o \left(1 - \frac{1}{\cosh L/\Re}\right) \quad , \tag{6}$$

at the wall $y = 0$ to the point $y = L$ at which it intersects the surface (see figure 2) forming a front. The maximum velocity

$$u = (g'h_o)^{1/2} \tanh L/\Re \quad , \tag{7}$$

occurs at the front.

The width of the current is determined by conservation of cross-sectional area of the upper layer,

$$\int_0^L \eta(y)dy = h_o L_o \quad . \tag{8}$$

Then, from (6), we find

$$\frac{L}{\Re} - \tanh \frac{L}{\Re} = \frac{L_o}{\Re} \quad . \tag{9}$$

The values of the final width $L/\Re$ are plotted against the initial width $L_o/\Re$ on figure 3. When the initial width of the buoyant layer is small compared with the deformation radius $(L_o/\Re \ll 1)$, $L \sim (2\Re L_o)^{1/2}$, while for wide layers $(L_o/\Re \gg 1)$ the final current width is closely approximated by the straight line

$$L = L_o + \Re \quad , \tag{10}$$

showing that the width of the layer increases by about one deformation radius in this limit. In this case (5) shows that the motion is also confined to the region of width $0(\Re)$ near the front, and that all the interface slope is also restricted to this region.

The implications of these results in terms of the instabilities of fronts and, in particular, the contribution of barotropic and baroclinic energy sources are described later.

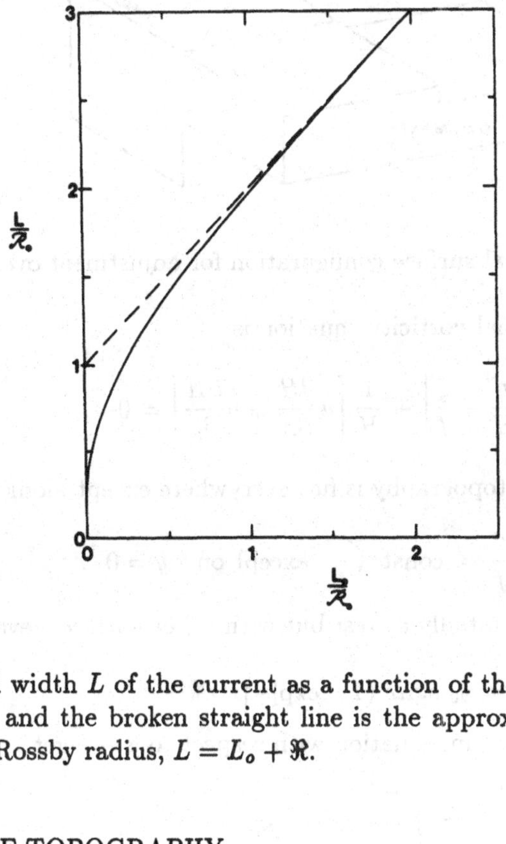

Fig. 3 The final width L of the current as a function of the initial width L_o. The solid line is (9) and the broken straight line is the approximation that the front spreads by one Rossby radius, $L = L_o + \Re$.

3. EFFECTS OF TOPOGRAPHY

The effect of depth variations in the downstream direction are discussed by Gill *et al* [3] and Allen [4]. The main effects are revealed by considering Rossby adjustment of one-layer fluid over an abrupt change in depth. The basic configuration is shown in figure 4, where the initial step in the free-surface elevation is perpendicular to the topographic change. As in the case above, the variation in the free-surface elevation is restricted to within a Rossby radius of the initial discontinuity. The important point to note in the present context is that the mass flux associated with the geostrophically adjusted flow away from the vicinity of the step is proportional to the depth of fluid (see (13) below). Consequently there is an imbalance in the mass fluxes on the two sides of the step which is accommodated by flow along the step.

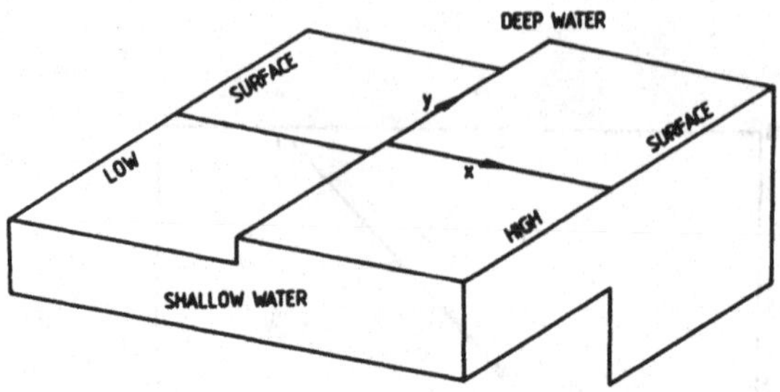

Fig. 4 Geometry and initial surface configuration for adjustment over a step.

The linearized potential vorticity equation is

$$\frac{\partial}{\partial t}\left[\frac{\eta}{H}-\frac{\zeta}{f}\right]+\frac{1}{H}\left[u\frac{\partial H}{\partial x}+v\frac{\partial H}{\partial y}\right]=0 \quad . \tag{11}$$

With a step at $y = 0$, the topography is flat everywhere except along this line, and so

$$\frac{\eta}{H}-\frac{\zeta}{f}=\text{const} \quad , \qquad \text{except on} \quad y=0 \ .$$

We treat the case of fluid initially at rest but with a free-surface elevation η_0 sgn$x$. The solution is then

$$\eta=\eta_0 \ \mathrm{sgn}x \ (1-\exp[-|x|/\Re]) \quad .$$

Integrating the momentum equation with respect to x, we get

$$f\int_{-\infty}^{\infty} v dx = 2g\eta_0 \quad , \tag{12}$$

and the volume flux is given by

$$H\int_{-\infty}^{\infty} v dx = 2gH\eta_0/f \quad . \tag{13}$$

Since H is the different on the two sides of the step, the flux is different as well. Clearly something must happen at the step.

The only available disturbance seems to be a double Kelvin wave which propagates along the topographic step with the shallow water on its right for $f > 0$

(Longuet-Higgins, [5]). Thus the double Kelvin wave can only affect the region $x > 0$ and it is this property that gives the problem its basic asymmetry.

The double Kelvin wave produces a surface elevation of the form

$$\eta = \eta_0 \operatorname{sgn} x - A(x,t) \exp[-|y|/\Re_\pm] \quad ,$$

where $\Re_\pm = (gH_\pm)^{1/2} f^{-1}$ is the Rossby radius associated with the appropriate depth on each side of the step. The along-step velocity is given by

$$u = \begin{cases} \frac{gA}{f\Re_-} \exp[y/\Re_-] \ , & y < 0, \\ \frac{-gA}{f\Re_+} \exp[-y/\Re_+] \ , & y > 0. \end{cases}$$

The flow is in opposite directions on the two sides of the step. The magnitude of A is determined by matching the fluxes and we find $A = 2\eta_0$. Thus all the flow is deflected at the step and it acts as a complete barrier to the approaching jet. A sketch of surface contours (streamlines) is shown in figure 5.

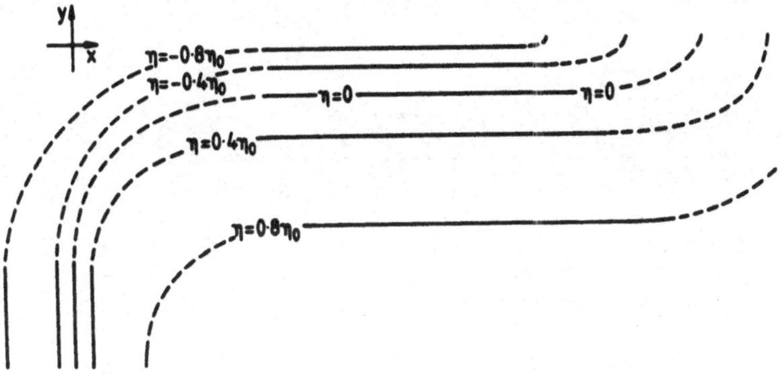

Fig. 5 A sketch of contours of surface displacement in the region $y < 0$ after the release of the initial configuration in figure 4. A double Kelvin wave is travelling in the $+x$ direction. Details of the flow in the vicinity of the dashed contours have not been calculated.

The effect of a coast on the flow depends on its position with respect to the direction of Kelvin wave propagation. In the northern hemisphere ($f > 0$) with the current flowing with the coast on its right looking downstream and approaching shallow water, the Kelvin wave propagates away from the coast producing a tongue of water offshore (figure 6). An example of this flow is shown in figure 6b. If the current is approaching deep water, the Kelvin wave propagates towards the shore and crosses the step in a thin boundary-layer region close to the coast, as shown in figure 6c.

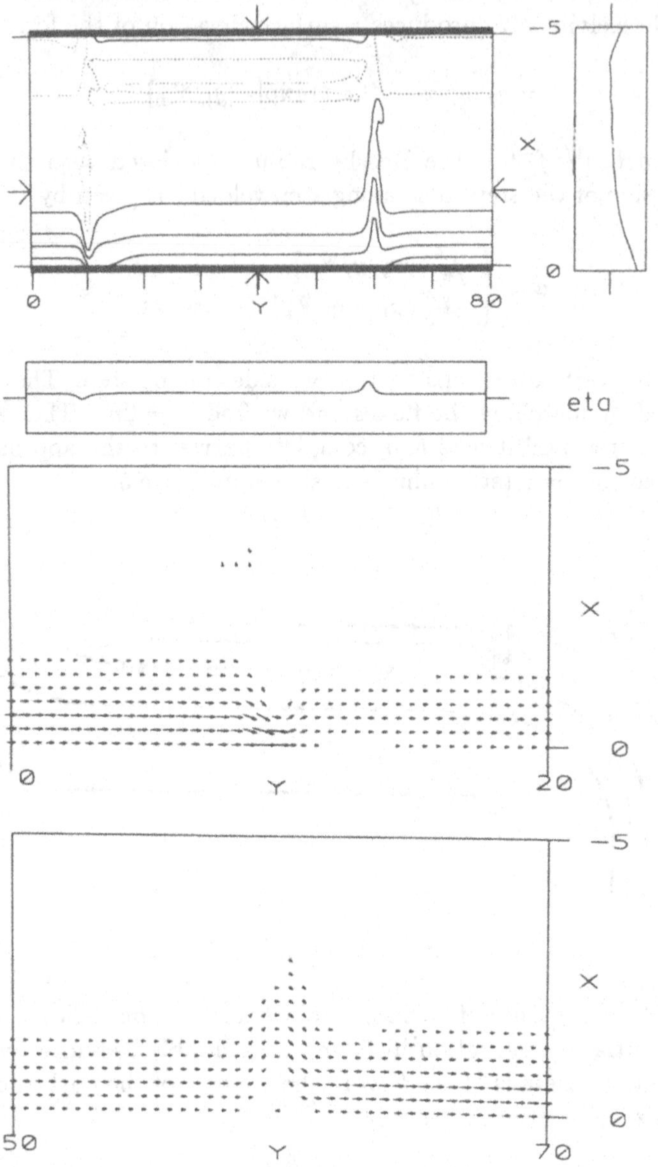

Fig. 6 Numerical calculations of flow approaching a step: a) surface contours for a step down at $y = 10$ and a step up at $y = 60$, b) flow in the vicinity of the step up, c) flow in the vicinity of the step down.

These results have been extended by Allen [4], who has examined the effects of sloping topography between two regions of uniform depth. Linear theory implies that the flow is stagnant over the slope and the motion perpendicular to the coast is concentrated into regions whose width scales on the Rossby radius, one at the top and one at the bottom of the slope. Nonlinear effects produce an asymmetrical flow with greatest velocities near the top of the slope. These effects are consistent with those observed in laboratory experiments, as shown in figure 7.

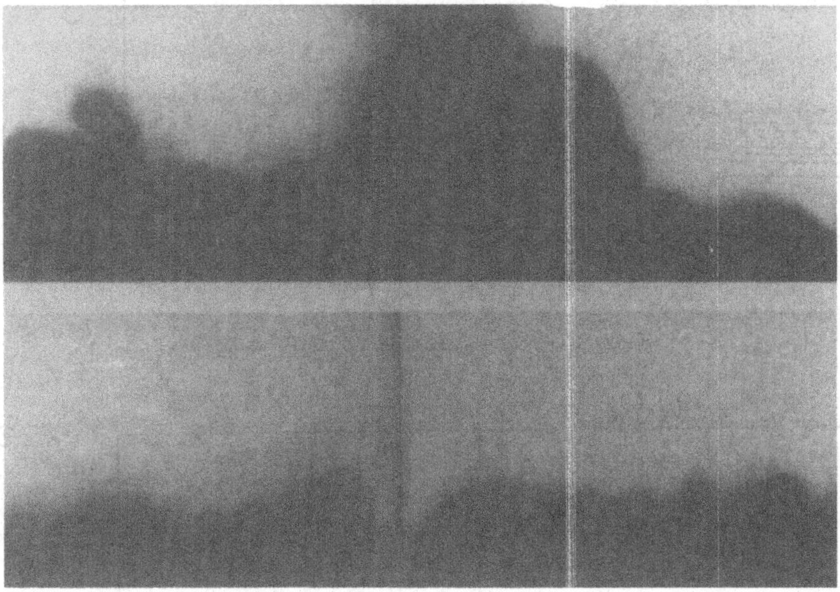

Fig. 7 The adjustment of a coastal current bounded by a front: a) the current of dyed fluid is propagating from left to right as it approaches the shallow end and is diverted across the tank over the step; b) the current approaches the deep end and is diverted towards the coast.

As an approximation to a baroclinic flow, Allen [4] has investigated the flow of a two-layer fluid over sloping topography. The response can be considered as a combination of the baroclinic and barotropic modes. The barotropic mode responds in much the same way as the one-layer flow with both layers travelling in the same direction and with the Kelvin wave having a unique propagation velocity determined by the topography. The baroclinic mode, on the other hand, has Kelvin waves propagating in both directions. The flow in the two layers is in opposite directions and an example is shown in figure 8.

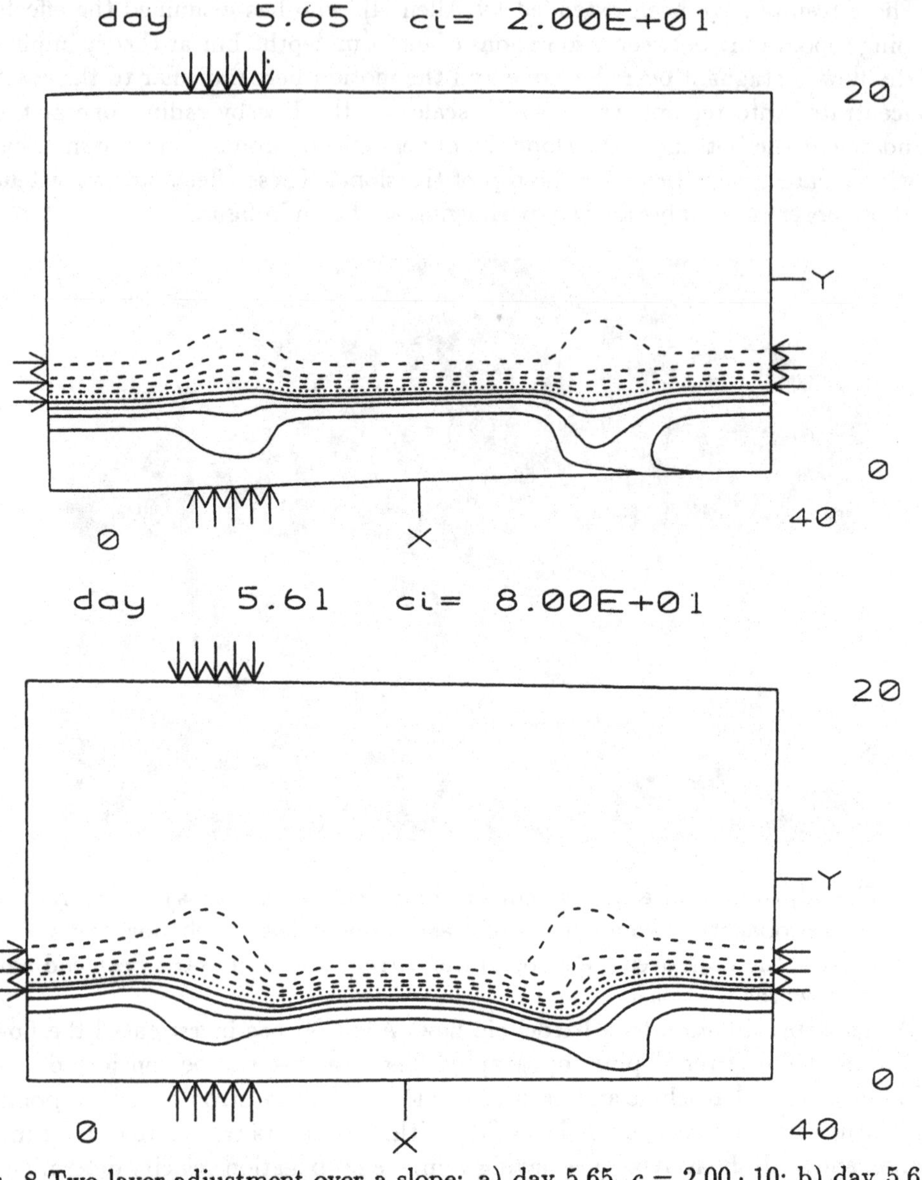

Fig. 8 Two-layer adjustment over a slope: a) day 5.65, $c = 2.00 \cdot 10$; b) day 5.61, $c_i = 8.00 \cdot 10$. The slopes are $10\Re$ wide and located at $x = 10$ (up) and $x = 30$ (down). The nonlinear run b) shows increased asymmetry of the tongue compared with the linear calculation a). The propagation of the double Kelvin waves in both directions is responsible for the broadening of the current over the slope. From Allen [4].

4. FRONTOGENESIS

The process whereby fronts are formed may be described by considering any scalar quantity S which is conserved during the motion. Then, restricting attention to one-dimension where $S = S(x,t)$

$$\frac{DS}{Dt} = S_t + uS_x = 0 \quad .$$

Differentiating this equation with respect to x we obtain the equation

$$\frac{DS_x}{Dt} = -u_x S_x \quad , \tag{14}$$

for the variation in the horizontal gradient of S. From (14) we see that increases in S_x are produced by regions of convergence in the flow ($u_x < 0$).

Hoskins & Bretherton [6] wrote a seminal paper showing how a convergent deformation field leads to infinite spatial gradients in a finite time. However, this and subsequent papers studied the frontogenesis which occurs when the deformation field is imposed. In a more recent study Simpson & Linden [7] discuss how a gravity driven flow will advect the density field to produce a front.

Consider the flow generated from rest in an unbounded fluid containing a uniform horizontal density gradient $\rho = \overline{\rho}(1-\alpha x)$. The governing equations for inviscid, Boussinesq flow are (with $w = 0$)

$$\overline{\rho} u_t = -p_x, \; g\rho = -p_z, \; \rho_t = -u\rho_x, \; u_x = 0 \quad . \tag{15}$$

where the velocity $\mathbf{u} = (u, 0, w)$ and p is the pressure.

Elimination of the pressure using the hydrostatic relations and differentiation of the resulting equation with respect to x, shows that for all times

$$\rho_{xx} = 0 \quad . \tag{16}$$

Therefore, if the initial horizontal density gradient is constant, then (16) shows that it remains constant throughout the motion and frontogenesis cannot occur. We conclude, therefore, that frontogenesis is associated with the presence of nonuniformities in the horizontal density gradient, and the generation of vertical velocities and a transverse circulation. The time evolution of the density gradient is determined by the sign of $(u\rho_x)_x$. Frontogenesis is produced by curvature in the density profile, and will only occur when the horizontal motion in the region of stronger density gradient is towards the region of weaker density gradient. Recently, frontogenesis in a quadratic density field has been studied by Jacquin [8] and Kay [9] and they observe very similar features and the production of a front in the density field.

The full solution to the problem (15) is

$$u = -gxzt \quad ,$$

$$\rho = \overline{\rho}(1 - \alpha x) - \frac{1}{2} g \overline{\rho} x^2 z t^2 \quad . \tag{17}$$

A uniform shear is established which accelerates at a constant rate. The isopycnals remain straight and rotate towards the horizontal. The gradient Richardson number Ri $= g\rho_z/\overline{\rho}u_z^2 = \frac{1}{2}$, and is constant throughout the motion. This result suggests that the flow will be linearly stable. (Note also that the solution (17) also satisfies the viscous equation of motion).

The solution for a piecewise constant density gradient is shown in figure 9. The front is formed adjacent to the lower boundary at the isopycnal which originally divides the two gradient regions. Laboratory experiments which confirm these theoretical results are shown in figure 10.

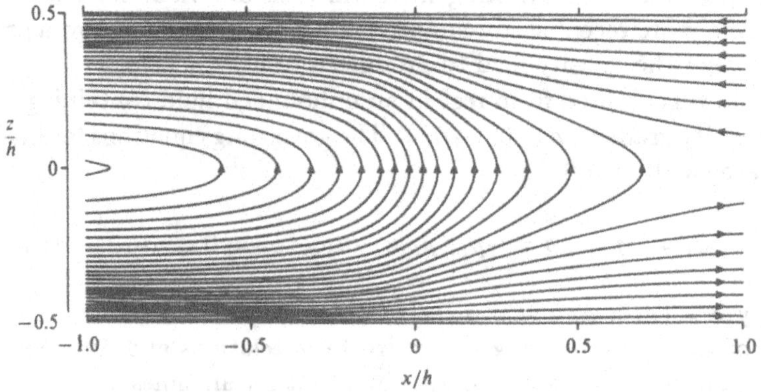

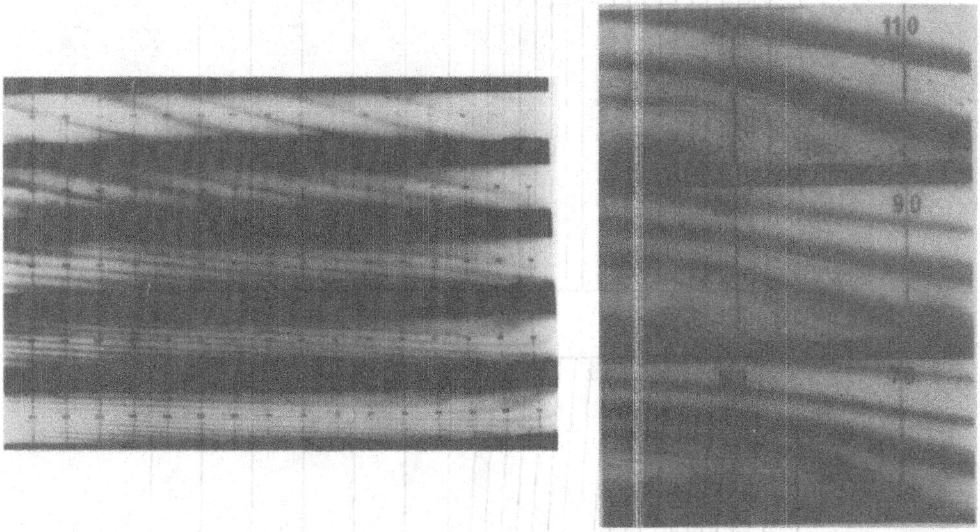

Fig. 9 Laboratory observations of the isopycnals during the adjustment of fluid containing a horizontal density gradient: a) uniform gradient - the isopycnals remain straight ($t = 0$); b) piecewise constant gradient - a front forms on the lower boundary ($t = 1$).

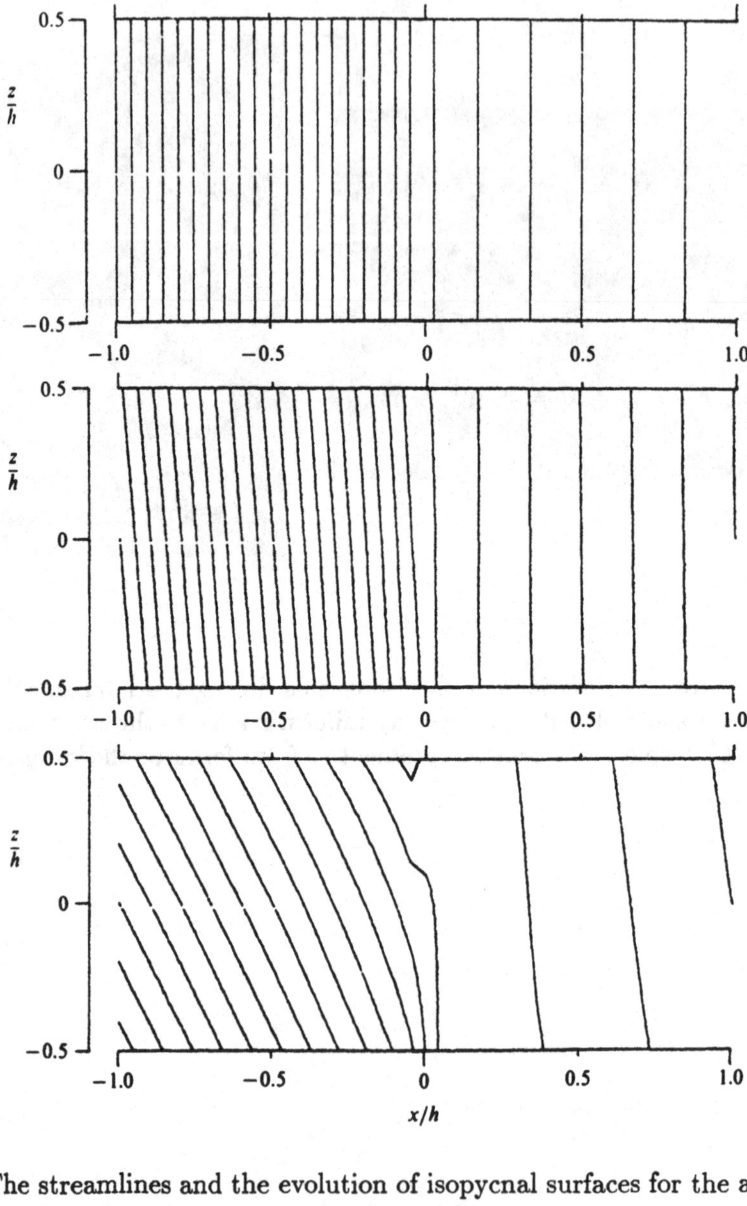

Fig. 10 The streamlines and the evolution of isopycnal surfaces for the adjustment of a fluid with a piecewise constant density gradient.

5. FRONTAL INSTABILITY

Fronts are observed to be almost always unstable. Generally waves grow on fronts at a range of scales up to several times the Rossby deformation radius. As was discussed earlier, QG theory suggests that there is a Rayleigh criterion for these flows which requires that the horizontal gradient of PV change sign if instability is to occur. Instability may release energy from the horizontal shear of the horizontal velocity (barotropic instability) or it may release potential energy (baroclinic instability) or it may result from a combination of the two.

One way in which these two processes may be separated is to consider a current bounded on one side by a front and on the other by a coast as in §2. Then the baroclinicity of the flow is set by the width of the current compared with the deformation radius. When the current is wide ($L_o/\Re \gg 1$) the flow is essentially baroclinic, but narrow currents ($L_o/\Re \ll 1$) are primarily barotropic.

While insight may be obtained using QG theory, near the front where the depth of the upper layer reduces to zero, conservation of PV means that the relative vorticity $\zeta \to -f$, the local Rossby number is 0(1), and it is necessary to consider ageostrophic effects. Killworth and Stern [10] and Killworth *et al* [11] examined the stability of a two-layer flow in which the interface intersects the surface forming a front. They found that the flow is unstable with perturbations receiving energy from the kinetic energy of the mean flow by the action of Reynolds stresses and from the potential energy of the mean flow.

5.1 Laboratory studies

The instability of a front on a coastal current has been studied by Griffiths & Linden [12], [13]. It was observed that waves grew on the front and that, at finite amplitude, the waves break forming eddies with closed streamlines. The shear in the current is anticyclonic and the upper layer eddies that result are anticyclones which form at the crests of the waves. In the troughs of the waves the lower layer deepens and cyclonic vorticity is produced, and these regions close to form cyclonic eddies. Thus the instability leads to the production of anticyclone-cyclone vortex pairs. An example of the growth to large amplitudes is shown in figure 11. The dipoles propagate under the influence of their own vorticity fields, thereby providing a mechanism for cross-front transport.

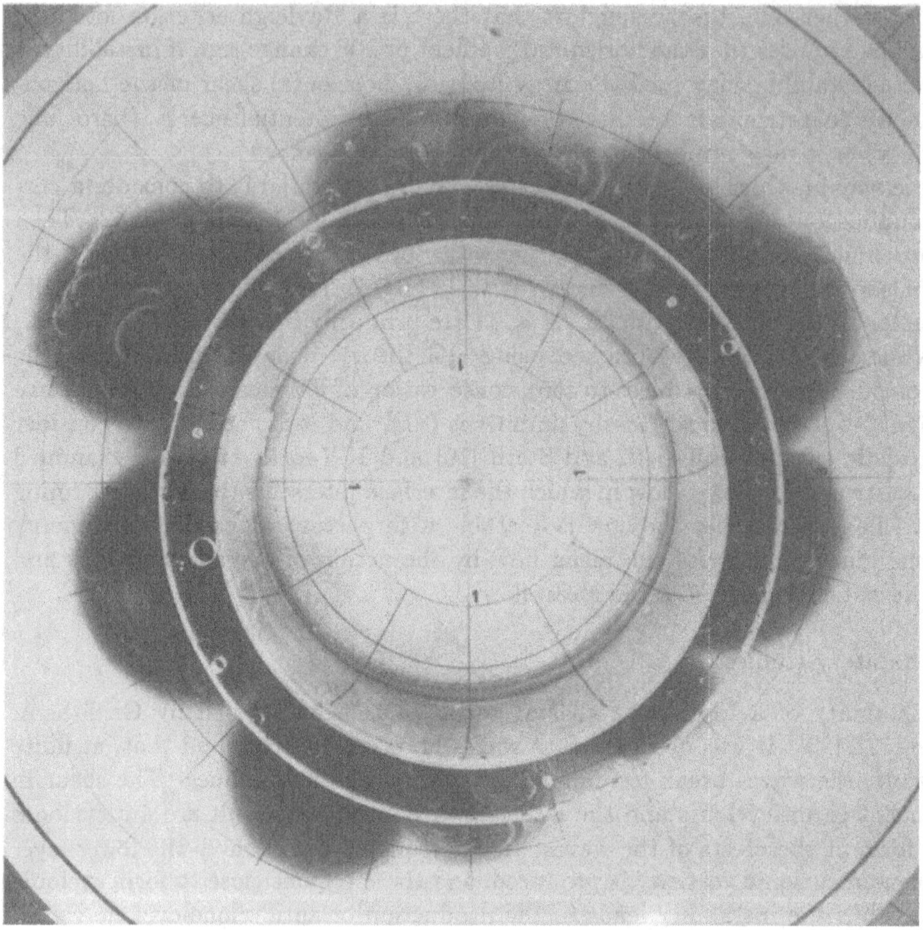

Fig. 11 The production of baroclinic dipoles by the instability of a coastal current.

The waves were observed to grow within a few rotation periods, and their wavelength λ non-dimensionalised by $\Re$, the Rossby deformation radius, is shown in figure 12, plotted against the Froude number $F = (L_o/\Re)^2$. For wide currents, $F > 10$, the wavelength is proportional to the deformation scale with $\lambda/2\pi\Re = 1.1 \pm 0.3$ in agreement with the calculations of Killworth *et al* [8]. For narrow currents, $F < 10$ the wavelength is reduced and scales on the width L_o of the current (see figure 13). The available potential energy increases as the width of the current exceeds the Rossby deformation radius. These results are, therefore, consistent with the release of energy being primarily baroclinic for wide currents, and primarily barotropic for narrow currents.

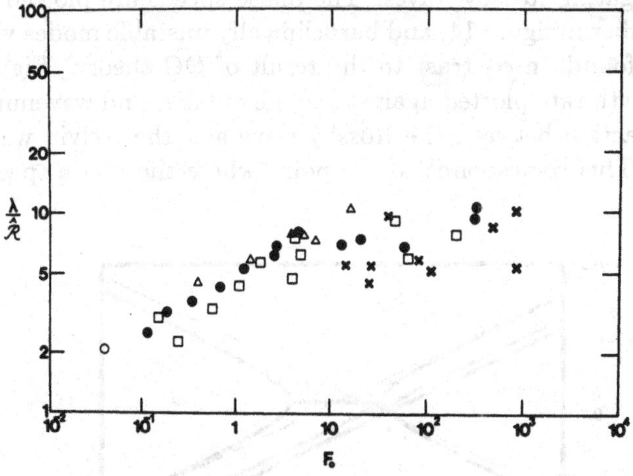

Fig. 12 The large-amplitude length scale λ normalized by the Rossby radius plotted as a function of the Froude number F.

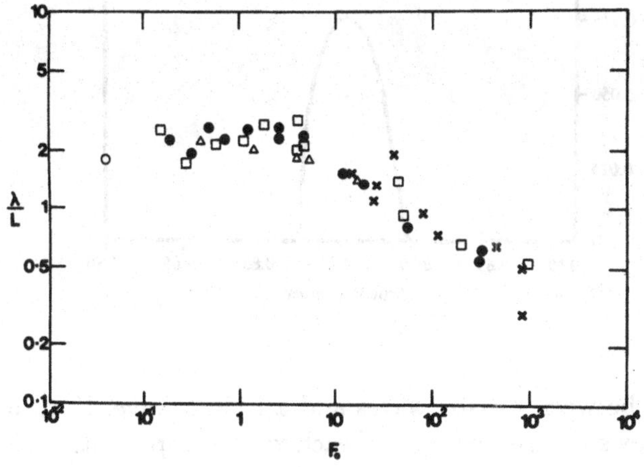

Fig. 13 The large-amplitude λ normalized by the current width L plotted as a function of the Froude number F.

5.2 Theory

The same formulation as described in the previous chapter is applied by Sakai [4] to examine the instability of fronts. Here the interaction is between a Kelvin wave in the upper layer and Rossby waves in the lower layer. A Kelvin wave travelling upstream against the current near the front has a small phase speed and can interact with slowly propagating Rossby waves. The phase speeds are plotted as functions of the Froude number in figure 14, and baroclinically unstable modes with non-zero phase speeds are found, in contrast to the result of QG theory. Figure 15 shows the maximum growth rate plotted against Froude number and wavenumber and we see that the interaction between the Rossby wave and the Kelvin wave occurs at around $F \simeq 0.7$. This corresponds to the point where the phase speeds match in figure 14.

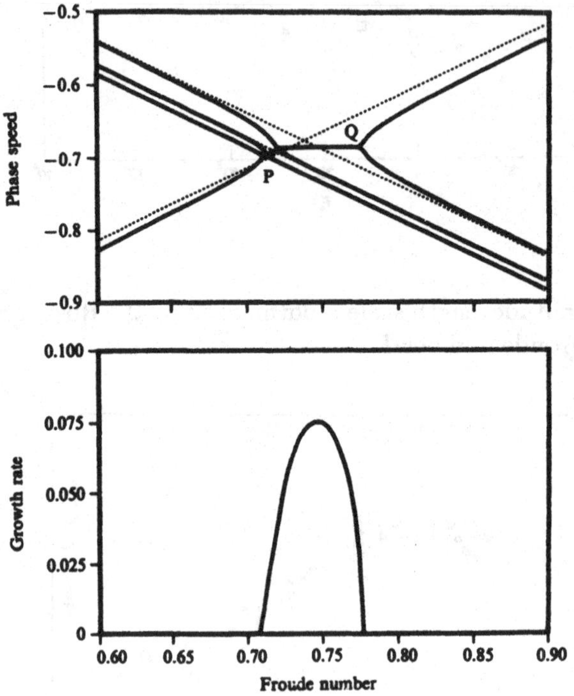

Fig. 14 Resonance between Rossby waves and a Kelvin wave, $Y_{\max} = 1/\sqrt{2}$, $k = 2.5$. Dotted line shows the phase speed of each wave component.

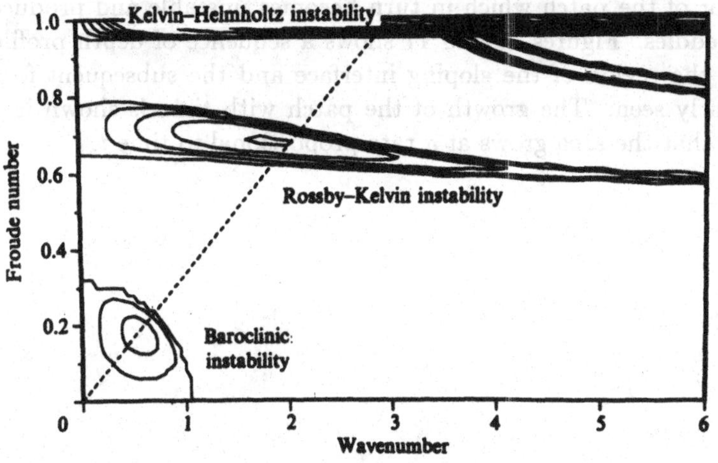

Fig. 15 Maximum growth rate for $\Delta H \equiv SY_{\max} = 0.5$. Contour interval is 0.02.

Sakai's results explain a number of features of the observed frontal instability, in particular that the growing frontal waves are stationary. In addition, Paldor [15] showed that the front was stable if the lower layer was infinitely deep which is consistent with the fact that Rossby waves cannot propagate in this case.

A front may be unstable even though the lower layer is infinitely deep provided the flow has a positive PV gradient towards the front (Killworth & Stern, [10]; Kubokawa, [16]). This instability results from the interaction between a front trapped Kelvin wave and a Rossby wave in the upper layer, and the interaction is horizontal rather than vertical.

6. HORIZONTAL SPREADING OF OCEAN FRONTS

The frontal instabilities described above lead to a breakdown of the geostrophic constraint and cause the buoyant fluid to spread laterally. The rate at which this lateral spreading occurred was studied by Griffiths & Hopfinger [17] in an experiment where a circular pool of fresh water was placed on a saline layer, both layers being initially at rest in a rotating frame of reference. The patch was released and the formation of eddies caused by instabilities on the geostrophically adjusted front and the subsequent increase in radius of the patch was measured.

In accordance with (5) the initial motion is restricted to within one Rossby radius $\Re$ of the front, and the eddies form cyclone-anticyclone pairs with diameters scaling on $\Re$. These pairs propagate away from the front and the patch spreads and reduces in depth. This reduction in depth induces further anticyclonic motion in the interior of the patch which in turn becomes unstable and produces further anticyclonic eddies. Figures 16 and 17 shows a sequence of depth profiles of such a patch and the region of the sloping interface and the subsequent formation of eddies is clearly seen. The growth of the patch with time is shown in figure 18, which shows that the area grows at a rate proportional to time t.

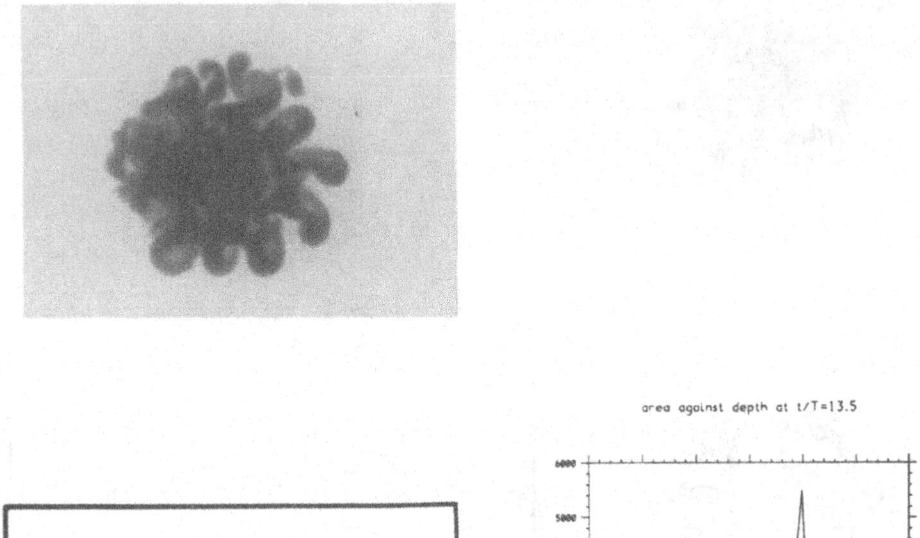

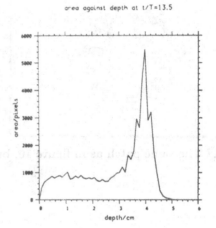

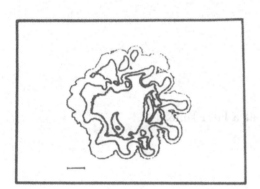

Fig. 16 The shape of the surface patch (a) the corresponding depth profiles (b) and the histogram of the depth (c). Note the steepening of the gradients near the fronts of the growing waves and, at early times, the central region of the patch is unchanged.

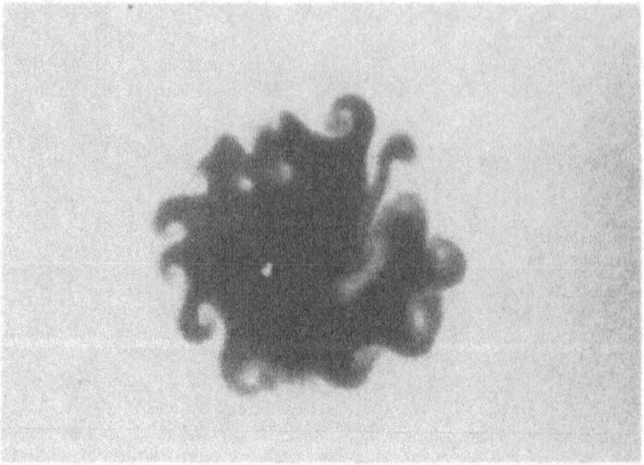

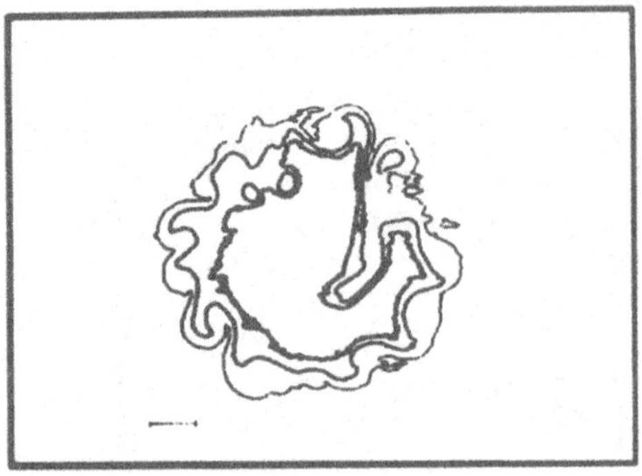

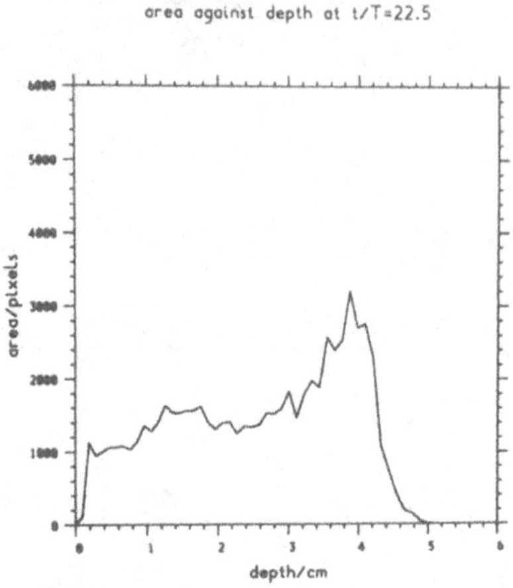

Fig. 17 The same patch as in figure 16, but at a later time.

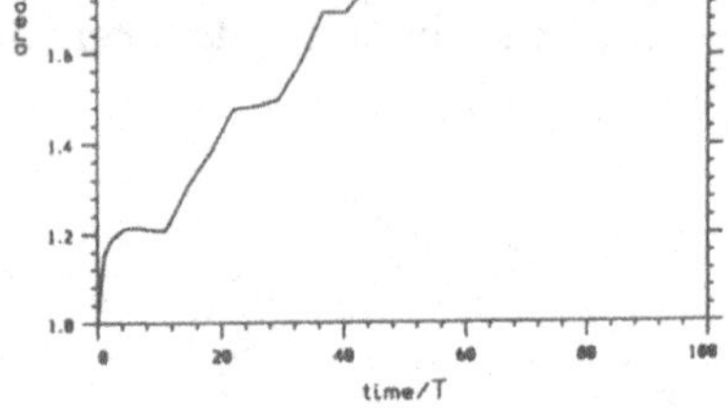

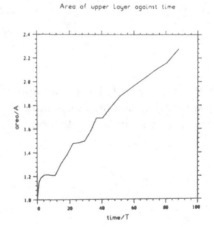

Fig. 18 The growth of the area of the surface patch.

Given a patch with initial area A_0, depth H_0 and reduced gravity g_0', then it is possible to derive the rate of spreading from dimensional analysis. In the absence of friction, a non-rotating patch will spread axisymmetrically depending only on its initial buoyancy $B_0 = g_0' H_0 A_0$. Then

$$A(t) = c(g_0' H_0 A_0)^{1/2} t \ , \tag{18}$$

where c is a dimensionless constant. Experiments (Linden & Simpson, [18]) suggest $c \simeq 1.2$. In a rotating system we expect (18) to be modified in the form

$$\frac{A}{A_0} = c \left(\frac{g_0' H_0}{A_0} \right)^{1/2} t f(F) \quad , \tag{19}$$

where $f(F)$ is a dimensionless function of the Froude number $F = \frac{f^2 A_0}{g_0' H_0}$, with $f(F) \to 1$ as $F \to 0$.

The front first adjusts on the scale $\Re$ and eddies form on the timescale

$$\tau \simeq \frac{\sqrt{g_0' H_0}}{f} \cdot \frac{1}{\sqrt{g' H_0}} \simeq \frac{1}{f} \quad .$$

As this process is repeated the rate of advance of the front is

$$\frac{\Re}{\tau} \sim f \Re \quad .$$

Thus at radius r

$$\frac{dA}{dt} \sim 2\pi r f \Re \quad ,$$

and hence

$$A \sim (g_0' H_0 A_0)^{1/2} t \quad .$$

This rate of growth corresponds to the low F limit of (19) and requires a continuous increase in the number of eddies with time.

At later times when the patch is full of eddies of scale $l \sim \sqrt{g_0' H_0}/f$, moving with velocity $u \sim \sqrt{g_0' H_0}$, we can approximate the growth by a turbulent diffusion with a diffusion coefficient

$$\tau \sim Ul \sim \frac{g_0' H_0}{f} \sim \frac{dr^2}{dt} \quad .$$

Then

$$\frac{A}{A_0} \sim \frac{g_0' H_0 t}{f A_0} \quad , \tag{20}$$

and comparison of (20) with (21) shows that

$$f(F) \propto F^{-1/2} \text{ for } F \gg 1 \quad . \tag{21}$$

The rate of growth of the patch is thereby reduced by an amount proportional to $\frac{\Re}{R_0}$, the ratio of the deformation radius to the size of the patch.

ACKNOWLEDGEMENT

I am grateful to Joanne Holford for allowing me to publish the data shown in figures 16, 17 and 18.

REFERENCES

1. Federov K.N. 1987. The physical nature and structure of oceanic fronts. *Spring-Verlag, Berlin.*
2. Ruddick, B.R. & Turner J.S. 1979. The vertical length scale of double-deffusive intrusions. *Deep-Sea Res.* **26**, 903-913.
3. Gill, A.E. Davey, M.K., Johnson E.R. & Linden, P.F. 1986. Rossby adjustment over a step. *Journal of Marine Research*, **44**, 713-738.
4. Allen, S.E. 1988. Rossby adjustment over a slope *PhD thesis University of Cambridge.*
5. Longuet-Higgins, M.S. 1968. On the trapping of waves along a discontinuity of depth in a rotating ocean. *J. Fluid Mech.* **31**, 417-434.
6. Hoskins, B.J. & Bretherton F.P. 1972. Atmospheric frontogenesis models: mathematical formulation and solution. *J. Atmos. Sci.* **29**, 11-37.
7. Simpson, J.E. & Linden, P.F. 1989. Frontogenesis in a fluid with horizontal density gradients. *J. Fluid Mech.* **202**, 1-16.
8. Jacquin, D., 1991. Frontogenesis driven by horizontally quadratic distribution of density. *J. Fluid Mech. to appear.*
9. Kay, A. 1991. Frontogenesis in gravity-driven flows with non-uniform density gradients. *J. Fluid Mech. submitted.*
10. Killworth, P.D. & Stern, M.E. 1982. Instabilities on density-driven boundary currents and fronts. *Geophys. Astrophys. Fluid Dyn.* **22**, 1-28.
11. Killworth, P.D., Paldor, N. & Stern, M.E.. 1984. Wave propagation and growth on a surface front in a two-layer geostrophic current. *J. Mar. Res.* **42**, 761-785.
12. Griffiths, R.W. & Linden P.F. 1981. The stability of buoyancy driven coastal currents. *Dyn. Atmos. Oceans.* **5**, 281-306.
13. Griffiths, R.W. & Linden P.F. 1982. The influence of a side wall on rotating flow over bottom topography. *Geophys. Astrophys. Fluid Dyn.* **27**, 1-33.
14. Sakai, S. 1989. Rossby-Kelvin instability: a new type of ageostrophic instability caused by a resonance between Rossby waves and gravity waves. *J. Fluid Mech.* **202**, 149-176.
15. Paldor, N. 1983. Linear stability and stable modes of geostrophic fronts. *Geophys. Astrophys. Fluid Dyn.* **27**, 217-228.
16. Kubokawa, A. 1985 Instability of a geostrophic front and its energetics. *Geophys. Astrophys. Fluid Dyn.* **33**, 223-257.
17. Griffiths, R.W. & Hopfinger, E.J., 1984. The structure of mesoscale turbulence and horizontal spreading at ocean fronts. *Deep Sea Res.* **31**, 245-269.

18. Linden, P.F. & Simpson J.E. 1990. Continuous two-dimensional releases from an elevated source. *J. Loss Prev. Process Ind.* **3**, 82-87.
19. Griffiths, R.W. & Pearce, A.F. 1985 Instability and eddy pairs on the Leeuwin Current south of Australia. *Deep-Sea Res.* **32**, 1511-1534.

PART IV

WAVES

Chapter IV.1

WAVE MOTIONS IN A ROTATING AND/OR STRATIFIED FLUID

T. Maxworthy
University of Southern California, Los Angeles, CA, USA

1 Introduction

When considering wave motions in natural systems one immediately realizes that a number of important external effects need to be taken into account. In particular these motions take place in a fluid that is compressible, stratified and rotating on a spherical surface. Thus any general theory is necessarily complex and any simplifying concepts lost in the overwhelming mass of mathematical details. It therefore behooves us to make rational assumptions concerning the relative magnitudes of these effects and then study their consequences in isolation. Firstly we ignore the effects of the sphericity of the system and assume the motion takes place on a plane surface rotating at constant angular frequency. The details of this assumption are treated later but essentially restrict both the lateral or horizontal and the vertical extent of the motion. As a first approximation we ignore, also, the compressibility of the fluid, essentially requiring that the "scale-height" of the system be much larger than the vertical extent of the fluid. This is a very good approximation in the ocean but obviously much weaker in the atmosphere, where in fact we should, finally, consider the effects of compressibility in order to be completely realistic.

This leaves us with two remaining effects to be considered, rotation and stratification. As discussed in the opening lectures by Professor Hopfinger [1] the major effects of these two constraints are contained in two dimensionless parameters that measure the relative magnitude of an inertial force of the motion to both the Coriolis and buoyancy restoring forces. Thus:

$$\text{The Rossby No.} = R_0 = \frac{U}{2L_h\Omega} \equiv \frac{\text{Inertial force}}{\text{Coriolis force}}$$

$$\text{and the Richardson No.} = L_v^2 N^2/U^2 = \frac{\text{Buoyancy force}}{\text{Inertial force}},$$

where U, L_h and L_v are characteristic velocity, horizontal and vertical length scales of the motion, respectively, Ω is the rotation frequency and $N = [(g/\rho)/(\partial\rho/\partial z)]^{1/2}$ the intrinsic or Brunt-Väisälä frequency, discussed in detail later. Here g is the acceleration of gravity and $|\partial\rho/\partial z|$ the distribution of density (ρ) with height or depth, depending on the system under consideration. Thus one measure of the relative importance of buoyancy and rotation is given by the Burgers number $(B) = R_0^2 R_i = [N^2/(2\Omega)^2][L_v^2/L_h^2]$ as we will see in detail later. For the moment one can crudely think of L_h as the distance traveled by an internal wave during one period of rotation. Typically, $N \sim 10^{-2}$ s^{-1} $f = 2\Omega \sim 10^{-4}$ s^{-1} so that when $L_h/L_v \lesssim 100$ buoyancy dominates the effect of rotation. This criterion is very easily exceeded by many phenomena in both the atmosphere and oceans. In particular only in the largest scale systems is rotation important in the study of stratified flow over mountains, for example.

2 Internal waves

This very simple order-of-magnitude analysis then justifies our initial concentration on the effects of stratification alone and in particular in the properties of internal waves. The classical example of such a wave is given in Lamb (1945), among others, where he considers the motion of an internal interface between two immiscible fluids of density ρ_1 and ρ_0. Since the fluid are incompressible and inviscid, Laplace's equation for the velocity potential must be solved in each layer separately and then matched across an interface of form

$$\eta = a \cos kx \cdot e^{i\sigma t}$$

where η is the interface displacement, a its amplitude, k the wavenumber, σ the frequency and t time. By matching the particle displacements and pressure across the interface one obtains the disperson relationship between σ and k:

$$c^2 = \frac{\sigma^2}{k^2} = \frac{g}{k}\frac{(\rho_1 - \rho_0)}{(\rho_1 + \rho_0)} \tag{1}$$

where c is the wave speed. If the layers are of finite thicknesses h_1 and h_0 this becomes:

$$c^2 = \frac{g}{k}(\rho_1 - \rho_0)[\rho_0 \coth kh_0 + \rho \coth kh_1]^{-1} \tag{2}$$

If the upper surface is free the internal interface is no longer of importance since, in general, $(\rho_1 - \rho_0)/\rho_0 \ll 1$, and the surface wave speed (c_1) becomes:

$$c_1^2 = \frac{g}{k} \tanh k(h_0 + h_1) \tag{3}$$

the so-called barotropic mode.

One interesting consequence of these results is that if we take the long-wave limit of (2) and superimpose two waves traveling in opposite directions we obtain a standing wave of period:

$$T = 2L \left[\frac{g(\rho_1 - \rho_0)}{\rho_0} \frac{h_0 h_1}{h_0 + h_1} \right]^{-1/2},$$

where L is the length of the rectangular basin in which the wave propagates. This result has great consequences for the calculation of the simplest response generated by the relaxation of the wind set-up of the closely-two-layer stratification of most lakes and reservoirs.

As one approach to the description of waves in the continuously stratified environments of most cases of interest it is possible to extend the treatment given above to include a large number of layers, although the algebra required becomes tedious even for only a small number of layers (Roberts [2]).

A simpler mathematical approach considers a continuous stratification from the start, however before embarking on this endeavour it is interesting to discuss a simple physical model of the basic particle force-balance in internal waves and stratified flows in general.

Consider, as in figure 1a, a basic, static density distribution $(\rho(z))$. A particle of volume dv, displaced, without mixing or density loss, from the point A to point B finds itself with a density deficit $(\rho_0 - \rho)$ with respect to its surroundings. The force tending to restore it to its original location is $(\rho_0 - \rho)gdv$ or its equation of motion is in this approximation, $\rho_0 \ddot{z} dv = -(\rho_0 - \rho)gdv$; but $\rho = \rho_0 + (\partial\rho/\partial z)z$ so that $\ddot{z} = -\left[(g/\rho_0)(\partial\rho/\partial z)\right] z$, representing a simple harmonic motion with frequency

$$N = \left[\frac{-g}{\rho_0} \frac{\partial \rho}{\partial z} \right]^{1/2}$$

This then is the Brunt-Väisälä frequency introduced earlier, but now its significance is somewhat clearer.

As a next step assume that the particle dv oscillates at an angle rather than vertically (figure 1b). Then the effective density gradient is $(\partial\rho/\partial z)\cos\theta$ and effective gravity $g\cos\theta$. Hence the frequency of oscillation becomes $\left[(g/\rho_0)(-\partial\rho/\partial z)\cos^2\theta\right]^{1/2} \equiv N\cos\theta$. Thus oscillations in the vertical have the maximum frequency, forcing at any other lower frequency (σ) gives oscillation at an angle $\theta = \cos^{-1}\frac{\sigma}{N}$. As we will see

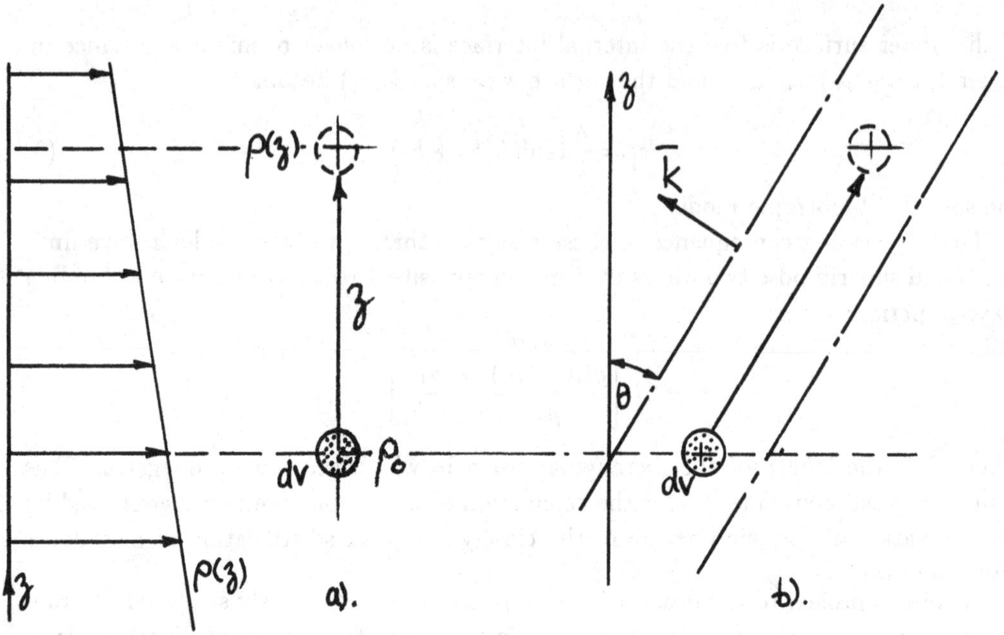

Figure 1: a) Vertical particle displacement and oscillation. b) ditto at an angle theta.

forcing at these higher frequency leads to an exponential decay of the motion away from the source. Note also that the particle motions are at right angles to the direction of wave propagation and that θ is the angle of the phase fronts to the vertical or wave vector to the horizontal.

In order to describe these motions explictly we start with the general equations of motion for an incompressible, inviscid, non-heat conducting fluid namely:

$$\text{Momentum equation: } \rho\left[\frac{\partial \bar{u}}{\partial t} + \bar{u}\nabla\bar{u}\right] = -\nabla p + \rho\bar{g} \tag{4}$$

$$\text{Continuity equation: } \nabla\bar{u} = 0 \tag{5}$$

$$\text{The particle density is conserved as it moves: } \frac{D\rho}{Dt} = 0 \tag{6}$$

Here $\bar{u}$ is the velocity vector, p the pressure and $\frac{D}{Dt}$ the convective derivative.

Let $p = p_0 + p'$ and $\rho = \rho_0 + \rho'$ in (4) where the primed quantities are deviations from the mean state. After subtracting out this mean state: $-\nabla p_0 + \rho_0 g = 0$, we obtain:

$$\rho_0\left(\frac{D\bar{u}}{Dt}\right) = -\nabla p' + \rho'\bar{g} \tag{7}$$

indicating that only deviations from the mean state are important in determining, $\bar{u}$.

If we further assume that $\bar{u}$ is a small quantity so that we can linearize the left-hand side of equation (7) we obtain:

$$\left[1+\frac{\rho'}{\rho_0}\right]\frac{\partial \bar{u}}{\partial t} = -\frac{1}{\rho_0}\nabla p' + \frac{\rho'}{\rho_0}\bar{g}$$

At this stage we can introduce the so-called, Boussinesq approximation in which we ignore the density variations in the inertial term and only retain them in the buoyancy term $g' = (\rho'/\rho_0)g$, often called the reduced gravity.

We obtain

$$\frac{\partial \bar{u}}{\partial t} = -\frac{1}{\rho_0}\nabla p' + \bar{g}' \tag{8}$$

We can linearize equation (6), also, to obtain

$$\frac{\partial \rho'}{\partial t} + w'\frac{\partial \rho_0}{\partial z} = 0$$

where w' is the perturbation velocity in the vertical direction, z. We see from this result that the change in density at any fixed point is simply the result of convecting the basic density field vertically with a velocity w'.

The full set of equations to be solved in two dimensions (x, z) become:

$$\frac{\partial u'}{\partial t} + \frac{1}{\rho_0}\frac{\partial p'}{\partial x} = 0; \quad \frac{\partial w'}{\partial t} + \frac{1}{\rho_0}\frac{\partial p'}{\partial z} + g' = 0$$

$$\frac{\partial u'}{\partial x} + \frac{\partial w'}{\partial z} = 0; \quad \frac{\partial g'}{\partial t} - N^2 w' = 0, \tag{9}$$

upon substituting for $\partial\rho_0/\partial z$.

Eliminating the pressure from the upper two equations gives:

$$\frac{\partial}{\partial t}\left[\frac{\partial u'}{\partial z} - \frac{\partial w'}{\partial x}\right] = \frac{\partial g'}{\partial x} \tag{10}$$

This gives the interesting physical observation that horizontal gradients of the buoyancy (g') result in generation of horizontal vorticity, the bracketed term of equation (10).

Eliminating g' and u' from this equation using the lower two equations of (9) leads to

$$\frac{\partial^2}{\partial t^2}\left[\frac{\partial^2 w'}{\partial x^2} + \frac{\partial^2 w'}{\partial z^2}\right] + N^2\frac{\partial^2 w'}{\partial x^2} = 0 \tag{11}$$

We assume now a plane wave solution of the form:

$$w' = \hat{w}(z)e^{i(kx-\sigma t)}$$

which upon substitution into (11) gives:

$$\frac{d^2\hat{w}}{dz^2} + \left[\frac{N^2}{\sigma^2} - 1\right] k^2\hat{w} = 0 \tag{12}$$

An equation that is only hyperbolic (i.e., can support a wave motion) if $\sigma < N$ with N as the upper limit on σ, and giving an elliptic equation with exponentially decaying solutions when $\sigma > N$.

We can also note a number of interesting possibilities. For example if N is variable such that $N^2/\sigma^2 = 1$ somewhere in the domain, then the waves will be trapped in the layer for which N excees σ.

In order to study this case further consider a fluid for which N = constant i.e. a linear density gradient in our approximation. Then the solution to the vertical structure equation (12) requires $\hat{w} = \bar{w}e^{imz}$, whence:

$$-m^2 + \left(\frac{N^2}{\sigma^2} - 1\right) k^2 = 0 \text{ or } \sigma = N\left[\frac{k^2}{k^2+m^2}\right]^{1/2} \tag{13}$$

or taking account of the geometry of the wavenumber vectors (figure 2b)

$$\sigma = N\cos\theta$$

and the phase speed $c_p = \sigma/k = N/(k^2+m^2)^{1/2}$.

In general

$$\sigma^2 = N^2\left[\frac{l^2+k^2}{l^2+k^2+m^2}\right]^{1/2} \tag{14}$$

where l is the wave-number in the y direction.

Furthermore writing $\bar{u} = \hat{\bar{u}}e^{i(\bar{k}_0\bar{x}-\sigma t)}$ and substituting into the continuity equation gives $i\bar{k}_0\hat{\bar{u}} = 0$ or that the particle motion is perpendicular to the wave-number vector, as assumed in our simple physical model.

If the fluid is contained within horizontal planes on which $\hat{w}$ must be zero, then only discrete modes can be excited, and mH must be an integer (n) times π in order to satisfy the boundary condition, hence:

$$\sigma = \frac{k}{\left(k^2 + \frac{n^2\pi^2}{H^2}\right)^{1/2}} \cdot N \tag{15}$$

In wave motions such as the one discussed above there is considerable interest in determining the sign and magnitude of the energy flux, since in the application to mountain waves, in particular, the information is of great value in determining the tilt of the surfaces of constant phase.

Consider the kinetic energy contained in a slab of fluid of thickness $\lambda = 2\pi/k$ perpendicular to the wave vector $\bar{k}$ (figure 2a)

$$E|_{\text{unit area}} = \int_0^\lambda \frac{1}{2}\rho(v'^2 + v'^2 + w'^2)d\zeta$$

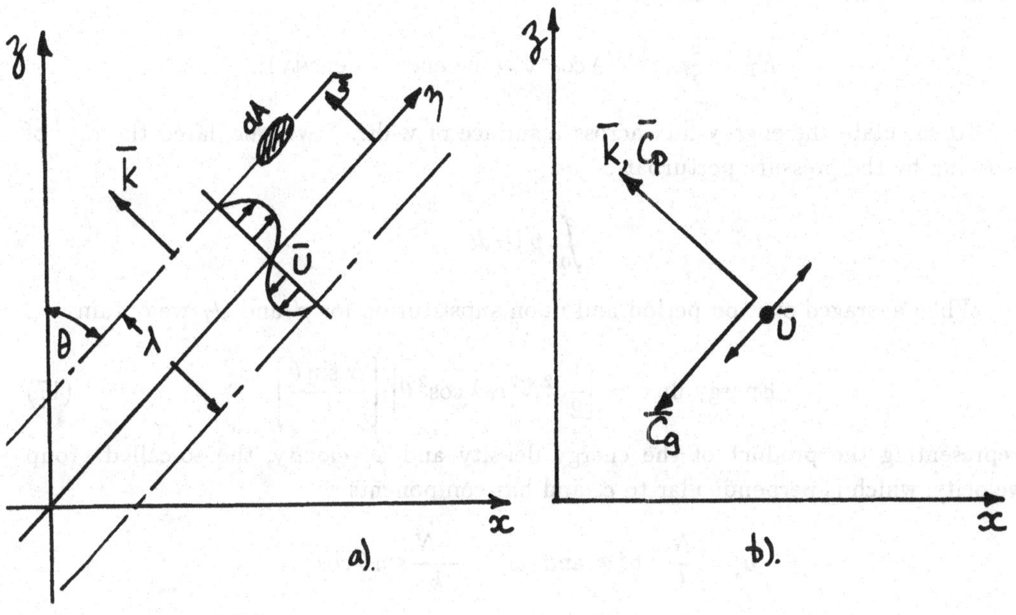

Figure 2: a) Sketch for the calculation of the energy flux. b) Direction of wave velocities w.r.t the wavevector.

Averaged over one period of the oscillation this gives

$$\bar{E}|_{\text{unit area}}|_{\text{period}} = \frac{\sigma}{2\pi}\int_0^{\frac{2\pi}{\sigma}}\int_0^\lambda \frac{1}{2}\rho U_T^2 d\zeta dt$$

where U_T is the absolute particle velocity.

However it can be shown that if the particle displacement is given by $-a\sin(\bar{k}\cdot\bar{x}-\sigma t)$ then

$$U_T = aN\frac{k_z}{k}\cos(\bar{k}\cdot\bar{x} - \sigma t)$$

and

$$P' = \frac{aN^2}{k}\rho_0\frac{k_x k_z}{k^2}\cos(\bar{k}\bar{x} - \sigma t)$$

So that

$$\bar{E} = \frac{1}{4}\rho_0 a^2 N^2 \lambda \cos^2\theta \tag{16}$$

It can be shown, also, that the averaged KE and potential energy (PE) are equal so that the total averaged energy is:

$$\bar{E_T} = \frac{1}{2}\rho_0 a^2 N^2 \lambda \cos^2\theta, \text{ (the energy density)}$$

To calculate the energy flux across a surface of width λ we calculated the rate of working by the pressure perturbation as:

$$\int_0^\lambda p' U_T d\xi$$

When averaged over on period and upon substitution for p' and U_T we obtain

$$\text{Energy flux} = \left[\frac{1}{2}a^2 N^2 \rho_0 \lambda \cos^2\theta\right]\left[\frac{N\sin\theta}{k}\right] \tag{17}$$

representing the product of the energy density and a velocity, the so-called group velocity, which is perpendicular to c_p and has components

$$u_g = \frac{N}{k}\sin^2\theta \text{ and } w_g = \frac{-N}{k}\sin\theta\cos\theta.$$

The relative orientations of the velocities are shown in figure 2b. In particular one can see immediately if the sources of the disturbances are at the bottom of the atmosphere, the energy must propagate upwards and the phase downwards. Furthermore one can generalize the expression for the group velocity, c_g to give:

$$c_{g_x} = \frac{\partial\sigma(\bar{k})}{\partial k}, \quad c_{g_y} = \frac{\partial\sigma(\bar{k})}{\partial l}, \quad c_{g_z} = \frac{\partial\sigma(\bar{k})}{\partial m}$$

in three dimensions.

One interesting consequence of the derivations given above is that wave reflection from a solid surface has some interesting properties. Since the wave angle depends only on the ratio σ/N then these waves cannot, in general, obey Snell's Law and in fact the magnitude of the wave vector must change during such an interaction (figure 3).

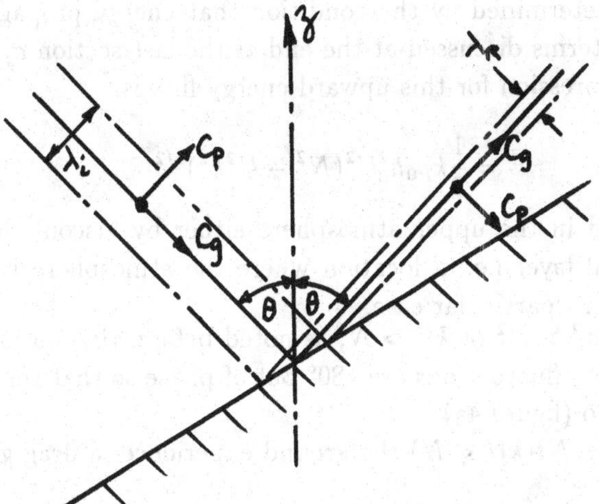

Figure 3: Angled wave reflection.

3 Mountain (Lee) Waves in a Stratified Fluid

In order to show, clearly, the relationship between the background we have developed so far and the subject at hand we consider a simple model due to Gill [3], in which a range of sinusoidal "mountains" is towed through a stratified fluid ($N =$ constant) at a constant speed U. In the frame of reference of the fluid the mountain shape is given by

$$h = h_0 \sin[k(x + Ut)]$$

so that the frequency of the motion produced is $\sigma = -Uk$ i.e. $\sigma/2\pi$ is the frequency at which surface air-parcels encounter crests. The linearized boundary condition on $z = 0$ gives:

$$w' = U\frac{dh}{dx} = Ukh_0 \cos(kx - \sigma t)$$

The solution to (11) in this case is:

$$w' = Ukh_0 \cos(kx + mz - \sigma t)$$

where

$$m^2 = k^2(N^2 - \sigma^2)/\sigma^2 = \left(\frac{N}{U}\right)^2 - k^2 \tag{18}$$

The sign of m is determined by the condition that energy propagates away from the source, so that in terms discussed at the end of the last section c_g is upwards and c_p downwards. The expression for this upward energy flux is

$$\bar{E} = \frac{1}{2} k \rho_0 h_0^2 U^2 (N^2 - U^2 k^2)^{1/2} \tag{19}$$

and must be deposited in the upper atmosphere either by viscous dissipation or by absorption at a critical layer i.e. a location where the atmosphere is moving locally with a velocity U, in this particular case.

For the case when $\sigma^2 > N^2$ or $kU > N$, as noted before, the motion is evanescent, the pressure and velocity fluctuations are 180° out of phase so that the rate of working by the boundary is zero (figure 4a).

However, when $\sigma^2 < N^2 (kU < N)$ the ground experiences a drag given by

$$\tau\big|_{\text{unitarea}} = -\rho_0 \overline{u'w'} = \frac{\bar{E}}{U} = \frac{1}{2} k \rho_0 h_0^2 \left[\left(\frac{N^2}{U^2} - k^2 \right) \right]^{1/2} U^2, \tag{20}$$

since high pressure is found where the particles are moving upwards.

Gill ([3], p. 145) gives estimates for the magnitudes of these effects from a prior paper by Bretherton in which he finds a wave drag of 0.4 N/m^2 with absorption at the 20 km level, here $U = 15$ m/s so that $\bar{E} \sim 6$ watts/m^2, for a model that also includes the effects of the variation of U and N with height.

According to the arguments given before, the angle between the wave crests and the vertical (θ) is given by $\cos\theta = Uk/N$ where N/U can be thought of as the critical wavenumber which divides the two types of solution. Thus as shown in figure 4 the group velocity is directed along the wave crests with magnitude $c_g = c_{g_x} \sin\theta = (\partial\sigma/\partial k)\sin\theta = U\sin\theta$. Thus the group velocity relative to the ground has a magnitude $U\cos\theta$ (see figure 4). The vertical component is $U\sin\theta\cos\theta$ with a maximum value of $\frac{1}{2}U$ at $\theta = \pi/4$.

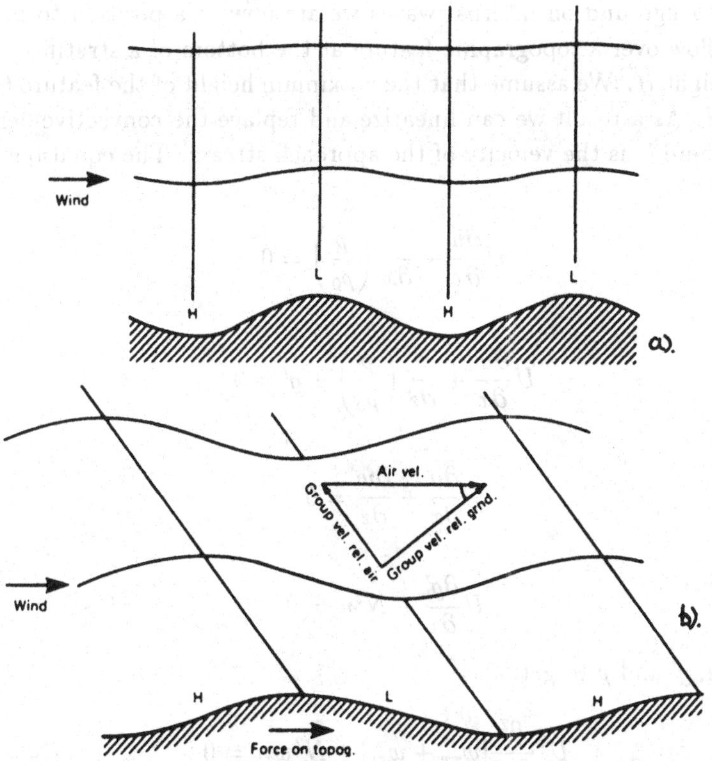

Figure 4: The motion produced by uniform flow of a uniformly stratified fluid over sinusoidal topography of small amplitude. The sinuous lines indicate the displacement of isopycnal surfaces whose equilibrium configurations are horizontal, and the straight lines join crests and troughs. (a) For small-wavelength topography, i.e., wavenumber $k > N/U$, where N is buoyancy frequency and Y is a fluid velocity relative to the ground (a typical value of U/N for the atmosphere is 1 km). The drawing is for $kU = 1.25N$. Note the decay of amplitude with height, showing that energy is trapped near the ground. H and L indicate positions of maximum and minimum pressure perturbation, respectively, i.e., there is suction over crests. when the lower half plane is fluid, this can lead to instability (Kelvin-Helmholtz) when the relative velocity between fluids is great enough for the suction to overcome gravity. (b) The response to large-wavelength topography, i.e., $k < N/U$ (the drawing is for $kU = 0.8N$). Now the displacement of isopycnals is uniform with height, but wave crests move upstream with height, i.e., phase lines are tilted as shown. The group velocity relative to teh air is along these phase lines, but the group velocity relative to the ground is at right angles, i.e., upward and in the downstream direction. High and low pressures are now at the nodes, so there is a net force on the topography in the direction of flow.

(Reproduced from Gill [3], with permission)

With this background on internal waves we are now in a position to discuss their generation by flow over a topographic feature at the bottom of a stratified tank ($N =$ constant) of height H. We assume that the maximum height of the feature ($\hat{h}$) is small compared to H. As a result we can linearize and replace the convective derivative $\frac{D}{Dt}$ by $\frac{\partial}{\partial t} + U\frac{\partial}{\partial x}$ where U is the velocity of the approach stream. The equations of motion become:

$$U\frac{\partial u}{\partial x} + \frac{\partial}{\partial x}\left(\frac{p}{\rho_0}\right) = 0$$

$$U\frac{\partial w}{\partial x} + \frac{\partial}{\partial z}\left(\frac{p}{\rho_0}\right) + g' = 0$$

$$\frac{\partial u}{\partial x} + \frac{\partial w}{\partial z} = 0$$

$$U\frac{\partial g'}{\partial x} - N^2 w = 0$$

Eliminate u, g' and p to get:

$$U^2\frac{\partial^2}{\partial x^2}[w_{xx} + w_{zz}] + N^2 w_{xx} = 0 \tag{21}$$

with boundary conditions $w = 0$ at $z = H$, $w \approx U(\partial h/\partial x)$ at $z = 0$, and $w \to 0$ as $x \to -\infty$.

The latter condition is important being a result of the observation that energy is swept downstream, because under most circumstances $c_g < c_p$ and $c_p = -U_1$ in order for the wave to be stationary with respect to the obstacle.

Using Fourier Transform methods for which:

$$w(x,z) = \int_{-\infty}^{\infty} \hat{w}(k,z)e^{ikx}dk$$

so that (21) becomes, on substitution, $\hat{w}_{zz} + \left(\frac{N^2}{U^2} - k^2\right)\hat{w} = 0$ with boundary conditions $\hat{w} = 0$ at $z = H$, $\hat{w} = ikU\hat{h}(k)$ on $z = 0$. Hence

$$\hat{w} = \frac{-ikU\hat{h}}{\sin(k_0^2 - k^2)^{1/2}H}\sin\{(k_0^2 - k^2)^{1/2}(z - H)\} \text{ for } |k| < k_0$$

and

$$\hat{w} = -ikU\hat{h}\frac{\sinh\{(k^2 - k_0^2)^{1/2}(z - H)\}}{\sinh\{(k^2 - k_0^2)^{1/2}H} \text{ for } |k| > k_0$$

where $k_0 = \frac{N}{U}$.

The poles at $(k_0^2 - k^2)H$ are integral multiples of π i.e. $n\pi$

$$k_n^2 H^2 = k_0^2 H^2 - n^2\pi^2$$

Calculating the residue at each pole leads to the following upon performing the inverse Fourier Transform

$$w(x,z) = -\sum_n \frac{4\pi^2 nU}{H^2}\hat{h}(k)\sin\frac{n\pi z}{H}\cdot\cos\left[k_0^2 - \frac{n^2\pi^2}{H^2}\right]^{1/2} x$$

if $\hat{h}(k)$ is symmetric and a similar expression with $\sin\left[k_0^2 - \frac{n^2\pi^2}{H^2}\right] x$ terms when the mountain is asymmetric.

Solutions to flows of this type are to be found in papers by Long [4] and Davis [5] for example. An example from the latter paper for a value of the relevant dimensionless parameter, the Froude Number, $Fr = U/NH = 0.80$ is given in figure 5a together with a sketch of the flow field found in a parallel experimental study (figure 5b). The effect of separation and/or internal hydraulic jump under the first lee-wave is evident, as are the very large velocities on the lee-side of the "mountain". Such effects are characteristic of flows over a wide range of Fr and are also observed in natural flows e.g. in the lee of the Rockies, where winds can reach hurricane magnitude, and in the Owens Valley in the lee of the Sierra Nevada, where such waves are used by glider pilots to reach record-breaking altitudes and where the downstream separated region is a serious environmental hazard. Streamlines for more realistic atmospheric applications are given in Gossard and Hooke [6].

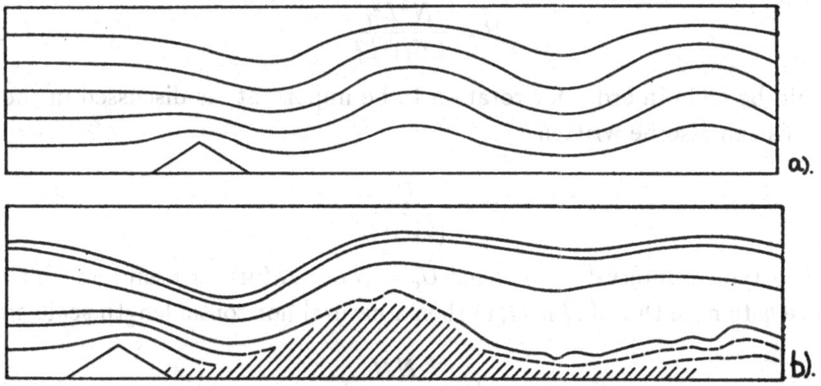

Figure 5: (a) Calculated flow over triangular obstacle for $k = 1.55$. (b) Observed flow over triangular obstacle for $k = 1.55$.

(Reproduced from Davis [5], with permission)

4 The Combined Effects of Rotation and Stratification

A simple argument to show when rotation is important in natural flows can be constructed using, at first, only the dispersion relationship for internal waves, equation (13). Typically in these flows $2\pi/N \sim 10^3$ s while $2\pi.\Omega \sim 10^5$ s. In equation (13) if $k/m \sim O(1)$ then $\sigma \sim O(N)$ and rotation is not important. However if the wave is propagating almost vertically i.e. $k \ll m$, then equation (13) gives $\sigma \sim Nk/m$ so that when $k/m \sim O(10^{-2})$, $\sigma \sim N(k/m) \sim \Omega$ and rotation is important.

More formally one can derive the exact dispersion relationship from the equations (9) by adding Coriolis force terms to the upper two equations. Namesly $-fv'$ and fu' where $f = 2\Omega$. By following procedure similar to that used to derive (13), as found in Gill ([3], p. 256) for example, one obtains the final dispersion relationship:

$$\sigma^2 = \frac{N^2k^2}{k^2+m^2} + (2\Omega)^2\frac{m^2}{k^2+m^2} \tag{22}$$

here if k/m is small:

$$\sigma^2 \to N^2\frac{k^2}{m^2} + (2\Omega)^2$$

From this one can deduce immediately that a critical parameter in determining the importance of rotation is, as shown earlier, the Burgers Number:

$$B = \frac{N^2k^2}{(2\Omega)^2m^2} \tag{23}$$

or using the length scales L_v and L_h

$$B = \frac{N^2L_H^2}{(2\Omega)^2L_v^2} \tag{24}$$

which should be $O(1)$ in order for rotation to be important, as discussed in the introduction. This can also be written:

$$B = \frac{U_c^2}{(2\Omega)^2L_v^2} \tag{25}$$

where U_c is a typical interval wave speed $U_c = NL_h$. A further useful and interesting consequence is to note that if B is $O(1)$ then a natural horizontal length scale emerges:

$$L_h = \frac{NH}{2\Omega} \tag{26}$$

often called the Internal Rossby Radius of Deformation, this being the natural scale for the evolution of disturbances in these systems e.g. see the section on Baroclinic Instability.

Based on the simple arguments given above one can then imagine simplifying the basic equations in order to remove terms that are small under the approximation $L_h \gg L_v$, leading to the so-called Hydrostatic Equations.

Consider the equations of motion including rotation at a latitude, ϕ, on a rotating sphere, and for simplicity consider only the x and z momentum equations

$$\frac{Du}{Dt} + 2\Omega \cos\phi w - 2\Omega \sin\phi v + \frac{1}{\rho_0}\frac{\partial p}{\partial x} = 0 \tag{27}$$

$$\frac{Dw}{Dt} - 2\Omega \cos\phi u + g' + \frac{1}{\rho_0}\frac{\partial p}{\partial z} = 0 \tag{28}$$

Under the assumptions stated initially one would expect certain of these terms to be small. In particular if U is a characteristic horizontal speed then:

$$w \sim U\frac{L_v}{L_h} \ll 1$$

and

$$\frac{Dw}{Dt} \sim \frac{L_v}{L_h}\frac{Du}{Dt}$$

From their definition $(p_z/p_x) \sim (L_V/L_h)$ where the subscripts, z and x refer to differention with respect to that subscript. Then, assuming initially that Du/Dt is at least of order $(1/\rho_0)(\partial p/\partial x)$, we find:

$$\frac{Dw}{Dt}/\frac{1}{\rho_0}p_z \sim \frac{L_v^2}{L_h^2}\frac{Du/Dt}{(1/\rho_0)p_x} \sim \frac{L_v^2}{L_h^2} \ll 1$$

Also in (27):

$$\frac{(2\Omega\cos\phi)w}{(2\Omega\sin\phi)v} \sim \cot\phi\frac{L_v}{L_h} \ll 1$$

and in (28):

$$\frac{(2\Omega\cos\phi)u}{\frac{1}{\rho_0}p_z} \sim \frac{L_v}{L_h}\frac{(2\Omega\cos\phi)u}{\frac{1}{\rho_0}p_z} \sim \frac{L_v}{L_h}\frac{(2\Omega\cos\phi)u}{(2\Omega\sin\phi)v} \sim \frac{L_v}{L_h}\cot\phi \ll 1$$

However, at the equator $\cot\phi \to \infty$ and these last two assumptions cannot be taken for granted. However they are still valid if the Rossby Number is large, thus:

$$\frac{2\Omega\cos\phi w}{\frac{Du}{Dt}} \sim \Omega T\frac{L_v}{L_h} \sim \frac{\Omega L_v}{U} \ll 1$$

where $T \sim L_h/u$ and

$$\frac{2\Omega \cos\phi u}{\frac{1}{\rho_0}p_z} \sim \frac{L_v}{L_h}\frac{2\Omega\cos\phi u}{\frac{1}{\rho_0}p_x} \sim \frac{L_v}{L_h}\frac{2\Omega\cos\phi u}{\frac{Du}{Dt}} \sim \Omega\frac{L_v}{L_h}T \sim \frac{\Omega L_v}{U} \ll 1.$$

So near the equator we require both $\frac{L_v}{L_h} \ll 1$ and $\frac{\Omega L_v}{U} \ll 1$.

Typically, at mid-latitudes, $\Omega L_v \sim 2.5$ km/hr in the atmosphere and 1 km/hr in the ocean.

Therefore in this approximation the linearized, Boussinesq equation become upon introducing $N^2(z)$ and $f = 2\Omega \sin\phi$:

$$u_t - fv + \frac{p_x}{\rho_0} = 0; \quad v_t + fu + \frac{1}{\rho_0}p_y = 0$$

$$g' + \frac{p_z}{\rho_0} = 0; \quad u_x + v_y + w_z = 0 \tag{29}$$

$$g'_t + N^2(z)w = 0$$

If we assume solutions of the form:

$$w = w_n(z)w^*(x,y,t) \text{ etc.}$$

we obtain equations of the form:

$$u_n u_t^* - f v_n v^* + p_n p_x^* = 0 \text{ etc.}$$

which can be reduced to one equation for $w_n w^*$. This can in turn be separated into two independent equations both equal to a constant eigenvalue say c_n^2, to give a vertical structure equation for the eigenfunction, w_n:

$$\frac{d^2 w_n}{dz^2} + \frac{N^2}{c_n^2}w_n = 0 \tag{30}$$

which is to be compared to equation (12) for the full vertical structure equation which, of course, reduces to (30) as $k \to 0$.

The equations for the starred quantities become:

$$u_t^* - fv^* + p_x^* = 0$$

$$v_t^* + fv^* + p_y^* = 0 \tag{31}$$

$$u_x^* + v_y^* + \frac{1}{c_n^2}p_t^* = 0$$

If $f = 0$ these reduce to:

$$\left[\frac{\partial^2}{\partial t^2} - c_n^2 \nabla^2\right]\left[\frac{\partial u^*}{\partial x} + \frac{\partial v^*}{\partial y}\right] = 0 \tag{32}$$

the shallow water wave equation which if f had been retained would have been recognized as the tidal equations.

As a further simplification assume that motion is restricted to the neighborhood of the equator so that to a first approximation $f = \beta y$ where y is the distance from the equator.

Let $v^* = \hat{v}(y)e^{i(kx-\sigma t)}$ etc. Eliminating $\hat{u}$ and $\hat{p}$ leads to:

$$\frac{d^2\hat{v}}{dy^2} + \left[\frac{\sigma^2}{c_n^2} - \beta\frac{k}{\sigma} - k^2 - \frac{\beta^2 y^2}{c_n^2}\right]\hat{v} = 0. \tag{33}$$

It is clear that if the bracketed term is positive near $y = 0$ an oscillatory solution exists in y, which must always tend to a decaying solution as y becomes large. The, so-called, β effect traps these ROSSBY WAVES at the equator. An inspection of (33) shows that there is a natural length scale $\left[(c_n/\beta)^{1/2}\right]$ and frequency $[(c_n\beta)^{1/2}]$ for this problem so that upon substituting $\sigma = \tilde{\sigma}(\beta c_n)^{1/2}$, $y = \tilde{y}(c_n/\beta)^{1/2}$, and $k = \tilde{k}(\beta/c_n)^{1/2}$ into (33) we obtain:

$$\frac{d^2\hat{v}}{d\tilde{y}^2} + \left(\tilde{\sigma} - \tilde{k}^2 - \frac{\tilde{k}}{\tilde{\sigma}} - \tilde{y}^2\right)\hat{v} = 0 \tag{34}$$

which has solutions of the form:

$$\hat{v}(\tilde{y}) = AH_m[\tilde{y}]\exp\left[-\frac{1}{2}\tilde{y}^2\right] \tag{35}$$

where H_m are the Hermite polynomials of order m, and $2m + 1 = \tilde{\sigma}^2 - (\tilde{k}/\tilde{\sigma}) - \tilde{k}^2$. Equation (34) is a cubic equation with three roots in $\tilde{\sigma}$ when $\tilde{k}$ and m are specified. We show a typical solution in figure 6. These three roots turn out to correspond to two inertio-gravity wave traveling east and west and one Rossby wave traveling west.

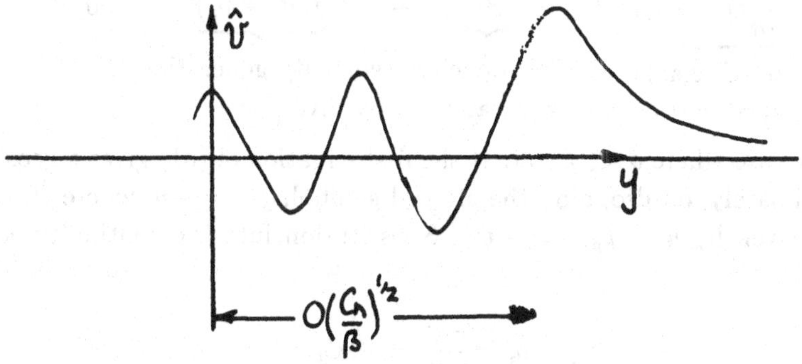

Figure 6: N.-S. eigenfunction for waves trapped at the equator.

These results have an interesting interpretation if one assumes a locally sinusoidal solution in $\tilde{y}$ i.e. allow $m \to \infty$.

Since $2m + 1 + \tilde{k}^2 = \tilde{\sigma}^2 - (\tilde{k}/\tilde{\sigma})$ this means that either $\tilde{\sigma}^2$ or $\tilde{k}/\tilde{\sigma}$ is large. If the former is true then:

$$\tilde{\sigma}^2 \approx \tilde{k}^2 + 2m + 1$$

that is we have ignored the $\beta k/\sigma$ term in the original equation or have assumed that f does not vary over one wavelength. This leads to a dispersion relationship:

$$\sigma^2 = f^2 + (k^2 + l^2)c_n^2 \quad \text{[these are the inertio-gravity waves]}$$

Where if $f^2 \ll c_n^2(k^2+l^2)$ we obtain pure gravity waves or if $f^2 \gg c_n^2(k^2+l^2)$ we obtain inertial oscillations with a frequency independent of wavenumber. Such motions are often detected in the ocean where their source is often a subject of intense speculation. These waves are still trapped at the equator since σ and k are constant, so as f^2 increases l^2 must decrease and eventually become zero, and $\bar{k}$ turns into the E-W direction.

A second possibility occurs when $\tilde{k}/\tilde{\sigma} \gg 1$ in this case we have ignored the σ^2/c_n^2 term and the dispersion relationship becomes:

$$\sigma = \frac{-\beta k}{k^2 + l^2 + \frac{f^2}{c_n^2}} \tag{36}$$

the expression for divergent Rossby Waves i.e. waves for which variations in surface height are dynamically important.

This result can be derived in a different manner that points up the important dynamics more clearly. One can derive a vorticity equation by differentiating the original horizontal momentum equations and combining to eliminate the pressure, the result is:

$$\underbrace{\frac{\partial}{\partial t}(u_y - v_x)}_{\substack{\text{Rate of change of}\\ \text{vertical vorticity}}} \quad - \quad \underbrace{\beta v}_{\substack{\beta\text{ vorticity}\\ \text{generation}}} \quad - \quad \underbrace{f(u_x + v_y)}_{\substack{\text{vorticity generation}\\ \text{by divergence}}} \quad = 0 \tag{37}$$

Consider the case where $\sigma \ll f$ so that the first equation of (31) gives a geostrophic balance approximately, on dropping the starred symbol, $f\psi = -p$, where the stream function, ψ, is given by $u = \psi_y, v = -\psi_x$. Substitution into the continuity equation gives:

$$u_x + v_y = \frac{-p_t}{c_n^2} = \frac{f}{c_n^2}\psi_t$$

Thus the vorticity equation (37) becomes:

$$\frac{\partial}{\partial t}\nabla_H^2\psi + \beta\psi_x - \frac{f^2}{c_n^2}\psi_t = 0 \qquad (38)$$

Using the Fourier transform leads to the recovery of the relationship (35). Now it can be seen that the term f^2/c_n^2 is a result of vortex stretching that opposes the basic Rossby wave mechanism. As shown in figure 7a as the Rossby wave propagation forces a fluid column northwards, for example, a relative anticyclonic flow is generated. This corresponds to high pressure or a raised surface which in turn induced a stretching of the vortex column which opposes the original contraction. For the sake of completeness in figure 7b we reproduce the whole physical argument that can be used to understand the propagation of Rossby waves, in this case ignoring the effects of divergence. If a particle at A is displaced northwards, to B, an anticyclonic flow is induced which in turn forces a particle at C toward a new location at D, at which loçation cyclonic motion is generated. This motion, in turn acts to prevent the particle at A from moving and constitutes a restoring force on the displaced particle. In figure 7c, a similar displaced particle argument can be used to explain the westward propagation of Rossby waves.

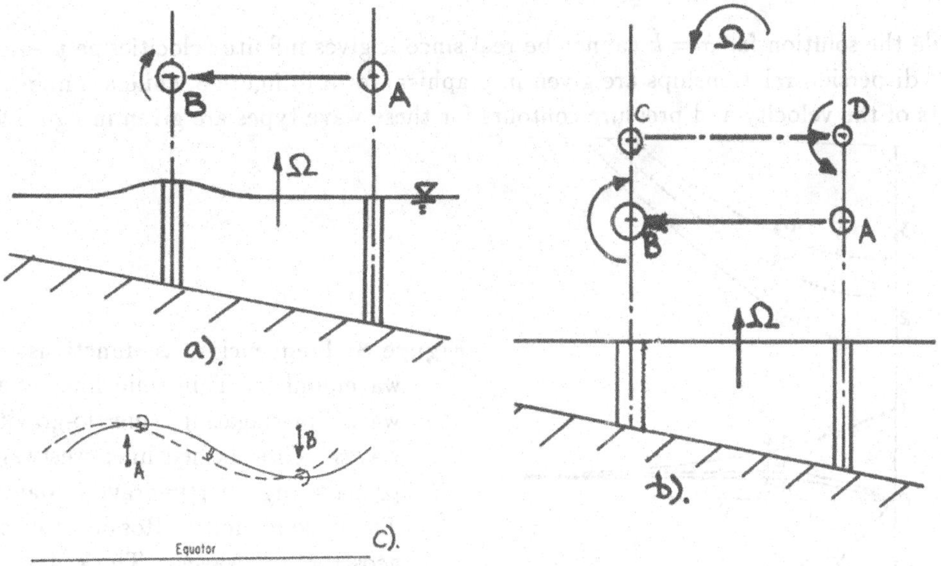

Figure 7: (a) Effect of surface divergence on vortex stretching and wave propagation. (b) Physical description Rossby wave restoring motion caused by particle displacement from A to B. (c) The mechanism of planetary wave propagation. Because potential vorticity is conserved, a particle displaced equatorward acquires cyclonic vorticity relative to its surroundings (as indicated by the arrow), whereas particles displaced poleward acquire anticyclonic vorticity. The motion induced by this relative vorticity distribution is indicated by the broad arrow, and is such as to produce westward propagation of the wave.

If a particle at location A is displaced northwards and that at B southwards then the particle at A acquires anticyclonic vorticity and that at B cyclonic vorticity. The net effect at C, for example, is to move particles westward so that the whole wave-form propagates.

Of course in (35) if $f = 0$ i.e. at the equator, or if the surface deformation is suppressed we recover the usual expression for the non-divergent Rossby wave:

$$\sigma = \frac{-\beta k}{k^2 + l^2}$$

The totality of possible cases is discussed in detail by Matsuno [7]. One special case is of interest however corresponding to $n = -1$, which cannot be extracted from the analysis above is the KELVIN wave for which it is necessary to find a solution for the case when $\hat{v} = 0$. From the original equations we deduce:

$$(\tilde{\sigma} + \tilde{k})(\tilde{\sigma} - \tilde{k}) = 0$$

and solution for $\tilde{w} = -\tilde{k}$ is:

$$u = ce^{-1/2\tilde{y}^2}$$

while the solution for $\tilde{\sigma} = \tilde{k}$ cannot be real since it gives infinite velocities as $\tilde{y} \to \infty$. The dispersion relationships are given in graphical form in figure 8 while a number of plots of the velocity and pressure contours for these wave types are given in figure 9.

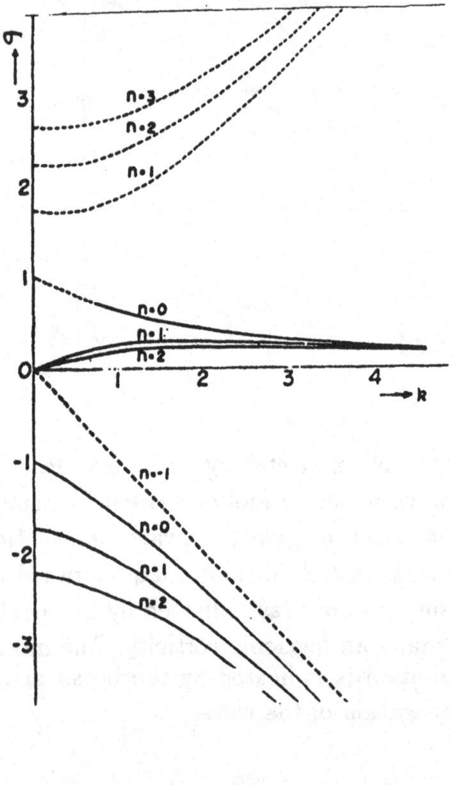

Figure 8: Frequencies as functions of wavenumber. Thin solid line: eastward propagating inertio-gravity waves. Thin dashed line: westward propagating inertio-gravity waves. Thick solid line: Rossby (quasi-geostrophic) waves. Thick dashed line: The Kelvin wave-like wave.

(Reproduced from Matsuno [7], with permission)

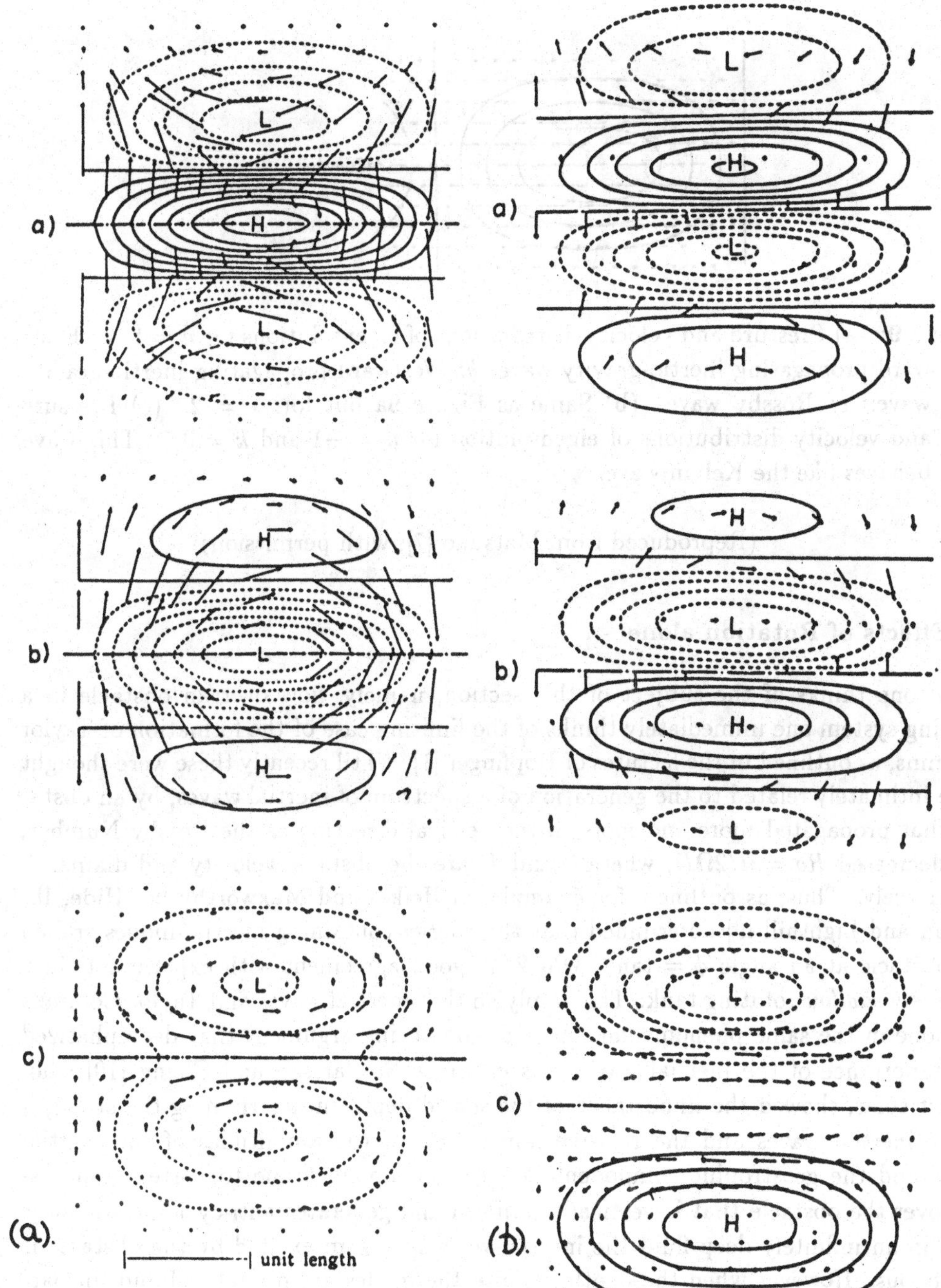

For caption see next page.

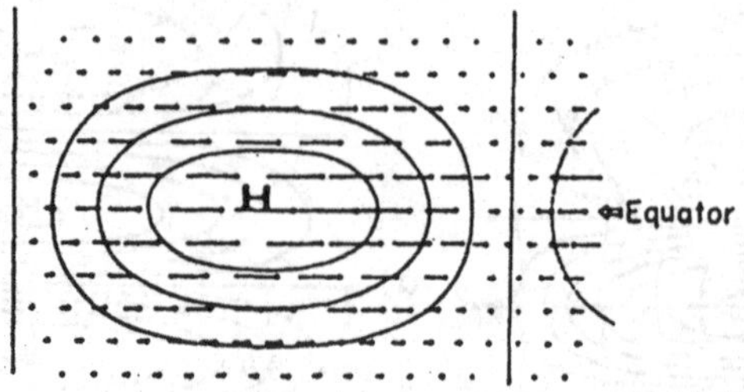

Figure 9: (a) Pressure and velocity distributions of eigensolutions for $n = 1$. *a:* Eastward propagating inertio-gravity wave. *b:* Westward propagating inertio-gravity wave. *c:* Rossby wave. (b) Same as Figure 9a but for $N = 2$. (c) Pressure and velocity distributions of eigensolution for $n = -1$ and $k = 0.5$. This wave behaves like the Kelvin wave.

(Reproduced from Matsuno [7], with permission)

5 Effects of Rotation alone

When one thinks of the subject of this section, namely, flow over an obstacle in a rotating system one immediately thinks of the limiting case of the formation of Taylor Columns, as outlined in the lectures of Hopfinger [1]. Until recently these were thought to be intimately related to the generation of a spectrum of inertial waves, by an obstacle, that propagated more and more in the vertical direction as the Rossby Number, *Ro*, decreased $Ro = u/2\Omega D$, where u and D are the obstacle velocity and diameter, respectively. Thus, as outlined, for example, in Heikes and Maxworthy [8], Hide, Ibbotson and Lighthill [9] determined that the energy-containing inertial-modes trailed the obstacle at an angle $\phi = \tan^{-1} 3Ro/2$, in good agreement with experiments in a relatively *shallow* rotating tank. This implyied that inertial waves and Taylor Columns were one in the same phenomenon. A more convincing argument that deemphasized the importance of the inertial waves was given by Stewartson and Cheng [10] who, in particular, showed the importance of the scaled depth in determining the strength of the inertial waves and the relative importance of the amplitudes of the inertial waves and the geostrophic component i.e. that component due to vortex compression over the obstacle that is vertically uniform and generates anti-cyclonic vorticity. Thus in an infinitely deep fluid the inertial-wave spectrum excited by the obstacle is continuous. However when the depth is finite the modes are discretized and, in particular, the low wavelength modes which are, in general, of the largest amplitude, are eliminated and the contribution from inertial waves becomes weaker. The parameter that determines the relative importance of the geostrophic and inertial components is $H = Ro(L_v/h)$, where h is the obstacle height and L_v the fluid depth. Therefore two distinct cases present themselves; when $H \ll 1$ the geostrophic component prevails and closed streamlines appear above the obstacle and the flow is vertically uniform. Unpublished photographs of a number of cases are shown in figure 10 where flow over a segment of a sphere and a pyramid are shown for various Rossby numbers. The tendency to closed streamlines is apparent but no stagnation is actually observed even at the lowest value of $Ro \sim 0.01$ and $H \sim 5 \times 10^{-2}$.

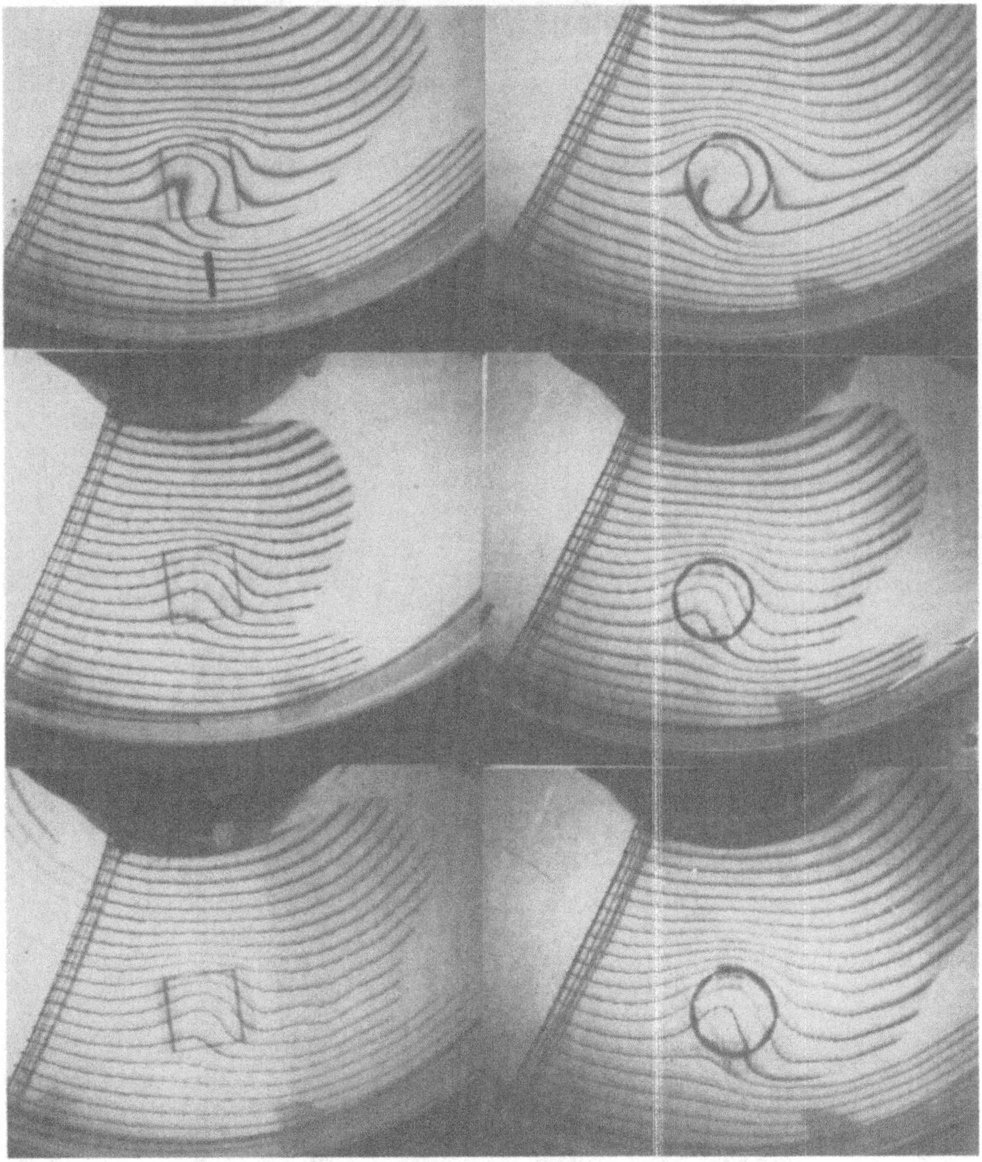

Figure 10: Rotating flow over two obstacle types for $h/L_v \sim 0.2$. Rossby number increasing from top to bottom.

Two other examples are shown in figure 11, where the flow over a slightly elongated obstacle is shown, and figure 12 where flow over a long ridge (simulating motion in the Antarctic Circumpolar Current) is presented. Here again partial stagnation at low R_0 is apparent. In figure 13 we show the inertial waves generated downstream of a transverse ridge for small values of h/L_v

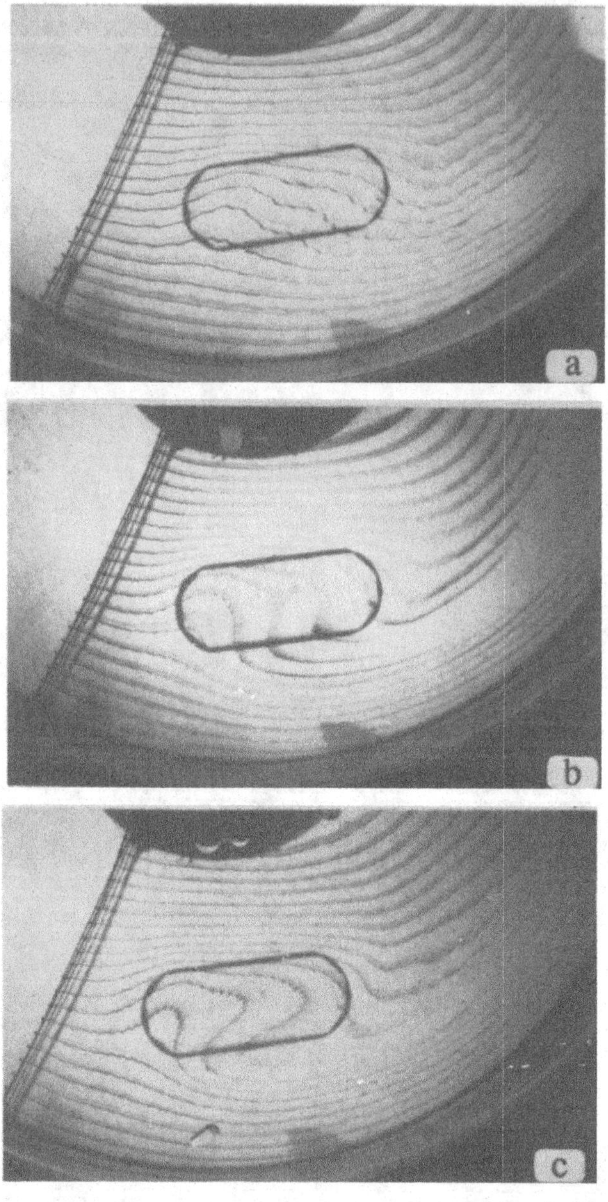

For caption see next page.

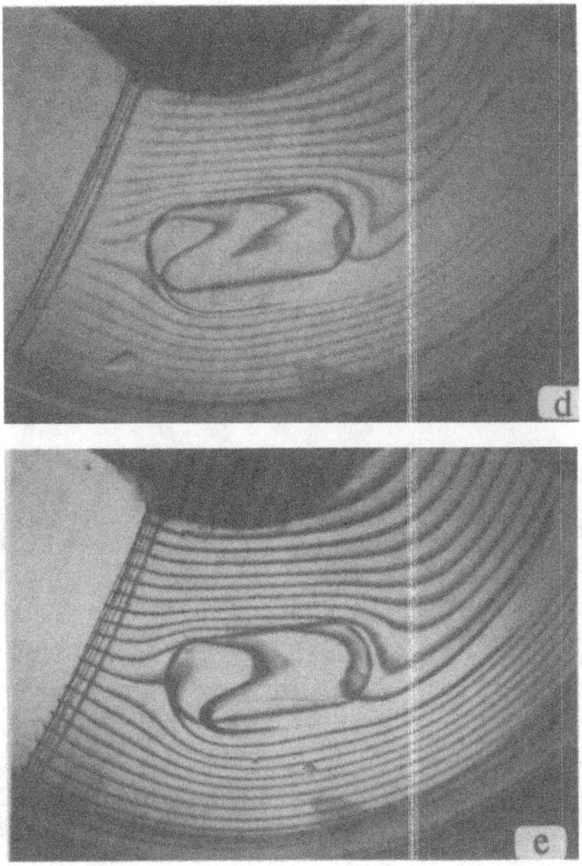

Figure 11: Rotating flow over an elongated obstacle. *Ro* decreasing from (a) to (e).

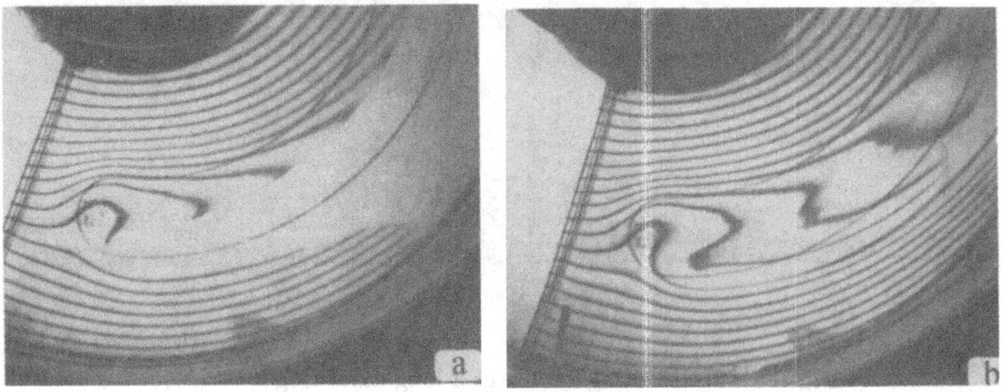

For caption see next page.

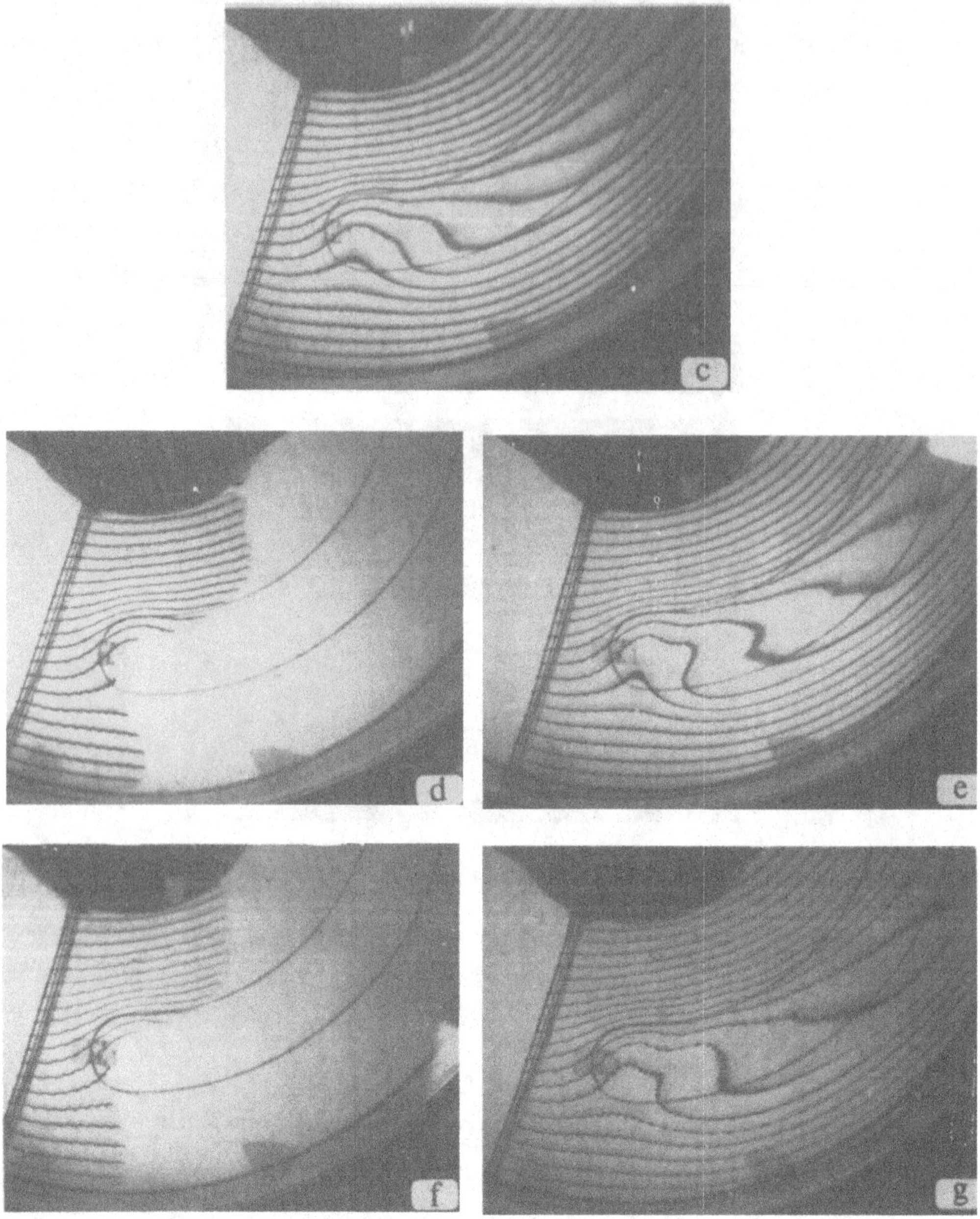

Figure 12: Rotating flow over a long axial ridge.

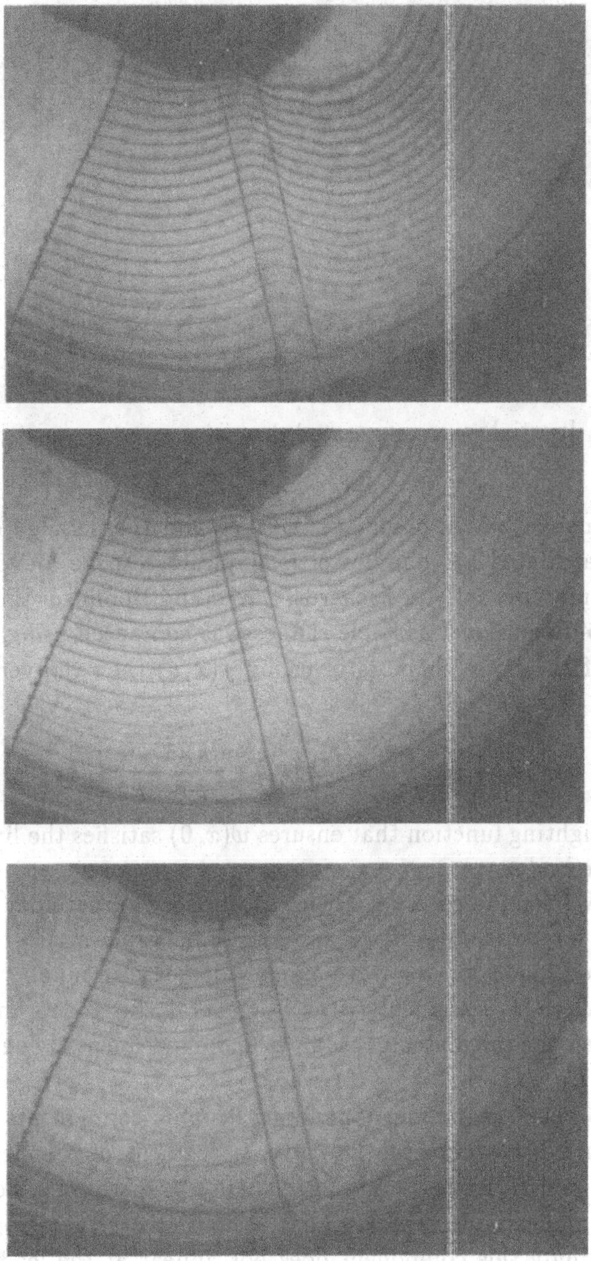

For caption see next page.

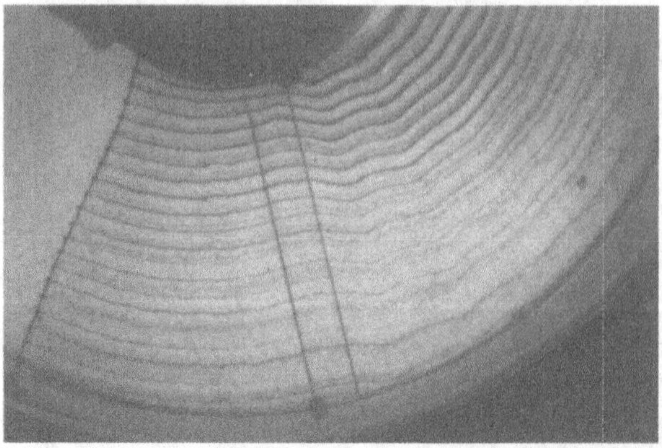

Figure 13: Inertial waves generated downstream of a transverse ridge.

When $H \gg 1$ inertial waves dominate the flow and the detailed structure of the flow field can be calculated. A number of examples are to be found in Heikes and Maxworthy [8] and more recently in Richards, Sneed and Hopfinger [11]. In the former paper flow over two-dimensional obstacle shapes was computed using Fourier integral techniques. Thus if the body shape is given by $f(x, y)$ then the vertical velocity is given by

$$w(x,z) = \frac{1}{2\pi}\int_{-\infty}^{\infty} e^{ik_x x} F(k_x) \frac{\sin[k_x(1-z)]}{\sin k_x} dk_x$$

where $F(k_x)$ is a weighting function that ensures $w(x, 0)$ satisfies the linearized boundary condition on the body.

The form of $F(k_x)$ for the cases of a cylindrical and a trapezoidal body shape are shown in figure 14 for two values of H and the same value of Ro. From this it is clear that the excited modes of low wave-number decrease in number and amplitude as H becomes smaller, in agreement with the general statements made previously. Comparisons between the calculated phase lines and experimentally determined values are shown in figure 15.

The relationship between the inertial and geostrophic components is made clearer by reference to figure 16. where the flow over a segment of a sphere is shown for various values of Ro and H. In particular in the figures the arrows indicate the relative positions of the two modes as Ro decreases, until eventually the geostrophic component dominates. Interestingly this component does not appear at the leading edge of the obstacle, but closer to the maximum height even for $Ro \sim 10^{-3}$, $H \sim 10^{-2}$, indicating that vortex compression does not have an effect immediately but is delayed somewhat. Note, also, in figures 16c and d that the inertial waves appear to be blocked or absorbed by the geostrophic flow, at its leading edge, in particular.

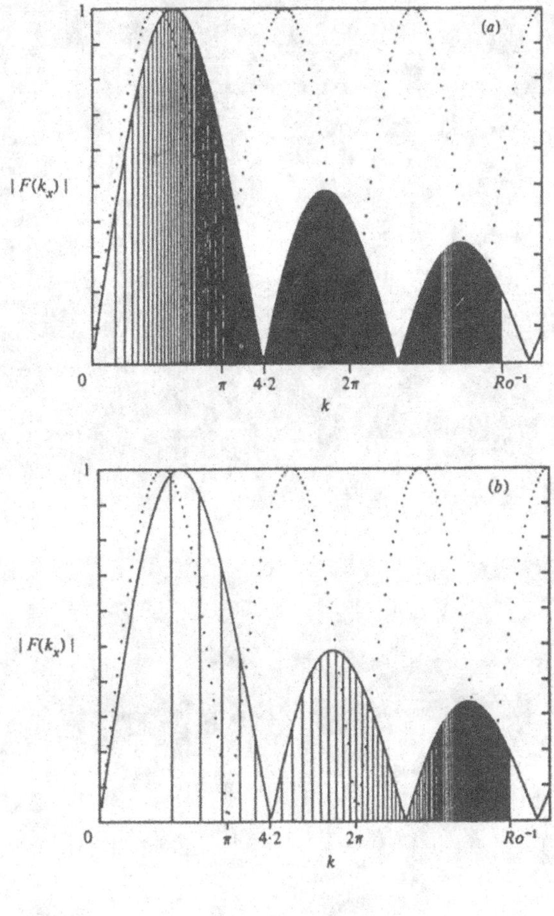

Figure 14: Normalized Fourier transforms, $|F_1(k_x)|/|F_{1max}|$ and $|F_2(k_x)|/|F_{2max}|$, of $partial f_i/\partial x$ for the cylindrical ridge $f_1(x)$ (solid) and the top-hat ridge $f_2(x)$ (dotted) respectively. Vertical lines denote $|F_1(k_x)|/|F_{1max}|$ for k_n^-, $n = 1, 2, 3, \ldots$, for (a) $Ro = 0.1$, $H = 10$ and (b) $Ro = 0.1$, $H = 1$.

(Reproduced from Heikes & Maxworthy [8], with permission)

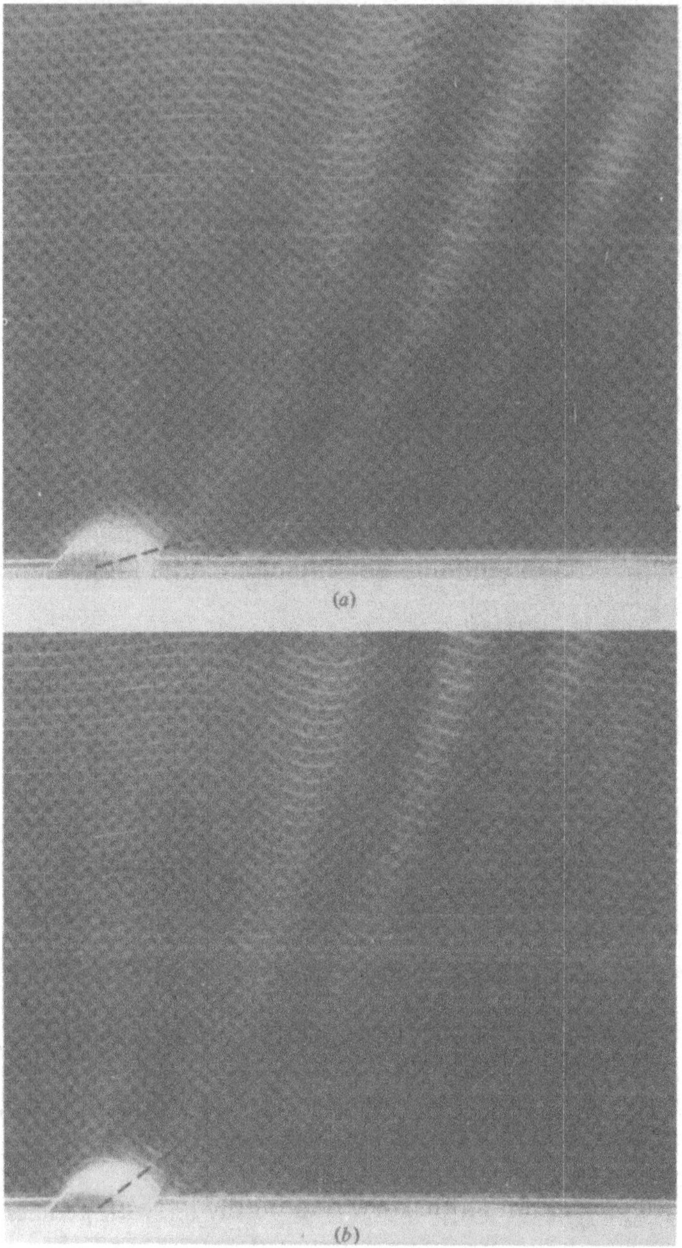

For caption see next page.

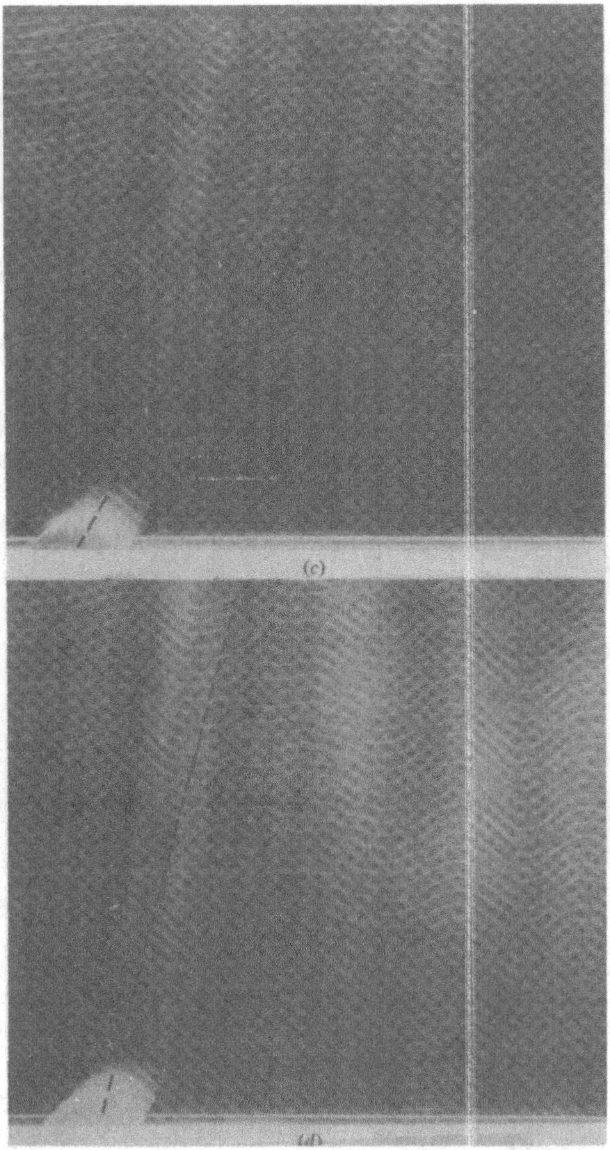

Figure 15: Flow over a cylindrical ridge $f_1(x)$ in the centerplane. Calculated streamlines are denoted by solid curves and group paths for $k_x = 4.20$ are denoted by dashed lines. The top of the photograph is at $z = 0.43$. In each case $L = 3.81$ cm, $D = 91.4$ cm, $h = 2.32$ cm. (a) $Ro = 0.179$, $H = 4.30$, $E = 3.13 \times 10^{-4}$ ($U = 1.37$ cm/s, $\Omega = 1.00$ s^{-1}); (b) $Ro = 0.121$, $H = 2.91$, $E = 3.13 \times 10^{-4}$ ($U = 0.974$ cm/s, $\Omega = 2.01$ s^{-1}); (c) $Ro = 0.0636$, $H = 1.53$, $E = 1.56 \times 10^{-4}$ ($U = 0.974$ cm/s, $\Omega = 2.01$ s^{-1}); (d) $Ro = 0.0310$, $H = 0.785 \times 10^4$ ($U = 0.946$ cm/s, $\Omega = 4.01$ s^{-1}). (Reproduced from Heikes & Maxworthy [8], with permission)

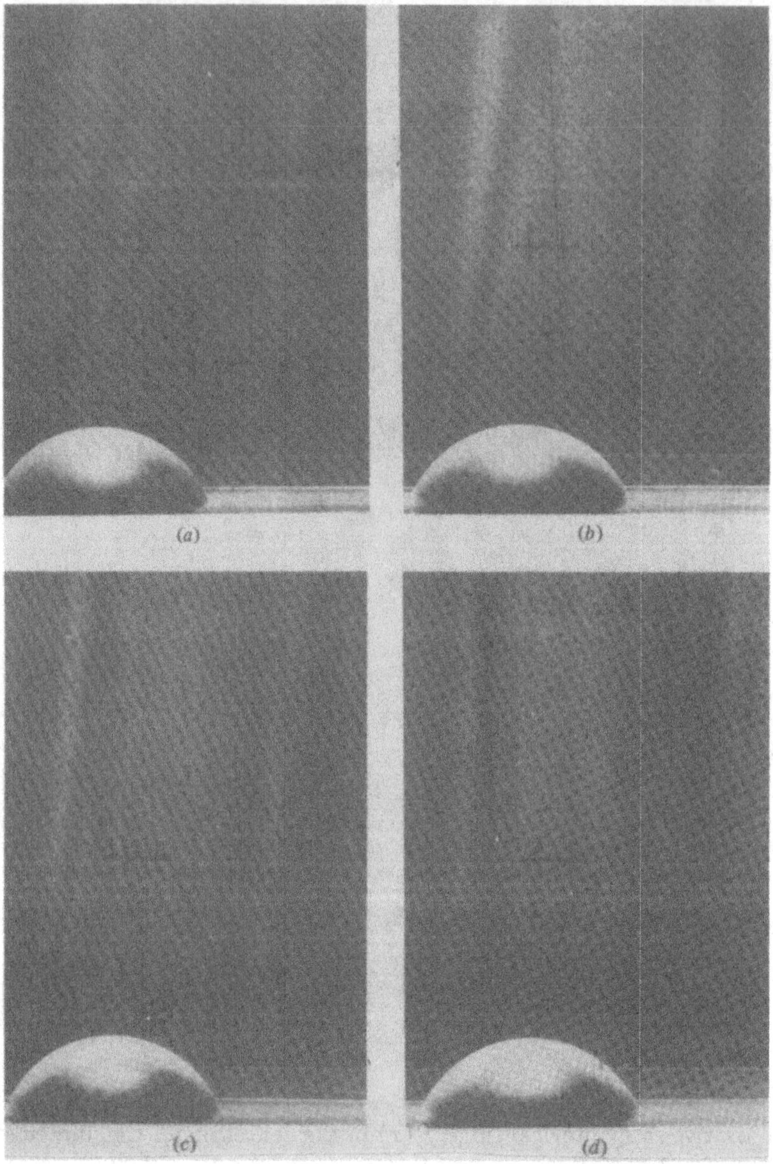

For caption see next page.

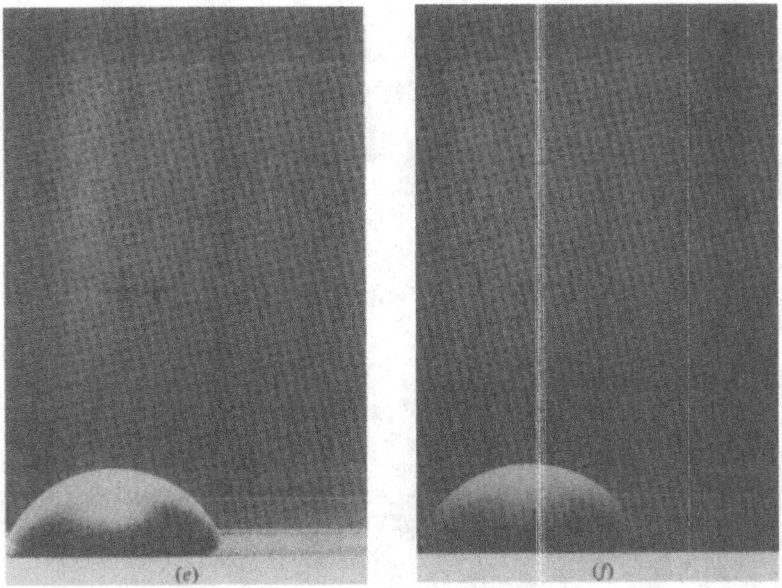

Figure 16: Flow over a spherical cap $f_1(r)$ in the plane $y = 0$. The top of the photograph is at $z = 0.52$. In each case $E = 0.111 \times 10^{-4}$ ($L = 10.2$ cm, $\Omega = 4.01$ s^{-1}, $D = 91.4$ cm, $h = 6.19$ cm). (a) $Ro = 0.0240$, $H = 0.216$ ($U = 1.95$ cm/s); (b) $Ro = 0.0178$, $H = 0.160$ ($U = 1.44$ cm/s); (c) $Ro = 0.0140$, $H = 0.126$ ($U = 1.14$ cm/s); (d) $Ro = 0.0093$, $H = 0.0838$ ($U = 0.758$ cm/s); (e) $Ro = 0.00329$, $H = 0.0296$ ($U = 0.270$ cm/s); (f) $Ro = 0.00150$, $H = 0.0135$ ($U = 0.122$ cm/s).

(Reproduced from Heikes & Maxworthy [8], with permission)

Richards et al. [11] used an identical linear theory to estimate a criterion for separation from compact, axisymmetric obstacles and found reasonable agreement between theory and experiment. In particular they note that when rotation is important separation takes the form of a single, trailing vortex, rotating clockwise when looking downstream (figure 17). In figures 18 and 19 we show a comparison between their calculated surface streamlines and one experiment, the separation criterion appears to be reasonable. Further comparison between theory and experiment is shown in figure 20 where for a hill with a $\cos^2$ shape the agreement is less impressive.

Figure 17: Rotating flow over a Gaussian hill ($h_0 = 10$ cm, $L = 12$ cm) with $R = 0.83$ and $D/L = 2.5$ ($U = 2.5$ cm s^{-1}, $f = 0.25$ rad s^{-1}): (a) plan view; (b) side view (flow from right to left).

(Reproduced from Richards et al. [11], with permission)

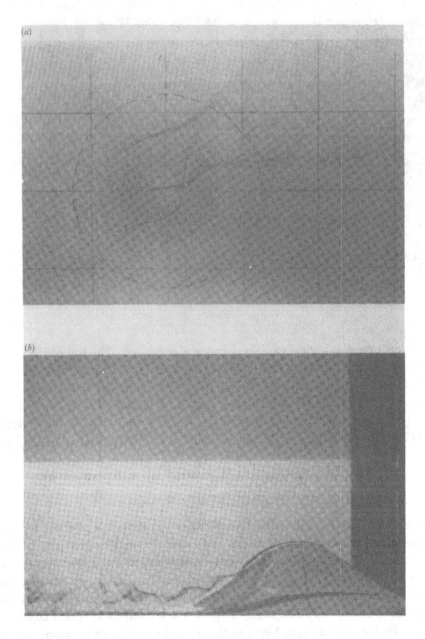

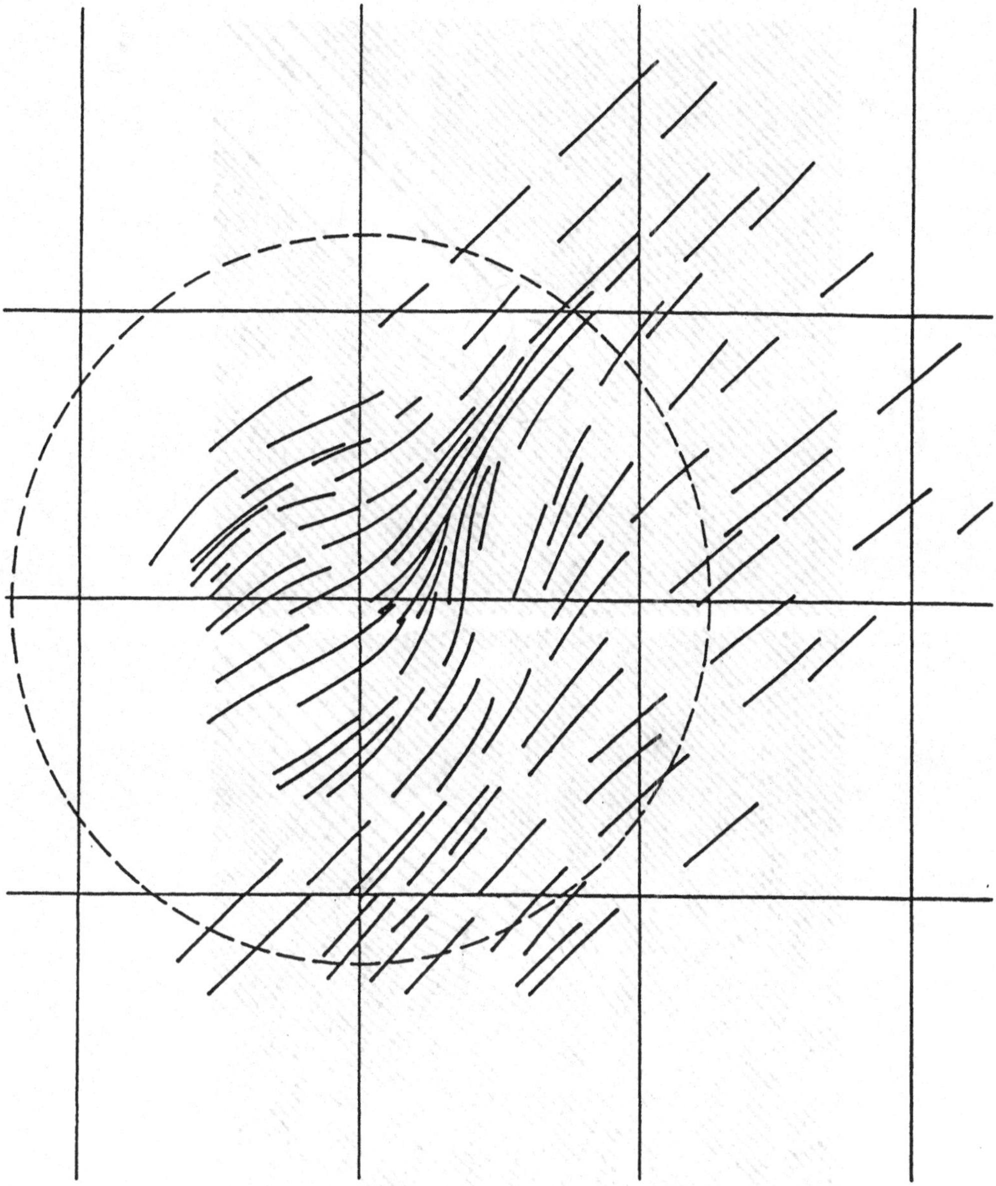

Figure 18: Surface stress field, visualized using potassium permanganate crystals, for rotating flow over a Gaussian hill ($h_0 = 5$ cm, $L = 12$ cm) with $R = 2.6$ and $D/L2.5$ ($U = 4$ cm s^{-1}, $f = 0.13$ rad s^{-1}).

(Reproduced from Richards et al. [11], with permission)

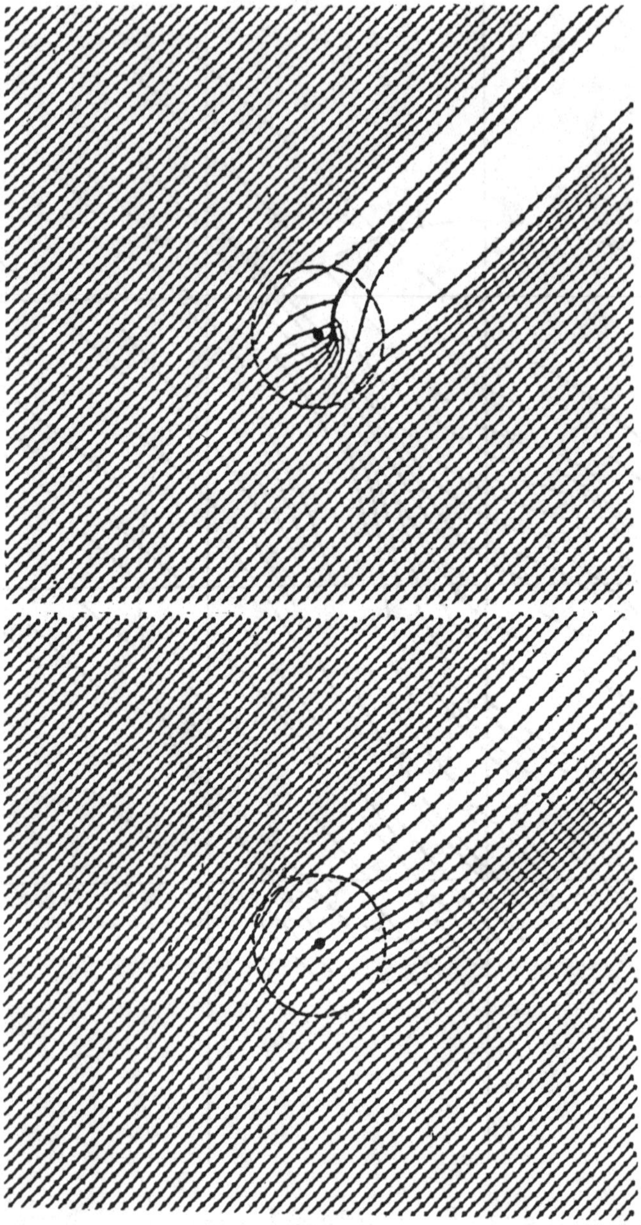

Figure 19: Calculated surface streamlines over a Gaussian hill: (a) $Ro = 0.5$, (b) $Ro = 2.0$.

(Reproduced from Richards et al. [11], with permission)

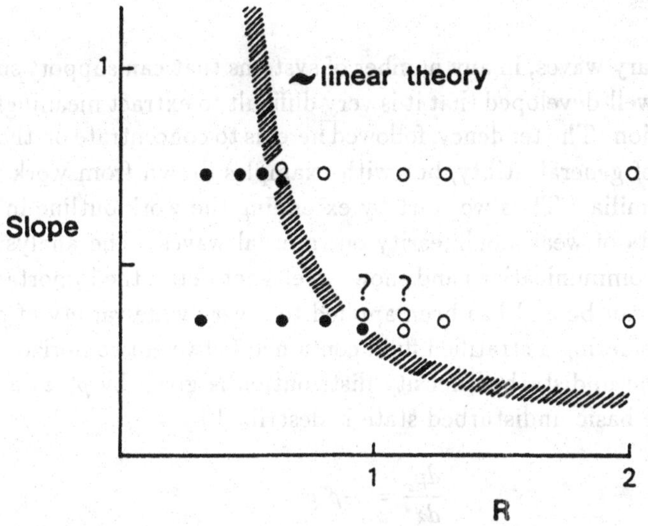

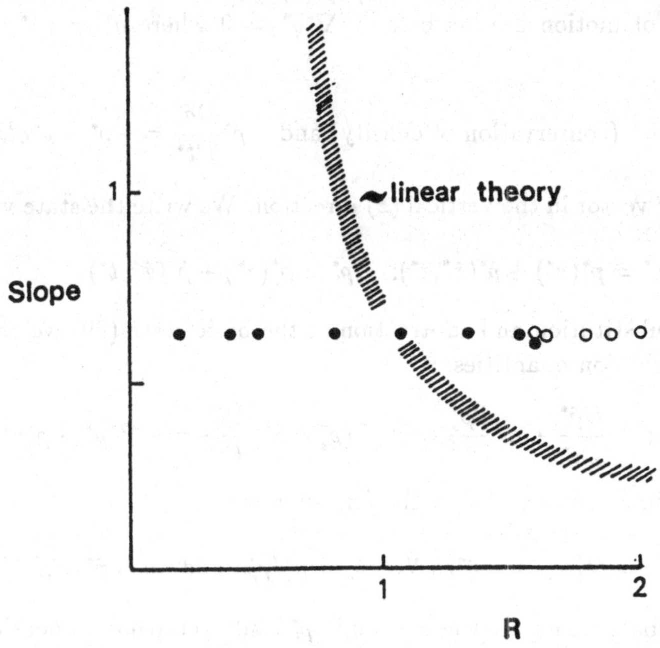

Figure 20: Comparison between linear-theory and experiment for the existence of separation as a function of Rossby number and maximum surface slope: (a) Gaussian hill, (b) $\cos^2$ hill.

(Reproduced from Richards et al. [11], with permission)

6 Solitary Waves

The subject of solitary waves, in any number of systems that can support such entities, is so extensive and well-developed that it is very difficult to extract meaningful material for such a short section. The tendency, followed here, is to concentrate on those subjects that appear to be of general utility, but with examples drawn from work with which one is intimately familiar. Thus we start by extending the work outline in section 1.2 to include the effects of weak nonlinearity on internal waves. The analysis is due to Redekopp (private communication) and shows in elegant detail the important elements of a technique that can be and has been applied to a very wide variety of problems.

We start by considering a stratified fluid contained between two horizontal surfaces at $z = 0$ and L. The undisturbed density distribution is given by $\rho_s^*(z)$ and pressure by $p_s^*(z)$ so that the basic undisturbed state is described by:

$$\frac{dp_x^*}{dz^*} = -\rho_s^* g \tag{39}$$

the undisturbed distribution of $N^{*2}(z^*) = (g/\rho_s^*)(\partial\rho_s^*/\partial z^*)$ with a maximum value N_0^{*2}.

The equations of motion are, as before: $\nabla^*\vec{q}^* = 0$ where $\vec{q}^* = (u^*, v^*, w^*)$ and $\vec{x}^* = (x^*, y^*, z^*)$.

$$\frac{D\rho^*}{Dt^*} = 0 \quad \text{(conservation of density) and} \quad \rho^*\frac{D\vec{q}^*}{Dt^*} = -p^* - \rho^* g\hat{l}_z$$

where $\hat{l}_z$ is the unit vector in the vertical (z) direction. We write the state variables as:

$$p^* = p_s^*(z^*) + \tilde{p}^*(\bar{x}^*, t^*); \qquad \rho^* = \rho_s^*(z^*) + \tilde{\rho}^*(\bar{x}^*, t^*)$$

So that upon substitution and subtraction of the basic state (39) we obtain equations for the perturbation quantities:

$$\nabla^* q^* = 0; \quad \frac{D\tilde{\rho}^*}{Dt^*} + w^*\frac{\partial\rho_s^*}{\partial z^*} = 0; \quad (\rho_s^* + \tilde{\rho}^*)\frac{D\bar{q}^*}{Dt^*} = -\nabla^*\tilde{p}^* - \tilde{\rho}^* g\hat{l}_z$$

the variables therein are now scaled in the following way

$$\bar{x} = \bar{x}^*/h; \; t = N_0^* t^*; \; \bar{q} = \bar{q}^*/hN_0^*; \; \rho_s = \rho_s^*/\rho_0^* \; \text{ and } \; \tilde{p} = \tilde{p}^*/\rho_0^* N_0^{*2} h^2.$$

Defining the perturbation buoyancy as $g'^* = g\tilde{\rho}^*/\rho_0^*$ leads to the non-dimensionalization $g' = g'^*/N_0^{*2}h$, where ρ_0^* is the average density.

Thus the non-dimensional equations of motion become:

$$
\begin{aligned}
u_x + v_y + w_z &= 0 \\
g'_t + ug'_x + vg'_y + wg'_z - w\mathcal{N}^2(z) &= 0 \\
(\rho_s + Fg')[u_t + uu_x + vu_y + wu_z] &= -p_z \\
(\rho_x + Fg')[v_t + uv_x + vv_y + wv_z] &= -p_y \\
(\rho_x + Fg')[w_t + uw_x + vw_y + ww_z] &= --p_z - g'
\end{aligned}
\tag{40}
$$

where

$$\mathcal{N}^2(z) = \frac{N^{*2}(z^*)}{N_0^{*2}} = \frac{-g\rho^*_{s_z}}{\rho_0^* N_0^{*2}} \tag{41}$$

and

$$F = \frac{N_0^{*2} h}{g} = \frac{-g\rho^*_{s_z}}{\rho_0^*}\frac{h}{g} \sim \frac{\Delta\rho^*}{\rho_0^*} \tag{42}$$

Note the Boussinesq limit is recovered by taking $F \rightarrow 0$. Consider the case of two-dimensional ($v = \partial/\partial y = 0$) long-waves, and introduce the appropriate Kortweg and DeVries (KdV) scaling: $\xi = \epsilon^{1/2}(x - c_0 t)$, $\tau = \epsilon^{3/2} t$ so that:

$$\frac{\partial}{\partial x} = \epsilon^{1/2}\frac{\partial}{\partial z} \text{ and } \frac{\partial}{\partial t} = -\epsilon^{1/2} c_0 \frac{\partial}{\partial \xi} + \epsilon^{3/2}\frac{\partial}{\partial \tau}$$

where ϵ is a perturbation parameter that measures the wave amplitude.

The transformed equations become:

$$
\begin{aligned}
\epsilon^{1/2} u_\xi + w_z &= 0 \\
\epsilon^{3/2} g'_\tau + \epsilon^{1/2}(u - c_0) g'_\xi + w g'_z - w\mathcal{N}^2 &= 0 \\
(\rho_s + Fg')[\epsilon^{3/2} u_\tau + \epsilon^{1/2}(u - c_0) u_\xi + w u_z] &= -\epsilon^{1/2} p_\xi \\
(\rho_s + Fg')[\epsilon^{3/2} w_\tau + \epsilon(u - c_0) w_\xi + w w_z] &= p_\xi - g'
\end{aligned}
\tag{43}
$$

We now expand the dependent variables in powers of ϵ:

$$u = \epsilon u^{(1)} + \epsilon^2 u^{(2)} + \cdots \quad \text{etc.}$$

with

$$w = \epsilon^{1/2}[\epsilon w^{(1)} + \epsilon^2 w^{(2)} + \cdots]$$

dictated by the continuity equation.

Substitution into (43) yields:

$$(u_\xi^{(0)} + w_z^{(1)}) + \epsilon(u_\xi^{(2)} + w_z^{(2)}) + \cdots = 0$$

$$-(c_0 g_\xi^{'(1)} + w^{(0)}\mathcal{N}^2) + \epsilon\{-(c_0 g_\xi^{'(2)} + w^{(2)}\mathcal{N}^2) + g_\tau^{(1)} + u^{(1)} g_\xi^{'(1)} + w^{(1)} g_z^{'(1)}\} + \cdots = 0$$

$$(44)$$

$$(p_z^{(1)} - c_0\rho_s u_\xi^{(1)}) + \epsilon\{p_\xi^{(2)} - c_0\rho_s u_\xi^{(3)} + \rho_s[u_\tau^{(1)} + u^{(1)}u_\xi^{(1)} + w^{(1)}u_z^{(1)}] + Fg'^{(1)}(-c_0 u_\xi^{(1)}\} + \cdots = 0$$

$$(p_z^{(1)} + g'^{(1)}) + \epsilon\{p_z^{(2)} + g'^{(2)} - \rho_s c_0 w^{(1)}\} + \cdots = 0$$

The order unity balance gives:

$$u_\xi^{(1)} = -w_z^{(1)} \text{—(a)};\; g_\xi^{'(1)} = -\frac{N^2}{c_0} w^{(1)} \text{—(b)}$$

$$(45)$$

$$p_\xi^{(1)} - c_0\rho_s u_\xi^{(1)} = 0 \text{—(c)};\; p_z^{(1)} + g'^{(1)} = 0 \text{—(d)}$$

Combining equations (45) gives

$$(\rho_s w_z^{(1)})_z + \frac{\mathcal{N}^2}{c_0^2} w^{(1)} = 0 \tag{46}$$

the vertical structure equation for long waves! Considering a separable solution of the form

$$w^{(1)}(\xi, z, \tau) = -\phi(z)\frac{\partial A(\xi,\tau)}{\partial \xi}$$

leads to the eigenvalue equation for the longwave speed (c_0)

$$(\rho_s\phi_z)_z + \frac{\mathcal{N}^2(z)}{c_0^2}\phi = 0 \tag{47}$$

with boundary conditions $\phi(0) = \phi(1) = 0$.

[Note: In the Boussinesq limit ($\rho_s = 1$) and for constant $N^2 = 1$ this equation becomes

$$\phi'' + \frac{1}{c_0^2}\phi = 0$$

with eigensolutions $\phi_n = \sin n\pi z$ and $c_0 = 1/n\pi$ or $c_0^* = N_0^* h/n\pi$]

The order ϵ balance equation can be evaluated in a similar fashion so that after considerable algebra one obtains a non-homogeneous equation for $w^{(2)}$:

$$(\rho_s w_z^{(2)})_z + \frac{\mathcal{N}^2 w^{(2)}}{c_0^2} = \frac{2\mathcal{N}^2}{c_0^3}\phi A_\tau - \rho_s\phi A_{\xi\xi\xi} - \frac{1}{c_0}\left\{\phi(\rho_{s_z}\phi_z)_z + \frac{2(\mathcal{N}^2)_z}{c_0^2}\phi^2 - F(\mathcal{N}^2\phi\phi_z)_z\right\} AA_\xi \tag{48}$$

The right hand side has the expected KdV terms. To obtain that equation we use a solvability condition by multiplying both sides by ϕ and integrating over the interval $h \geq z \geq 0$, to obtain:

$$A_\tau\left\langle\frac{2\mathcal{N}\phi^2}{c_0^3}\right\rangle-\frac{1}{c_0}\left\langle\phi^2(\rho_{sz}\phi_z)_z+\frac{2(\mathcal{N}^2)_z}{c_0^2}\phi^3-F(\mathcal{N}^2\phi\phi_z)_z\right\rangle AA_\xi-\langle\rho_s\phi^2\rangle A_{\xi\xi\xi}=0 \quad (49)$$

where $\langle\rangle \equiv \int_0^h(\ )dz$.

In the Boussinesq limit ($\rho_s = 1$, $F \to 0$) the coefficient of the nonlinear term becomes:

$$-\frac{2}{c_0^3}\langle(\mathcal{N}^2)_z\phi^3\rangle.$$

So that the nonlinear term vanishes when $\mathcal{N}^2 =$ constant, i.e. in this case the linear wave mode satisfies the full nonlinear equations with no restriction on wave amplitude. Also note that in this limit we can recover a nonlinear term if we allow for non-Boussinesq effects.

Equation (49) can be written in a more compact form namely:

$$A_\tau + R_0AA_\xi + S_0A_{\xi\xi\xi} = 0 \quad (50)$$

Where, obviously R_0 and S_0 are given by dividing the coefficients of AA_ξ and $A_{\xi\xi\xi}$ respectively by the coefficient of A_τ.

Equation (50) has a solitary wave solution in the scaled coordinate system moving with the zeroth order wave-speed (c_0)

$$A(\xi,\tau) = \text{sgn}\,(R_0S_0)\,\text{sech}^2\left\{\left|\frac{R_0}{12S_0}\right|^{1/2}(\xi - c_1\tau)\right\} \quad (51)$$

where $c_1 = -\frac{1}{3}|R_0|\ \text{sgn}S_0$.

This wave of permanent form then represents a balance between the tendency to spread due to wave dispersion and the tendency to steepen due to nonlinearity.

Under certain circumstances the term R_0 can vanish and then it is necessary to carry the expansion to one order higher in ϵ, at which level the modified KdV equation is obtained i.e. one with a nonlinear term of the form A^2A_ξ.

Equations of this type apply to a wide range of physical circumstances e.g. waves on a free surface, vortex core, rotating-stratified atmosphere with horizontal shear, liquid-bubble mixture, cold plasmas, nonlinear lumped circuits etc., etc.

In particular Redekopp [12] gives a very useful classification of different wave types that can be obtained in a simple, two-layer stratification. Here the connection between various types of wave equations, as the relative depths of the two layers is varied, can be seen very clearly.

Thus for two layers of depths h_1 and h_2 and density ρ_1 and ρ_2 contained between horizontal walls a distance h_1 and h_2 apart the full dispersion relationship is:

$$\sigma^2(k) = \frac{gk(\rho_2 - \rho_1)T_1T_2}{\rho_1 T_2 + \rho_2 T_1} \tag{52}$$

where $T_i = \tanh(kh_1)$

In the shallow water limit ($k \to 0$, h_1 and h_2 fixed) we obtain

$$\sigma^2 = c_0^2 k^2 - 2c_0\gamma k^4 + \ldots \tag{53}$$

where

$$c_0^2 = \frac{g(\rho_2 - \rho_1)h_1h_2}{\rho_1 h_2 + \rho_2 h_1}; \quad \gamma = \frac{1}{6}c_0 h_1 h_2 \frac{\rho_1 h_1 + \rho_2 h_2}{\rho_1 h_2 + \rho_2 h_1}$$

Considering only waves propagating to the right and retaining only the first dispersive term leads to $\sigma - c_0 k = -\gamma k^3$ which defines the relationship between the slow space and time scales for evolution in a frame moving with velocity c_0, $\xi = \mu(x - c_0 t)$, $\tau = \mu^3 t$, as used before. Here μ is a parameter measuring wavelength to wave-guide height. If the wave is also weakly nonlinear (i.e. $\epsilon = a/h_1 \ll 1$) and if $\mu \sim O(\epsilon^{1/2})$ then the wave equation for the interface shape (ζ) becomes:

$$\zeta_\tau + \gamma\zeta_{\xi\xi\xi} + \nu\zeta\zeta_\xi = 0 \tag{54}$$

where

$$\nu = \frac{3}{2}\frac{c_0}{h_1h_2}\frac{\rho_2 h_1^2 - \rho_1 h_2^2}{\rho_2 h_1 + \rho_1 h_2}$$

Note that when $\rho_2 h_1^2 = \rho_1 h_2^2$ one must expand to higher order to obtain the modified KdV equation, a consideration of importance when one considers waves propagating over a gently sloping bottom, e.g., the continental shelves of the world's oceans.

To study the deep water limit of (52) we take the limits: (1) $h_2 \to \infty$, k and h, fixed; (2) $k \to 0$ h, fixed, to obtain:

$$\sigma^2 = c_0^2 k^2 - 2\beta c_0 k^3 \operatorname{sgn} k \tag{55}$$

where $c_0^2 = \frac{\rho_2 - \rho_1}{\rho_1 g h_1}$ and $\beta = \frac{1}{2}\frac{\rho_2}{\rho_1}c_0 h_1$ So here again on considering propagation in one direction we obtain

$$\sigma - c_0 k = -\beta k|k| \tag{56}$$

suggesting a scaling $\xi = \mu(x - c_0 t)$, $\tau = \mu^2 t$. Assuming a weak nonlinearity leads to the Benjamin-Ono-(Davis) equation:

$$\zeta_\tau + \gamma\zeta\zeta_\xi + \beta\frac{\partial^2}{\partial\xi^2}H(\zeta) = 0 \tag{57}$$

Where $H(\zeta)$ is the Hilbert transform and $\gamma = \frac{3}{2}\frac{c_0}{h_1}$ Equation (57) has the solution

$$\zeta = \frac{a\lambda^2}{(x-ct)^2+\lambda^2}\,;\quad \frac{a\lambda}{h_1^2} = \frac{4}{3}\frac{\rho_2}{\rho_1}$$

with a wave speed correction $c = c_0(1+\frac{3}{4}\frac{a}{h_1})^{1/2}$

In his original derivation of this equation Benjamin [13] considered several cases but in particular the case when the density profile was given by $\rho(\eta) = \bar{\rho}(1-\bar{w}\tanh\alpha\eta)$ where $\bar{w} = (\rho_2-\rho_1)/(\rho_2+\rho_1)$ which gave a wave-speed with a linear correction term.

$$c = \frac{2g'}{\alpha}0.35\left(1+\frac{3}{10}a\alpha\right)$$

where a is the wave amplitude and η the vertical coordinate.

Maxworthy [14] has experimentally determined the validity of this latter result and found good agreement for wave amplitudes αa less than unity (figure 21).

This system also exhibits an intermediate depth limit in which one depth is much larger than the other such that $kh_2 = O(1)$ and $kh_1 = O(\mu)$ (Joseph [15]; Redekopp [12]).

We note that solitary waves of the types discussed here can interact in a very interesting way. As discussed by many authors two waves of different amplitude i.e.

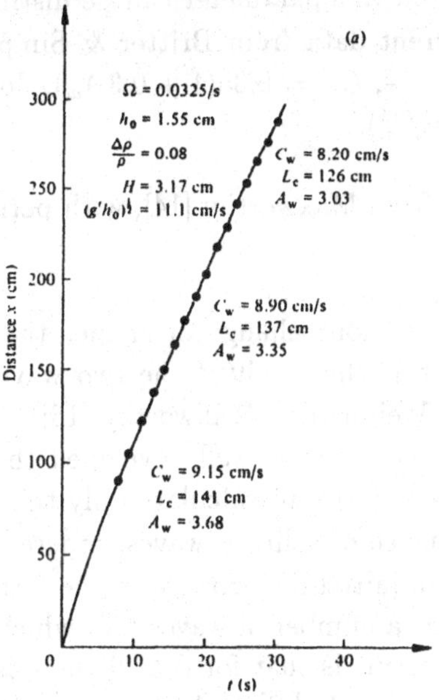

For caption see next page.

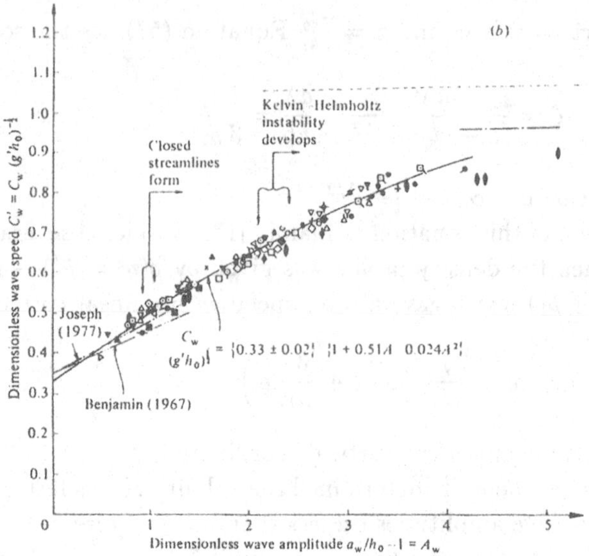

Figure 21: (a) Typical wave history, displacement x versus time t. Local values of wave velocity C_w, amplitude d_w and Rossby radius L_c are shown. (b) Non-dimensional local wave velocity $C_w/(g'h_s)$ versus non-dimensional wave amplitude $A_w = a_w/h_0 - 1$. These results represent values taken over a 20-fold range of L; no dependence on this parameter can be distinguished. $-\diamond-$, from Faust (1981). Gravity-current data from Britter & Simpson (1978, $\cdots$; 1981, -----); Benjamin (1967), $-\cdot\cdot-$, $C_w = 0.35(1 + 0.3A_w)$; Joseph (1977), $-\cdot\cdot\cdot-$ $C_w = 0.35(1 + 0.3A_w(1 + h_0/D)$.

(Reproduced from Maxworthy [14], with permission)

different waves speeds, interact without change of shape, the only memory of the interaction is a phase shift which is the result of the two waves exchanging fluid as they approach one another (e.g. Weidman & Maxworthy [16]).

Finally the theory of these systems is so well developed that it is possible using the so-called inverse-scattering transform, to calculate fully the evolution of an initial compact disturbance into a sequence of solitary waves, ordered by amplitude, and a dispersive wave-train. Thus for an initial disturbance in the form of a square pulse of length l, and height A one obtains a number of waves (N) which is the largest integer less than $S = lA^{1/2}/\pi + 1$. The result is that for $S > 0$ there is at least one solitary wave which, for small S, has an amplitude $\frac{1}{2}S^2A$. As S increases a second solitary wave appears when $S = \pi$ at which point the first wave has an amplitude $1.30A$ (Whitham [17]).

As mentioned previously numerous examples exist in nature a number of these will now be discussed briefly.

The most dramatic examples of internal solitary wave generation occur in regions of the world's oceans where barotropic tidal flow is large and the ambient is strongly stratified. A number of examples are given in Fu & Holt [18] where the surface signature of the waves can be seen clearly in the return from the Seasat SAR. In this case the surface convergence due to the internal wave motion modifies the short-surface-wave field which in turn modulate the reflected radar signal. In most cases the waves are generated by tidal flow over bottom topographic features e.g. ridges, the continental shelf-break etc. The basic mechanism of production has been outlined by Maxworthy [19] and [20] and Maxworthy et al. [21] based on prior work by Lee & Beardsley [22]. Under the simplest possible circumstances a two layer barotropic tidal flow over a ridge (figure 22) a downstream, standing lee-wave forms when the flow is supercritical. The phase velocity of this wave points upstream. As the tidal flow is reduced this wave can propagate upstream against the weaker on-coming flow. This wave or pulse can then evolve into a sequence of solitary waves since the basic stratification can support such entities which obey the KdV equation, as already dicussed. Under many natural circumstances the initial wave is of such a large amplitude that mixing can occur and this mixed fluid can in turn act as a wave-generating piston which enhances the wave production mechanism discussed above but also can generate waves propagating in the opposite direction, due to the mixed-region collapse.

Recently it has been realized that similar collapse phenomena can occur when intense convection (e.g. that due to intense storms) interacts with the tropopause and generates solitary waves. These waves then propagate away and can trigger squall-line activity far from their source. Similar mechanisms e.g. sea-breeze interactions, can generate waves some of which result in spectacular consequences, as in the Morning Glory of Northern Australia and the Southerly Buster along the Australian east coast.

Acknowledgments

Firstly I would like to thank the International Center for Mechanical Sciences in Udine and its director Professor G. Bianchi for inviting me to give these lectures at the kind suggestion of Professor E. Hopfinger, the course director.

I owe especial thanks to Dr. F. Bretherton who introduced me to many of the topics discussed here at the WHOI Summer School in 1965! Also the profound influence of Drs. A. Gill and J. Pedlosky is to be found in these notes, while my close colleagues E. J. Hopfinger, F. K. Browand, S. Narimousa, L. G. Redekopp among many others have strongly affected my attempts to think about the problems discussed here.

References

[1] Hopfinger, E.J., General concepts and examples of rotating fluids, mis volume (1992)

[2] Roberts, J., *Internal Gravity Waves in the Ocean*, Marcel Dekker, New York, 1975.

[3] Gill, A., *Atmosphere Ocean Dynamics*, Academic Press, New York, 1982.

[4] Long, R. R., "Some aspects of the flow of stratified fluids, I. A theoretical investigation", *Tellus*, **5**, 42, 1953.

[5] Davis, R. E., "Two-dimensional flow of a stratified fluid over an obstacle", *J. Fluid Mech.*, **36**, 127–143, 1969.

[6] Gossard, E. E. and Hooke, W., *Waves in the Atmosphere*, Elsevier, Amsterdam, 1979.

[7] Matsuno, T., "Quasi-geostrophic motions in the equatorial area", *J. Met. Soc. Japan*, **44**, 25–42, 1966.

[8] Heikes, K. and Maxworthy, T., "Observations of inertial waves in a homogeneous, rotating fluid", *J. Fluid Mech.*, **125**, 319–345, 1982.

[9] Hide, R., Ibbotson, A. and Lighthill, M. J., "On slow transverse flow past obstacles in a rapidly rotating fluid", *J. Fluid Mech.*, **67**, 397–412, 1968.

[10] Stewartson, K. and Cheng, H. K., "On the structure of inertial waves produced by an obstacle in a deep, rotating container", *J. Fluid Mech.*, **91**, 415–432, 1979.

[11] Richards, K. J., Smeed, D. A. and Hopfinger, E. J., "Boundary layer separation of rotating flows past surface-mounted obstacles", *J. Fluid Mech.*, in press, 1991.

[12] Redekopp, L. G., "Nonlinear waves in geophysics: Long internal waves", *Lectures in Appl. Math.*, **20**, 59–78, 1983.

[13] Benjamin, T. B., "Internal waves of finite amplitude and permanent form", *J. Fluid Mech.*, **25**, 241–270, 1966.

[14] Maxworthy, T., "Experiments on solitary internal Kelvin waves", *J. Fluid Mech.*, **129**, 365–383, 1983.

[15] Joseph, R. I., "Solitary waves in a finite depth fluid", *J. Phy. A: Math (Gen.)*, **10**, L255, 1977.

[16] Weidman, P. D. and Maxworthy, T., "Experiments on strong interactions between solitary waves", *J. Fluid Mech.*, **85**, 417, 1978.

[17] Whitham, G. B., *Linear and Non-Linear Waves*, Wiley, New York, 1974.

[18] Fu, L.-L., and Holt, B., "Seasat views oceans and sea-ice with synthetic-aperture radar", *Jet Propulsion Laboratory Publication 81–120*, Pasadena, 1982.

[19] Maxworthy, T., "A note on the internal solitary waves produced by tidal flow over a three-dimensional ridge", *JGR*, **84**, C.1, 338-346, 1979.

[20] Maxworthy, T., "A mechanism for the generation of internal solitary waves by tidal flow over submarine topography", *Ocean Modelling*, **14**, University of Cambridge, 1978 (Unpublished manuscript).

[21] Maxworthy, T., Chabert d'Hieres, G. and Didelle, H., "The generation and propagation of internal gravity waves in a rotating fluid", *JGR*, **89**, C4, 6383–6396, 1984.

[22] Lee, C. Y. and Beardsley, R. C., "The generation of long, nonlinear internal waves in a weakly stratified shear-flow", *JGR*, **79**, 453–462, 1974.

Chapter IV.2

INERTIAL WAVES

M. Mory
University J.F. and CNRS, Grenoble Cedex, France

Inertial waves are an essential feature of rotating fluids. They play a particularly important role in the process of two dimensionalisation by rotation of the flow field. The basic theory of linear inertial waves is presented here. Two limiting cases are then discussed in more detail: (i) when the frequency is just below the Coriolis frequency ("near-inertial waves"), and (ii) when the frequency of oscillations is small compared to the Coriolis frequency 2Ω. Lastly, experimental studies showing resonance phenomena of inertial waves are described.

1. THE THEORY OF LINEAR INERTIAL WAVES

11. The dispersion relationship

In order to examine the possibility of wave propagation in a rotating fluid, the Navier-Stokes equations are written in a coordinate system rotating around the Oz axis with angular velocity Ω. These equations are simplified under the following assumptions:

(i). the fluid is inviscid,

(ii). there is no mean flow component. The velocity components u,v and w are the only contributions of inertial waves,

(iii). the density is homogeneous (the combination of inertial and internal waves is treated in this volume by Maxworthy [1]),

(iv). small amplitude motions are considered so that the equations are linearised.

The Navier-Stokes equations are thus simplified into the form:

$$\frac{\partial u}{\partial t} - 2\Omega\, v = -\frac{1}{\rho}\frac{\partial p}{\partial x} \tag{1}$$

$$\frac{\partial v}{\partial t} + 2\Omega\, u = -\frac{1}{\rho}\frac{\partial p}{\partial y} \tag{2}$$

$$\frac{\partial w}{\partial t} = -\frac{1}{\rho}\frac{\partial p}{\partial z} \ . \tag{3}$$

The continuity equation is

$$\frac{\partial u}{\partial x} + \frac{\partial v}{\partial y} + \frac{\partial w}{\partial z} = 0 \ . \tag{4}$$

This equation set is assumed to have wavelike solutions:

$$u = U \exp\left[i\,(\vec{k}.\vec{x} - \omega t\,)\right]$$

$$v = V \exp\left[i\,(\vec{k}.\vec{x} - \omega t\,)\right]$$

$$w = W \exp\left[i\,(\vec{k}.\vec{x} - \omega t\,)\right]$$

$$p = P \exp\left[i\,(\vec{k}.\vec{x} - \omega t\,)\right] \tag{5}$$

ω is the wave frequency and $\vec{k}$ is the wave vector with wavenumber components k_x,k_y and k_z. Introducing this solution in (1) to (3) enables the velocity amplitudes U, V and W to be written in terms of the pressure amplitude P/ρ. This gives:

$$U = \frac{P}{\rho}\,\frac{-i\,2\Omega\, k_y - \omega\, k_x}{4\,\Omega^2 - \omega^2} \tag{6}$$

$$V = \frac{P}{\rho}\,\frac{i\,2\Omega\, k_x - \omega\, k_y}{4\,\Omega^2 - \omega^2} \tag{7}$$

$$W = \frac{P}{\rho}\frac{k_z}{\omega} \quad . \tag{8}$$

The dispersion relationship is deduced by introducing (6) to (8) into the continuity equation (4):

$$\omega^2 = 4\,\Omega^2 \, \frac{k_z^2}{k_x^2 + k_y^2 + k_z^2} \quad . \tag{9}$$

The first observation emerging from the dispersion relationship (9) is that inertial waves are low frequency waves. They propagate with a frequency lower than the Coriolis frequency 2Ω. Oscillatory motions with a frequency greater than 2Ω and wave number components k_x and k_y real require that the wavenumber component k_z be imaginary complex. This solution makes sense only when the fluid is bounded in the Oz direction; the perturbation decays exponentially in this direction.

12. The kinematics of inertial waves

The values obtained for the amplitude of the velocity components (eqs. 6 to 8) readily show that the motion of a fluid element takes place in the plane of constant phase in which the element is located because

$$k_x U + k_y V + k_z W = 0 \quad . \tag{10}$$

Furthermore it can be shown (Lighthill [2]) that fluid element trajectories produced by linear inertial waves are circles in the planes of constant phases.

The phase velocity of the wave is the vector with components

$$\left\{ \frac{2\Omega\, k_z}{k_x\sqrt{k_x^2+k_y^2+k_z^2}} \; , \; \frac{2\Omega\, k_z}{k_y\sqrt{k_x^2+k_y^2+k_z^2}} \; , \; \frac{2\Omega}{\sqrt{k_x^2+k_y^2+k_z^2}} \right\} \quad . \tag{11}$$

On the other hand, the group velocity of inertial waves is the vector with components

$$\left\{ \frac{-\,2\Omega\, k_x\, k_z}{\left(k_x^2 + k_y^2 + k_z^2\right)^{3/2}} \; , \; \frac{-\,2\Omega\, k_y\, k_z}{\left(k_x^2 + k_y^2 + k_z^2\right)^{3/2}} \; , \; \frac{2\Omega\left(k_x^2 + k_y^2\right)}{\left(k_x^2 + k_y^2 + k_z^2\right)^{3/2}} \right\} \quad . \tag{12}$$

The orientation of the group velocity is a characteristic property of the inertial wave; it is perpendicular to the wave vector $\vec{k}$.

13. The limiting case of waves with frequency approaching 2Ω

When the frequency ω approaches 2Ω, the wavenumber component k_z becomes much larger than the wavenumber components k_x and k_y. Thus

$$\frac{k_x^2 + k_y^2}{k_z^2} \approx \frac{2\Omega - \omega}{\Omega} \quad . \tag{13}$$

The wave vector $\vec{k}$ is almost parallel to the rotation axis.When the difference between ω and 2Ω decreases to zero, the circular trajectories of fluid elements tend to be in the planes (x,y) perpendicular to the rotation axis. The asymptotic values of the phase velocity and group velocity are at leading order

$$\vec{C}_\varphi \approx \left\{ \frac{2\Omega}{k_x} \ , \ \frac{2\Omega}{k_y} \ , \ 0 \right\} \tag{14}$$

$$\vec{C}_g \approx \left\{ \frac{-2\,(2\Omega - \omega)\,k_x}{k_x^2 + k_y^2} \ , \ \frac{-2\,(2\Omega - \omega)\,k_y}{k_x^2 + k_y^2} \ , \ 0 \right\} \quad . \tag{15}$$

The wave solution is singular when ω = 2Ω.The group velocity vanishes. Eqs. (1-4) show that the pressure no longer acts as a coupling factor of the velocity components u and v with the velocity component w, because $k_x = k_y = 0$. Inertial waves with a frequency approaching the Coriolis frequency have been much studied by oceanographers in the past few years. Such waves, which are often called near-inertial waves, have been observed in the mixed layer on top of the ocean (Weller [3]). They are thought to play a crucial role near fronts (Rubenstein and Roberts [4]) and to interact strongly with mean geostrophic shear flows (Kunze [5]). Though both the phase and group velocity components parallel to the rotation axis are small, near-inertial waves may transmit a significant amount of momentum to the deeper ocean in regions of negative vorticity, due to the trapping of inertial waves in these regions.

14. The limiting case of low frequency waves (ω << 2Ω)

The case in which the frequency of the wave is very small, i.e. ω << 2Ω, has a wide range of physical applications, because ω / 2Ω << 1 is equivalent to the fact that wave motion is characterised by a low Rossby number. Low frequency inertial waves are produced by large-scale geophysical motion. The dispersion relationship (9) implies that the wavenumber component k_z is small compared to the wavenumber components k_x and k_y. In other words, the lengthscale of the wave parallel to the rotation axis is very large compared to that in the direction perpendicular to the rotation.

The particular values taken by the wave vector, phase velocity and group velocity components when ω is very small are:

$$\vec{k} \approx \{ k_x \ , \ k_y \ , 0 \ \} \tag{16}$$

$$\vec{C}_\varphi \approx \vec{C}_g \approx \left\{ 0 \ , \ 0 \ , \ \frac{2\Omega}{\sqrt{k_x^2 + k_y^2}} \right\} \ . \tag{17}$$

These orders of magnitude imply that inertial waves of low frequency have wave fronts parallel to the rotation axis. The wave propagates energy parallel to the rotation axis. The wave is not dispersive at the leading order of ω/2Ω because the phase velocity and group velocity are equal. As shown more clearly in § 15, low frequency inertial waves are closely connected to the process of two-dimensionalisation of the flow field by rotation. This is consistent with the orientation of the group velocity parallel to the rotation axis.

15. Relationship with the Taylor-Proudman Theorem

A simple model, inspired by Stern ([6], p. 28-31), highlights the relationship between low-frequency waves and the Taylor-Proudman Theorem, according to which low Rossby number motions are two-dimensional. Consider the fluid layer contained between the plane z=H and a lower boundary which varies sinusoidally in the x direction with wavelength 2π/k around the mean position z=0 (Figure 1)

$$h(x) = \hat{h} \sin(kx) \ . \tag{18}$$

Flow is produced by pulling the lower boundary with constant velocity U in the Ox direction. The whole system rotates with angular velocity Ω around the Oz axis. The linear solution for motion inside the layer describes the flow produced by towing an obstacle at constant speed in a rotating coordinate system. Assuming that the height of the topography is small compared to the wavelength, the velocity component w forced by the topography in the mean plane z=0 is

$$w(x,y,z=0,t) = - U k \hat{h} \cos(kx-kUt) \quad . \tag{19}$$

The boundary condition on the upper boundary z=H is w(x,y,H,t)=0. The solution inside the layer (0<z<H) is found more easily when the linear set of motion equations (1-4) is reduced to a single equation with the unknown quantity w. One readily deduces from (1-4)

$$\left(\frac{\partial^2}{\partial t^2} + (2\Omega)^2\right)\frac{\partial^2 w}{\partial z^2} + \left(\frac{\partial^2}{\partial x^2} + \frac{\partial^2}{\partial y^2}\right)\frac{\partial^2 w}{\partial t^2} = 0 \quad . \tag{20}$$

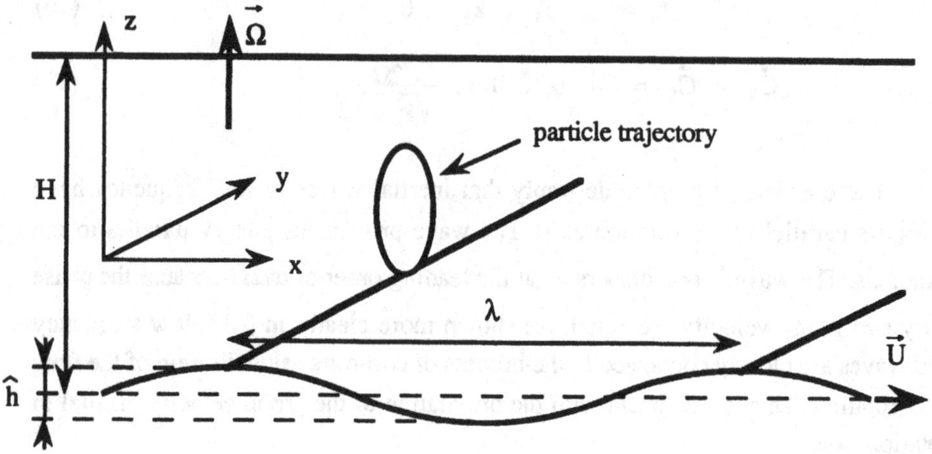

Figure 1: Flow produced by an obstacle moving with constant speed in a rotating shallow water layer.

For a solution of the form

$$w(x,y,z,t) = \tilde{w}(z) \cos(kx-kUt) \quad , \tag{21}$$

the dependence of the velocity component w with respect to z is determined from

$$\frac{\partial^2 \tilde{w}}{\partial z^2} + k^2 \frac{(Uk)^2}{(2\Omega)^2 - (Uk)^2} \tilde{w} = 0 \quad . \tag{22}$$

The solution depends on the wave number k_z defined as

$$k_z^2 = k^2 \frac{(Uk)^2}{(2\Omega)^2 - (Uk)^2} \quad . \tag{23}$$

When the frequency Uk is larger than the Coriolis frequency 2Ω, solution (21) decays exponentially with distance from the moving topography. The topography does not produce much motion inside the layer. On the other hand, when the frequency Uk is smaller than the Coriolis frequency 2Ω, w varies sinusoidally in the layer. The solution satisfying the required boundary conditions is

$$w(x,y,z,t) = - U k \hat{h} \frac{\sin(k_z H - k_z z)}{\sin(k_z H)} \cos(kx-kUt) \quad . \tag{24}$$

In the case when $k_z H << 1$, the velocity component w varies linearly inside the layer

$$w(x,y,z=0,t) \approx - U k \hat{h} \frac{H-z}{H} \cos(kx-kUt) \quad . \tag{25}$$

This is a valid solution when

$$(kH)^2 \frac{(Uk)^2}{(2\Omega)^2 - (Uk)^2} << 1 \quad , \tag{26}$$

i.e. when the velocity U of the topography is small ($Uk << \Omega$), and the depth of the layer not too large. Equations (6) and (7) imply that the velocity component perpendicular to Ω must be along Oy ($U = 0$ and $V = iPk/2\Omega\rho$ when $k_x=k$ and $k_y=0$). Introducing the value of u=0 and w (eq. 25) into the continuity equation readily shows that the velocity component v is independent of z. This concurs exactly with the Taylor-Proudman theorem, namely that material columns of fluid are rigid and remain vertical as they pass over an obstacle. Particle trajectories are also indicated in Figure 1.

2. OBSERVATIONS OF INERTIAL WAVES. RESONANCE

Inertial waves are very common, but it is difficult in practice to study them in a quantitative way and, for example, to test the dispersion relationships or to examine particle motion. Firstly, it is difficult to produce a single inertial wave. Secondly, it is more difficult to characterise a wave from fluid element motion than from the motion of an interface as for internal or surface waves. Experimental studies have characterised inertial waves from the observation of inertial wave resonance (Fultz [7]; McEwan [8]). The phenomenon of inertial wave resonance can be understood already by the simple model presented in § 15. In fact, the velocity component w(x,y,z,t) forced by the boundary motions must increase very much when $k_z H = n\pi$.

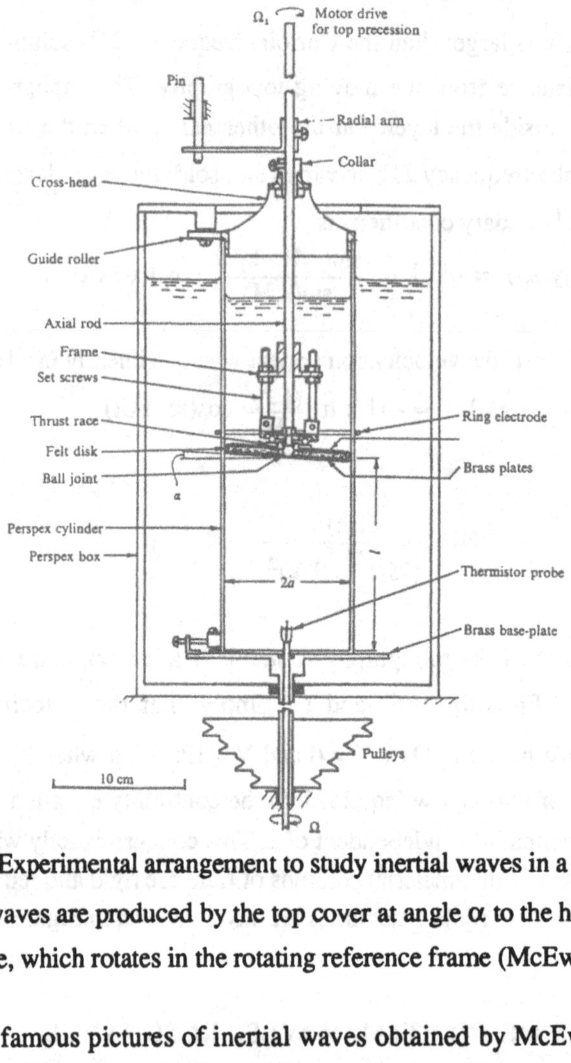

Figure 2: Experimental arrangement to study inertial waves in a rotating tank. The waves are produced by the top cover at angle α to the horizontal plane, which rotates in the rotating reference frame (McEwan [8]).

The famous pictures of inertial waves obtained by McEwan [8], which are reproduced in Greenspan's book [9], are discussed below. These inertial wave motions were generated in the experimental setup shown in Figure 2. It consists of a rotating cylindrical tank of radius a=4.73 cm and height l =30 cm. The waves are forced by the upper cover which makes a small angle α with the horizontal plane perpendicular to the rotation axis. The upper cover rotates with an angular speed Ω_1 that is slightly different from the basic rotation speed Ω of the tank. The frequency of the disturbances is thus $\sigma = (\Omega - \Omega_1)$.

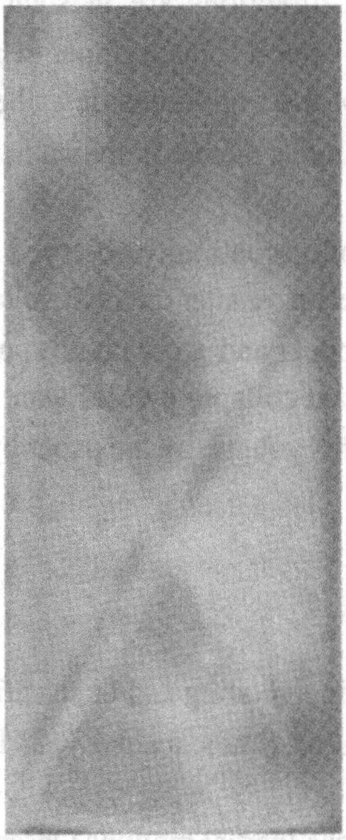

Figure 3: Characteristic surfaces of discontinuity produced by a resonant inertial wave in a rotating tank (McEwan [8]).

The flow produced within the tank is the combination of an infinite number of modes of wave motion having different wave vectors, but all of the same frequency σ. The wave motion becomes resonant for discrete values of the quantity

$$\frac{l\ \sigma}{a\sqrt{4\Omega^2 - \sigma^2}}\ , \tag{27}$$

for which case the amplitude of one of the modes becomes infinite (if viscosity is neglected). Characteristic surfaces then appear in the flow field across which the velocity gradient or the velocity are discontinuous, depending on the condition chosen. These surfaces were visualised by McEwan using some particular particles that adopt a preferred orientation relative to the viscous stress field. These particles mark regions of large stress by showing light or dark lines when the fluid is illuminated in a shallow sheet of light. The occurrence of resonant conditions is thus tested by the appearance of such lines. In the general case, this line has a wavy shape in the Oz direction, from which the wavenumber in the Oz direction is determined. In more particular cases, the wavenumber components in the Or and Oz directions are in such a ratio that the surfaces of discontinuity retrace themselves after a finite number of reflections on the walls of the tank. In this case, the surfaces of discontinuities describe regular diagonal lines in the illuminated sheet of light. This famous picture is shown in Figure 3.

The coherent surfaces of discontinuity observed in Figure 3 usually occur only during the early stages of the experiment after the relative rotation of the top cover begins. Later, oscillatory disturbances appear and grow rapidly, producing disordered motions. This phenomenon is called resonant collapse by McEwan. A complementary insight into resonant collapse is provided in this volume by the paper by Linden and Manasseh [10].

REFERENCES

1. Maxworthy T.: Wave motions in a rotating and/or stratified fluid, this volume.
2. Lighthill J.: Dynamics of rotating fluids: a survey, J. Fluid Mech., 26 (1966), 411-431.
3. Weller R.A.: The relation of Near-Inertial motions Observed in the Mixed Layer during the JASIN (1978) Experiment to the Local Stress and to the Quasi-Geostrophic Flow Field", J. Phys. Oceanogr., 12 (1982), 1122-1136.
4. Rubenstein D.M. & Roberts G.O.: Scattering of Inertial waves by an Ocean Front, J. Phys. Oceanogr., 16 (1986), 121-131.
5. Kunze E.: Near-Inertial Wave Propagation in Geostrophic Shear, J. Phys. Oceanogr., 15 (1985), 544-565.
6. Stern, M.E.: Ocean Circulation Physics, Academic Press, 1975.
7. Fultz D.: A note on overstability and the elastoid-inertia oscillations of Kelvin, Solberg and Bjerknes, J. Meteor., 16 (1959), 199-208.
8. McEwan A.D.: Inertial oscillations in a rotating fluid cylinder, J. Fluid Mech., 40 (1970), 603-640.
9. Greenspan G.P.: The theory of Rotating Fluids, Cambridge University Press, 1968.
10. Linden P.F. & Manasseh R.: Resonant collapse, this volume.

Chapter IV.3

ROTATING ANNULUS FLOWS AND BAROCLINIC WAVES

P.L. Read
University of Oxford, Oxford UK

ABSTRACT

Many aspects of the thermally-driven circulation in the atmosphere or oceans can be studied on the laboratory scale via the cylindrical rotating annulus. In this paper, we review the basic geophysical motivation for such experiments, and discuss all the principal flow regimes so far studied. Particular emphasis is placed (a) on the underlying dynamics of the steady axisymmetric flow, (b) the structure and stability of the baroclinic wave regime, and (c) the possible transition scenarios to chaotic and/or irregular flow studied to date.

1. GENERAL CIRCULATION STUDIES AND THE ROTATING ANNULUS

At its most basic level, the general circulation of the atmosphere is but one example of thermal convection due to impressed differential heating in the horizontal in a rotating fluid of low viscosity and thermal conductivity. Laboratory experiments investigating such a problem should therefore include at least these attributes, and be capable of satisfying scaling requirements for dynamical similarity to the relevant phenomena in the atmosphere. Such experimental systems may then be regarded [1] as representing the general circulation in the absence of various complexities associated, for example, with radiative transfer, atmospheric chemistry, boundary-layer turbulence, planetary curvature, topography, etc. (although some of the latter, such as planetary vorticity gradients and topography, can be included in a systematic way if required).

Experiments of this type are by no means a recent phenomenon, with examples published as long ago as the 19th century (e.g. [2], [3], and see [4] for a review of this early work). It was not until the late 1940s, however, that Fultz began a systematic series of experiments at the University of Chicago on rotating fluids subject to horizontal differential heating in an open cylinder (hence resulting in the obsolete term 'dishpan experiment'), and set the subject onto a firm footing. Independently and around the same time, Hide [5] began his first series of experiments at the University of Cambridge on flows in a heated rotating annulus, initially in the context of fluid motions in the Earth's liquid core. It is the latter system which is now considered in detail.

2. FLOW REGIMES AND TRANSITIONS

The typical construction of the annulus is illustrated schematically below, and consists of a working fluid (usually a viscous liquid, such as water or silicone oil) contained in the annular gap between two coaxial circular, thermally conducting cylinders, which can be rotated about their common

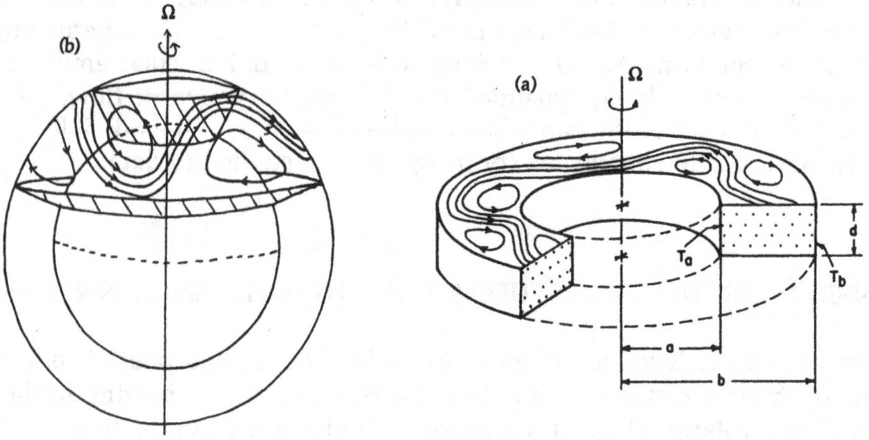

Figure 1: (a) Schematic diagram of a rotating annulus, (b) schematic equivalent configuration in a spherical fluid shell (cf an atmosphere).

(vertical) axis. The cylindrical sidewalls are maintained at constant but different temperatures, with a (usually horizontal) thermally insulating lower boundary and an upper boundary which is also thermally insulating and either rigid or free (i.e. without a lid).

Although a number of variations in these boundary conditions have been investigated experimentally, all such experiments are found to exhibit the same three main flow regimes, for example, as the rotation rate Ω varies. These consist of axisymmetric flow (in some respects analogous to Hadley flow in the Earth's tropics, and frequently referred to as the 'upper-symmetric regime'; see below) at very low Ω, regular waves at moderate Ω, and highly irregular, aperiodic flow at the highest values of Ω attainable. In addition,

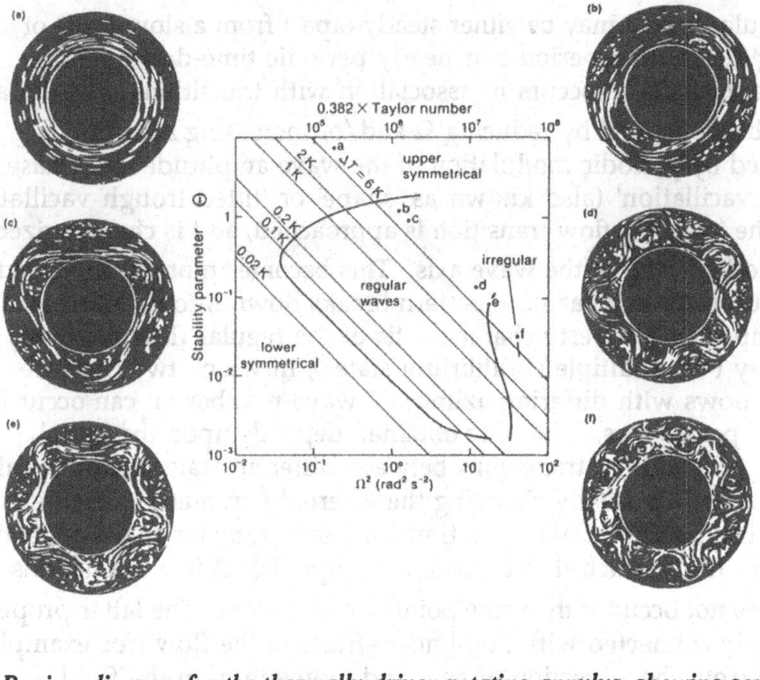

Figure 2. Regime diagram for the thermally-driven rotating annulus, showing some typical horizontal flow patterns at the top surface.

axisymmetric flows occur at all values of Ω at a sufficiently low temperature difference ΔT (a diffusively-dominated regime termed 'lower symmetric' to distinguish it from the physically distinct 'upper-symmetric regime' mentioned above). The location of these regimes are usually plotted an a 'regime diagram' with respect to the two most significant dimensionless parameters.

- a stability parameter or 'thermal Rossby number' $\Theta \equiv g\alpha\Delta T d/[\Omega(b-a)]^2$ providing a measure of the strength of buoyancy forces relative to Coriolis accelerations

- a Taylor number $T_a \equiv \Omega^2(b-a)^5/[\nu^2 d]$ measuring the strength of viscous dissipation relative to Coriolis accelerations

where g is the acceleration due to gravity, α the thermal expansion coefficient of the fluid, ν the kinematic viscosity and a, b and d are the dimensions indicated above. Fig. 2 shows a typical regime diagram with the locations of the main regimes indicated.

The regular waves may be either steady (apart from a slow drift) or 'vacillating' (i.e. with a periodic or nearly periodic time-dependence). 'Amplitude vacillation' occurs in association with transitions towards a lower wave number (obtained by reducing Ω and/or increasing ΔT), and is characterized by periodic modulation of the wave amplitude and phase speed. 'Structural vacillation' (also known as 'shape' or 'tilted-trough vacillation') occurs as the irregular flow transition is approached, and is characterized by a nearly periodic tilting of the wave axis. This becomes more pronounced as Ω is increased, until the regular flow pattern breaks down into fully irregular flow. Another important property characteristic of the regular flow regime is that of intransitivity (i.e. multiple equilibrium states), in which two or more alternative flows with differing azimuthal wave number m can occur for a given set of parameters. The state obtained depends upon the initial conditions. In addition, transitions between different states in the regular regime, achieved by slowly changing the external parameters, often exhibit hysteresis, in that the location of a transition in parameter space depends upon the direction from which that transition is approached (e.g. transitions from $m = 3 \rightarrow 4$ does not occur at the same point as $m = 4 \rightarrow 3$). The latter properties are intimately connected with non-linear effects in the flow (for example see [6]) arising from the advection of heat and momentum in the fluid.

From a consideration of the conditions under which waves occur in the annulus (especially the location in parameter space of the 'upper-symmetric' transition) and a comparison with the results of linear instability theory, it is concluded that the waves in the annulus are fully developed manifestations of baroclinic instability (often referred to as 'sloping convection' from the geometry of typical fluid trajectories, for example see [7]). Since these flows occur in the interior of the annulus (i.e. outside ageostrophic boundary layers) under conditions appropriate to quasi-geostrophic scaling, a dynamical similarity to the large-scale mid-latitude cyclones in the Earth's atmosphere is readily apparent, though with rather different boundary conditions. A more detailed discussion of the properties of these flows is given below and by [7]. Associated with this conclusion is the implication that the waves develop in order to assist in the transfer of heat both upwards (enhancing the static stability) and horizontally down the impressed thermal gradient (i.e. tending to reduce the impressed horizontal gradient).

3. AXISYMMETRIC FLOW & THE ROLE OF BAROCLINIC WAVES

The assertion that annulus waves develop from a baroclinic instability and serve primarily to transfer heat requires further justification and discussion, raising questions such as:

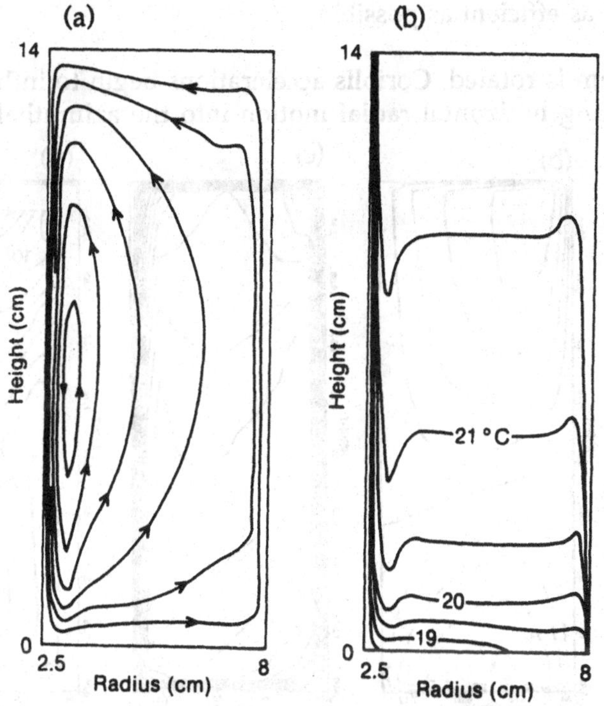

Figure 3. Cross-sections in the (r,z) plane of (a) streamfunction and (b) temperature in an annulus differentially-heated at the sidewalls with $\Omega = 0$ (see [8]).

- *why can't the axisymmetric flow component effect all heat transfer?*

- *why do isotherms/isopycnals slope when* $\Omega \neq 0$*?*

If the annulus were filled with a conducting solid instead of a fluid, heat would only be transferred by molecular conduction. Hence the isotherms would lie on cylindrical surfaces coaxial with the axis of symmetry. Given a fluid, heat is also transferred by advection, with hot fluid adjacent to heated sidewall rising and cold fluid adjacent to the cooled wall sinking. This motion takes place in thin buoyancy-dominated boundary layers with a return flow in the (quasi-inviscid) interior. The interior flow redistributes hot and cold fluid until it reaches a state of minimum potential energy - i.e. with as much as possible of the cold dense fluid at the bottom of the cavity. Hence, isotherms in the interior become virtually *horizontal* and stably-stratified in the vertical, adjusting to become vertical in thin conduction-dominated boundary layers adjacent to the sidewalls. A typical example of a non-rotating axisymmetric circulation is shown in Fig. 3. In this state, advective heat transport (by *laminar flow*) is apparently as efficient as possible..

When the system is rotated, Coriolis accelerations begin to influence the circulation, deflecting horizontal radial motion into the azimuthal direction.

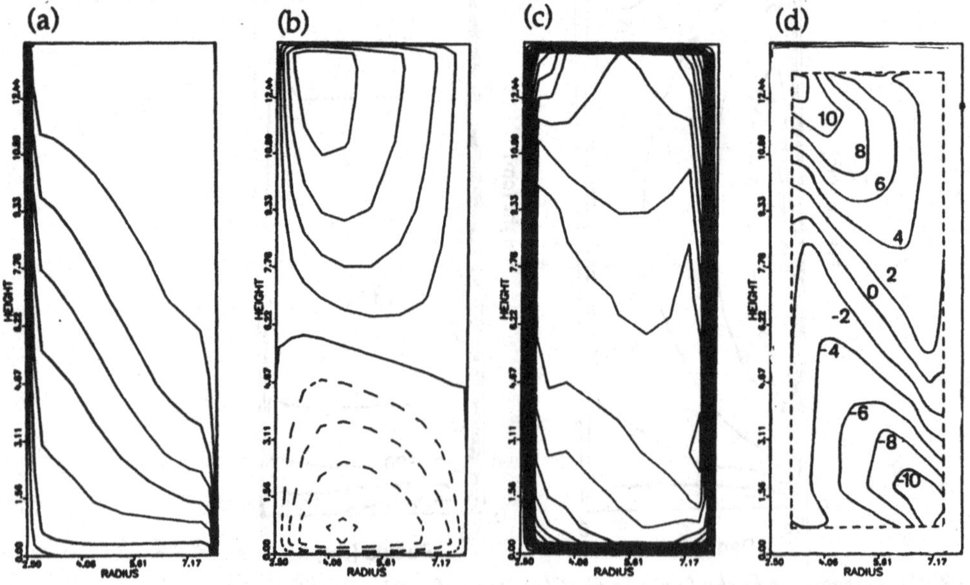

Figure 4: Cross-sections in the (r,z) plane of (a) temperature, (b) azimuthal velocity, (c) streamfunction and (d) radial gradient of QG potential vorticity (see [11]) in the axisymmetric regime of a rotating annulus subject to differential heating at the sidewalls.

The axisymmetric flow at low values of Ω is therefore similar to the non-rotating case, except (a) an azimuthal component of flow is induced, producing jets antisymmetric about mid-depth (for identical upper and lower boundary conditions), prograde at the top (where radial flow is inwards) and retrograde below (where radial flow is outwards), and (b) radial flow becomes largely confined to Ekman layers adjacent to the horizontal boundaries (for Ω sufficiently large, see §4 below). When Coriolis accelerations dominate the interior flow, any O(1) radial flow has to be geostrophic, requiring an azimuthal pressure gradient. Such a gradient cannot occur in an axisymmetric circulation unless a rigid meridional barrier is present.

Since the radial flow becomes confined to Ekman layers when $\Omega \neq 0$, the efficiency of advective heat transport is governed primarily by the *mass* transport which can be accommodated within an Ekman layer. For a given geostrophic zonal flow in the interior, the Ekman transport is proportional to the depth of the Ekman layer (i.e. proportional to $VE^{1/2}$, where E is the Ekman number, see below). Since E is inversely proportional to Ω, the efficiency of advective heat transport must decrease with Ω - given a constant ΔT, V is proportional to Ω^{-1} and Ekman transport is proportional to $\Omega^{-1}E^{1/2} \propto \Omega^{-3/2}$. Thus, as Ω increases, advective heat transport must decrease. Since conductive heat transport is always present and is unaffected by rotation, the resultant thermal field becomes increasingly dominated by conduction. Hence, the isotherm structure will tend towards the vertical alignment characteristic of the conductive state as Ω increases. It is reasonable, therefore, for waves to develop in such a flow provided they are able to transfer heat radially in the interior, which is possible in a geostrophically-balanced rotating fluid since waves are associated with a periodically varying azimuthal pressure gradient to balance a radial geostrophic flow. Laboratory experiments show that such wave-like disturbances with these properties will frequently develop under many circumstances, with the primary role of transferring heat and releasing 'available' potential energy.

Note that the above arguments will apply in a qualitative sense to a planetary atmosphere, save that the state to which the thermal structure relaxes in the absence of fluid motion is not one of conductive equilibrium, but of radiative(-convective?) equilibrium. In the case of the Earth, the latter would result in a temperature contrast between equator and poles of ~ 200K, instead of ~ 50K observed on average, indicating the dominant role of dynamical heat transports in the atmosphere.

4. REGIMES OF AXISYMMETRIC FLOW: HEAT & MOMENTUM TRANSPORT

Although the the above description of the axisymmetric flow gives a plausible explanation for the observed axisymmetric and wave regimes in the annulus, it is a highly simplified discussion which glosses over more subtle aspects of the problem. Previous chapters (e.g. [27]) have considered boundary layers in rotating flows, and discussions of baroclinic waves and instability typically make assumptions concerning the main force balances in the basic zonal flow state. In this section, we take a more quantitative view of the axisymmetric flow in the annulus, to put the above discussion on a stronger theoretical footing and as an illustration of the use of scale analysis and boundary layer theory.

We carry out the analysis for an incompressible Boussinesq fluid in a rotating annulus of vanishingly small relative curvature ($2[b - a]/[b + a] << 1$, so we may use Cartesian geometry), and neglect centrifugal accelerations. It is convenient to define a meridional streamfunction c such that

$$u = \partial\chi/\partial z; \quad v = -\partial\chi/\partial x \tag{1}$$

The steady state equations for momentum, continuity and heat then reduce to a zonal momentum equation

$$\nu\nabla^2 v = f\partial\chi/\partial z + J(v,\chi) \tag{2}$$

[where $f = 2\Omega$ and the Jacobian is defined as

$$J(c,d) = \partial c/\partial x.\partial d/\partial z - \partial c/\partial z.\partial d/\partial x \qquad (3)]$$

an azimuthal vorticity equation

$$\nu\nabla^4\chi = g\alpha\partial T/\partial x - f\partial v/\partial z - J(\chi,\nabla^2\chi) \tag{4}$$

[where T is the temperature, α the volumetric expansion coefficient, and vorticity ζ is defined as

$$\zeta = \partial u/\partial z - \partial v/\partial x = \nabla^2\chi \qquad (5)]$$

and the temperature equation

$$\kappa\nabla^2 T + J(\chi,T) = 0 \qquad (6)$$

We consider a container of unit aspect ratio (for simplicity) and apply boundary conditions

$$\chi = \partial\chi/\partial z = v = \partial T/\partial z = 0; \qquad z = 0,L \qquad (7a)$$

$$\chi = \partial\chi/\partial x = v = T - T_0 = 0; \qquad x = -L/2, +L/2 \qquad (7b)$$

We make use of dimensionless parameters such as the Ekman and Prandtl numbers E and σ defined by

$$E = \nu/\Omega L^2 \qquad (8a)$$

$$\sigma = \nu/\kappa \qquad (8b)$$

and the Rayleigh number

$$Ra = g\alpha.\Delta T L^3/(\kappa\nu) \qquad (8c)$$

It is also convenient to define a Nusselt number as measuring the ratio of total heat transport to that due to conduction, which we take to be

$$Nu = X/\kappa + 1 \qquad (9)$$

(where X is a characteristic scale for χ). We then carry out a scale analysis with the aim of deriving the dominant dynamical balances in the interior and principal boundary layers, and to obtain the dependence of Nu and the zonal velocity scale on external parameters over as wide a range as possible. Initial assumptions are restricted to:

(i) Single thickness scales assumed, l for the side and h for the horizontal boundary layers

(ii) Outside the boundary layers there is a distinct interior flow with length scale L, and $(l,h) \ll L$

(iii) Prandtl number $\sigma \gg 1$

(A) Non-rotating problem
We assume the flow to comprise an advective interior and thin sidewall boundary layers, and non-dimensionalise in the thin sidewall layer of thickness l using

$$\Delta x = l\Delta x^*,\ \Delta z = L\Delta z^*,\ T - T_0 = \Delta T\, T^*,\ \chi = X\chi^* \tag{10}$$

Thus (6) becomes

$$\nabla^2 T^* + Xl/\kappa L\, J(\chi^*, T^*) = 0 \tag{11}$$

For advective/diffusive balance, we require

$$X = \kappa L/l \tag{12}$$

For this case, (4) becomes

$$\nabla^4\chi^* = Ra(l/L)^4 \partial T^*/\partial x^* - \sigma^{-1} J(\chi^*, \nabla^2\chi^*) \tag{13}$$

If $\sigma >> 1$, we obtain a buoyancy/viscous balance in the sidewall boundary layer, implying that

$$l \doteq Ra^{-1/4} L \tag{14}$$

which was the result obtained in [27] for the principal boundary layer scale when $\sigma S >> E^{2/3}$. In this case, the Nusselt number is obtained from (12) as

$$Nu - 1 = X/\kappa = O(Ra^{1/4}) \tag{15}$$

(B) Effect of Rotation
In considering the relative impact of rotation on the circulation, we anticipate that the Ekman layer will be of importance. It is therefore convenient to follow an approach due to [9], in defining a parameter P measuring the (square of the) ratio of the thickness of the Ekman layer and sidewall buoyancy layer, thus

$$P = \mathrm{Ra}^{-1/2}\mathrm{E}^{-1} \tag{16}$$

which is proportional to Ω. Based on a consideration of the full range of P, we can effectively identify up to 6 distinct regimes of axisymmetric flow:

(i) No rotation, $P = 0$;
(ii) very weak rotation, $0 << P << \sigma^{-2}$;
(iii) weak rotation, $\sigma^{-2} << P << 1$;
(iv) moderate rotation, $P \approx 1$;
(v) strong rotation, $1 << P << \mathrm{Ra}^{1/6}$;
(vi) very strong rotation, $P >> \mathrm{Ra}^{1/6}$;

and briefly outline their characteristics as follows.

(i) No rotation: This has already been discussed above in (A), with consequent scales for l and Nu - 1. Note that we can obtain an estimate of isotherm slope $\gamma = \Delta T_H/\Delta T$ from a consideration of the balances in the zonal vorticity equation. Provided $\sigma >> 1$, a buoyancy/viscous balance holds in the interior, so that $g\alpha\partial T/\partial x$ $(= O(g\alpha\Delta T\gamma/L)) = \nu\nabla^4\chi$ $(= O(\nu X L^{-4}))$ - hence $\gamma < \mathrm{Ra}^{-3/4} << 1$, and isotherms are quasi-horizontal.

(ii) Very weak rotation: When f is no longer zero, (2) is coupled to (4) and gyroscopic torques render v non-zero. We obtain an estimate for the zonal velocity scale V by scaling (2) in the Ekman layer using $h = \mathrm{E}^{1/2}\mathrm{L}$ (see Lecture 5). Hence, (2) becomes

$$\sigma P^{1/2}\nabla^2 v^* = fL/V\partial\chi^*/\partial z^* + J(v^*,\chi^*) \tag{17}$$

Thus, for $P << \sigma^{-2}$ we have an inertial/Coriolis balance in the Ekman layer (i.e. there is no proper Ekman layer), and the entire flow is characterised by local conservation of angular momentum (hence $V = O(fL)$ which is proportional to P).

(iii) Weak rotation: For $P >> \sigma^{-2}$, the viscous term in (17) becomes dominant in the Ekman layer (i.e. normal Ekman layers exist), thus rescaling V to $O(\kappa \mathrm{Ra}^{1/2}P^{1/2}/\mathrm{L})$. This balance extends into the interior, while the previous balance in the sidewall layer is unchanged from (i). Despite the new scaling for V, the dominant balances (and scaling for Nu - 1) in (6) also remain unchanged from (i). The rescaling of V does, however, affect the interior balance in the azimuthal vorticity equation, from a buoyancy/viscous balance

to a buoyancy/Coriolis balance characteristic of the 'thermal wind' balance typical of geostrophic flow. The reason why the (now geostrophic) scale for V does not go as Ω^{-1} typical of a 'thermal wind' scale is because γ is now increasing rapidly with Ω ($\gamma = O(P^{3/2})$), which more than outweighs the P^{-1} dependence of V for constant γ. Note also the zonal Rossby number ($=V/fL$) = $O(\sigma^{-1}P^{-1/2})$ and is therefore << 1.

(iv) Moderate rotation: In this regime, the Ekman layer thickness is comparable with that of the sidewall buoyancy layer, and so is expected to begin to exert a strong influence on the meridional circulation and transport. Anticipating that V will eventually tend towards the 'thermal wind' scale $O(P^{-1})$, we identify this range of P as the regime where V reaches a maximum $V_o = O(\kappa Ra^{1/2}/L)$. If the Ekman layer exercises dominant control over the radial mass transport, X will be rescaled to $O(V_0 LE^{1/2}) = O(\kappa Ra^{1/4}P^{-1/2})$, implying a slow broadening of the sidewall advective/diffusive boundary layer from l to $lP^{1/2}$.

(v) Strong rotation: As P is increased beyond 1, the Ekman layers fully dominate the meridional circulation. By this point, the isotherm slope g has become O(1) and so cannot increase any further. V then rescales to the familiar 'thermal wind' scale $V = O(\kappa Ra^{1/2}L^{-1}P^{-1})$. The expansion of the advective/diffusive sidewall layer accelerates to $l' = O(Ra^{-1/4}LP^{3/2})$, extending

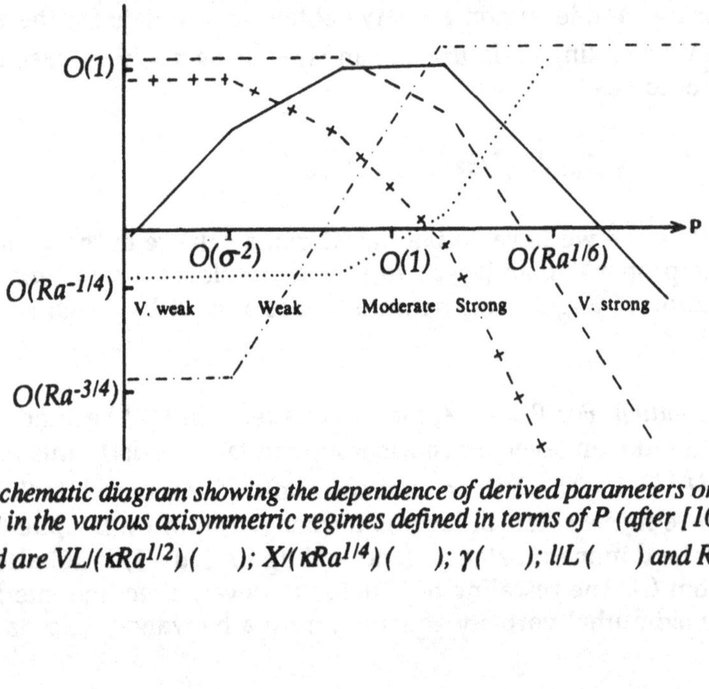

Figure 5: Schematic diagram showing the dependence of derived parameters on internal parameters in the various axisymmetric regimes defined in terms of P (after [10]). Quantities represented are $VL/(\kappa Ra^{1/2})$ ( ); $X/(\kappa Ra^{1/4})$ ( ); γ ( ); l/L ( ) and Ro ( ).

the influence of thermal diffusion further into the interior. The heat transport Nu - 1 is rescaled to $O(Ra^{1/4}P^{-3/2})$, though remains >> 1.

(vi) Very strong rotation: In the final regime, the diffusive thermal sidewall layer expands to fill the interior and no separate thermal boundary layer and interior can be distinguished (though Stewartson $E^{1/3}$ layers may exist in this limit) The critical value for *P* distinguishing regimes (v) and (vi) simply arises from equating *l'* (see above) with L. All other balances remain unchanged from (v) - i.e. the geostrophic interior and strong Ekman layers. Heat transport in this regime, however, is dominated by thermal conduction.

(C) Experimental Verification

The above regimes are capable (at least in principle) of existing in reality, given an experimental system operating in an appropriate parameter range. In practice, however, regimes (iv) - (vi) are not usually obtainable because of the development of non-axisymmetric baroclinic waves within regime (iv). This is

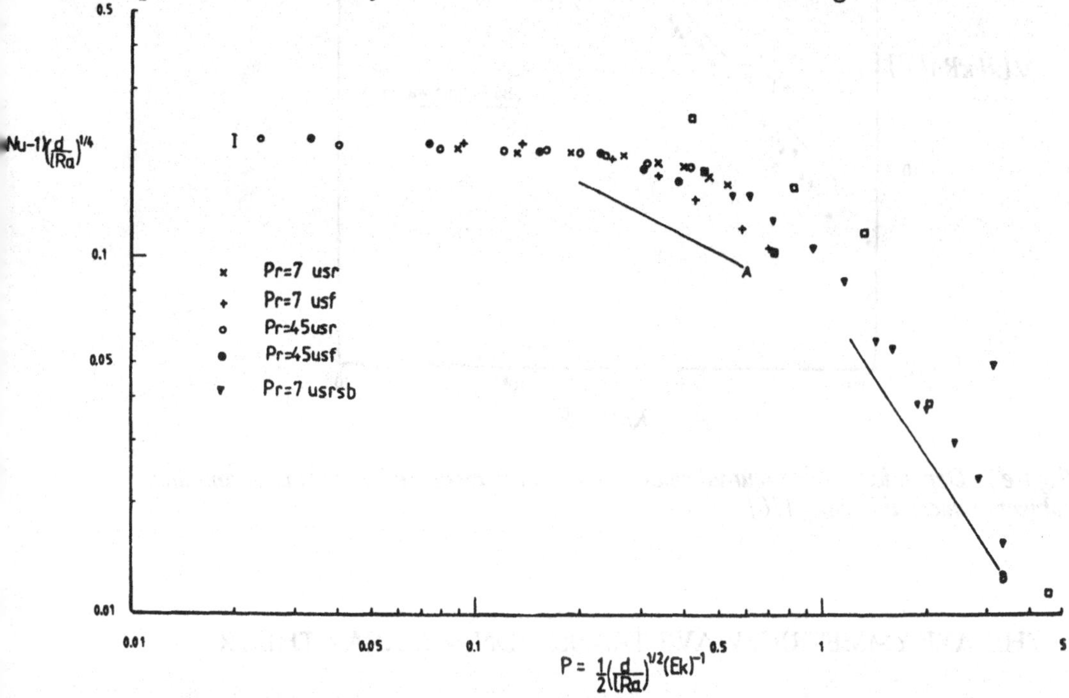

Figure 6: Scaled measurements of total heat transport in the axisymmetric regime of a rotating annulus as a function of P (from Hignett 1982).

consistent with the notion that baroclinic waves develop when Ekman layers begin to inhibit meridional heat transport.

An exception was provided by Hignett [11], who made heat transport measurements in a rotating annulus with parallel sloping upper and lower endwalls which sloped strongly in the same sense as the isotherms. Such a configuration tends to inhibit the development of baroclinic waves (by constraining fluid trajectories away from the 'wedge of instability'). As a result, Hignett was able to show the effect of almost the full range of behaviour from zero to very strong rotation on the total heat transport by axisymmetric flow in a rotating annulus. His results are shown above. The dependence of Nu and V on P in a rotating annulus subject to internal heating was also investigated by Read [10], which also confirmed the above analysis.

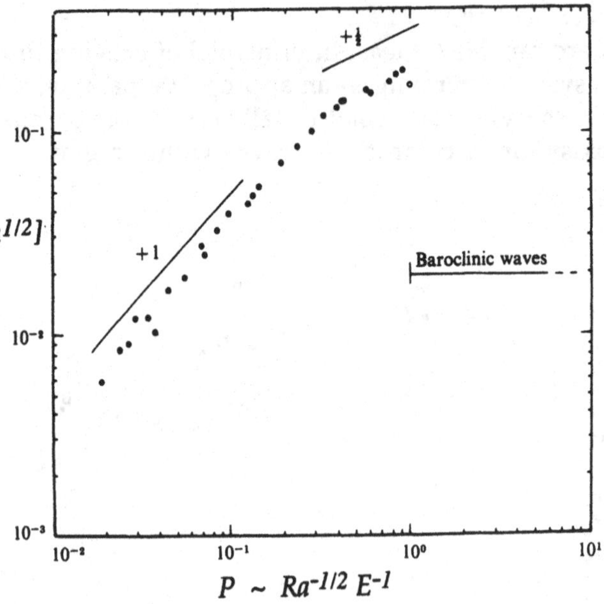

Figure 7: Dependence of azimuthal velocity scale on P, measured in a rotating annulus subject to internal heating [10].

6. THE AXISYMMETRIC/WAVE TRANSITION & LINEAR THEORY

The previous section indicated the conditions under which baroclinic waves occur in the annulus, and their role as a means of transferring heat upwards and against the horizontal temperature gradient. The Eady model of baroclinic instability has been invoked to account for the onset of waves from axisymmetric flow [7]. Although the Eady model is highly idealised, it does predict the location of the onset of waves remarkably close to the conditions actually observed - at least at high Taylor number (note that the Eady problem is inviscid). Apparent agreement can be made even closer if the Eady problem is modified to include Ekman boundary layers - by replacing the $w = 0$ boundary condition with the Ekman compatibility condition

$$w = (E/2)^{1/2}/Ro\ \nabla^2\psi; \qquad z=0,1 \qquad (18)$$

This naturally brings in the Taylor number familiar to experimentalists, and leads to a plausibly realistic envelope of instability at low Taylor number, confirming that the 'lower symmetric transition' is frictionally-dominated.

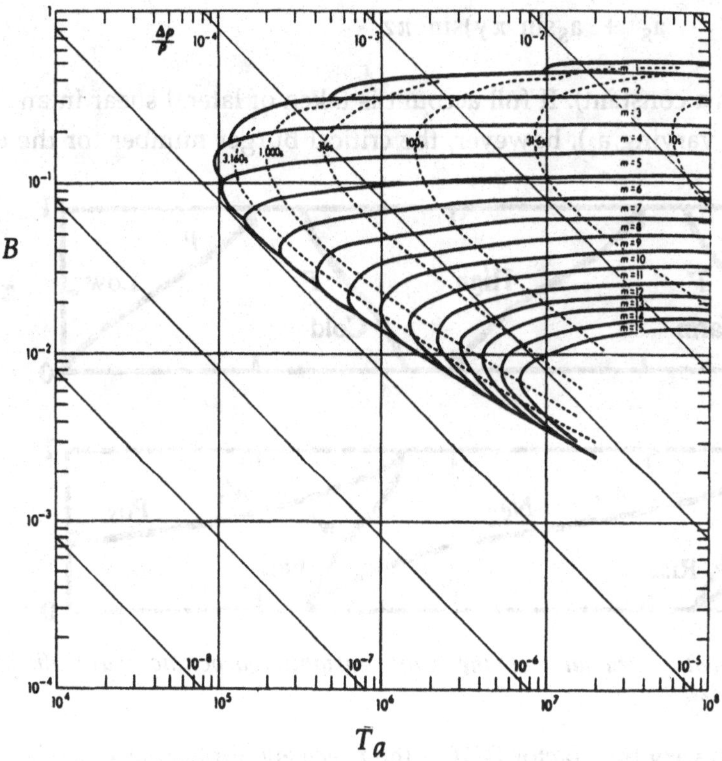

Figure 8: Regime diagram based on the extension of Eady's baroclinic instability theory to include Ekman layers (see [7]).The wavenumber of maximum instability transition curves and contours of e-folding time are given on a Burger no. (B ~ Θ; see [7]) against Ta plot.

The structure of the most rapidly growing instability has certain characteristic features in terms of phase tilts with height etc., (see Fig. 9). In the annulus, steady baroclinic waves are also seen to exhibit many of these features, as determined from experiment and numerical simulation. The extent to which Eady theory actually provides a complete theoretical description of the instability problem in the annulus, however, is still a matter for some controversy. The Eady model relies on the existence of horizontal temperature gradients on horizontal boundaries for the required change of sign in $\partial q/\partial y$ for instability. In practice, however, strong horizontal mass transports in the Ekman layers result in almost no horizontal temperature gradients at the boundaries - $\partial q/\partial y$ changes sign smoothly in the interior. Thus, instability of an *internal baroclinic jet* is a more appropriate starting point, preferably including consideration of lateral shears. Bell & White [13] have considered the stability of the form

$$U = 1/2(1 - a_S + a_S \sin \pi y) \sin \pi z \tag{19}$$

(where a_S is a constant). If full account is taken of lateral shear in an internal jet (by varying a_S), however, the critical Burger number for the onset

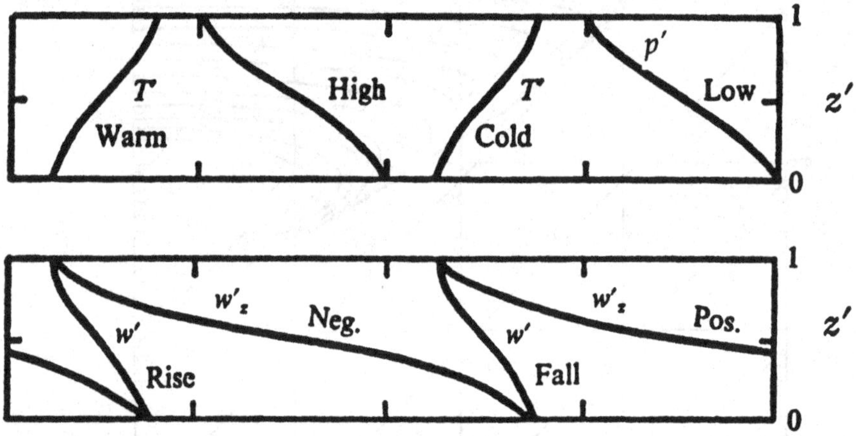

Figure 9: Schematic structure of the most unstable baroclinic wave disturbance in the Eady model.

of waves can vary by a factor O(10) - the precisely applicable value is probably dependent upon subtle details of the zonal flow shape and the lateral boundary conditions, since the true boundary conditions at the sides of the geostrophic interior ought really to take proper account of the complex viscous boundary layer structures (Stewartson layer etc.) instead of the cavalier use of impermeable boundaries as employed (for mathematical convenience) in studies to date.

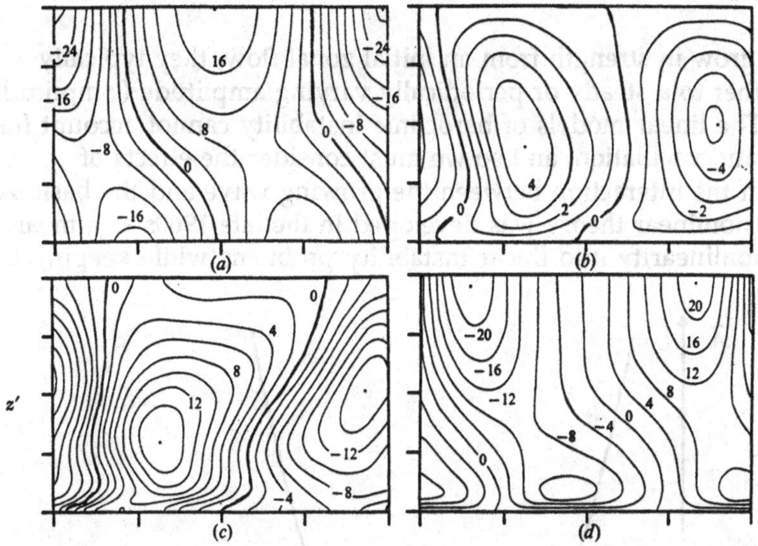

Figure 10: Cross-sections in the (θ,z) plane at mid-radius within a steady wave flow in a rotating annulus (cf [12]) of departures from the azimuthal mean fields of (a) pressure, (b) vertical velocity, (c) temperature and (c) azimuthal velocity.

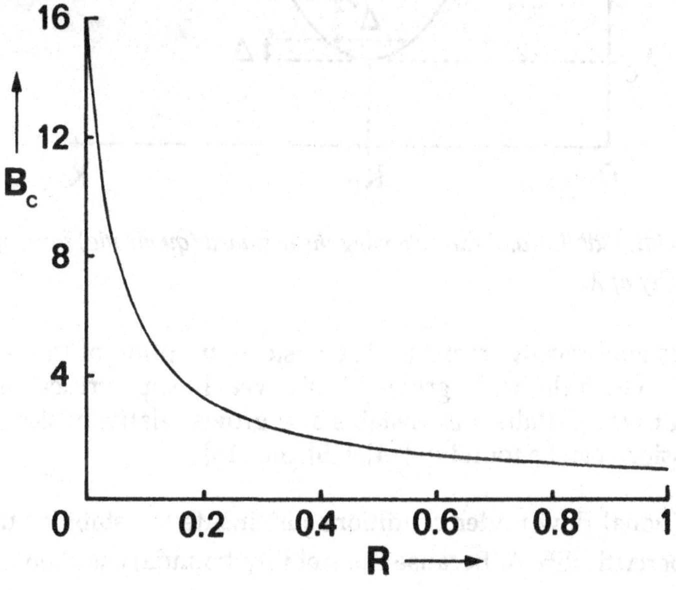

Figure 11: Variation of critical Burger number as a function of R (= 1 - a_s) in the model of laterally sheared baroclinic internal jet (see text and Eq (19)), after [13].

7. STEADY WAVES AND EQUILIBRATION: WEAKLY NONLINEAR THEORY

As waves grow in strength from an initial zonal flow, they typically equilibrate either to a steady or periodically varying amplitude ('amplitude vacillation'). The linear models of baroclinic instability cannot account for equilibration and vacillation, and so we must consider the effects of nonlinearity in the interaction between the growing wave and the basic zonal flow. Weakly nonlinear theory was developed in the late 1960s as a means of introducing nonlinearity into linear instability problems while keeping the

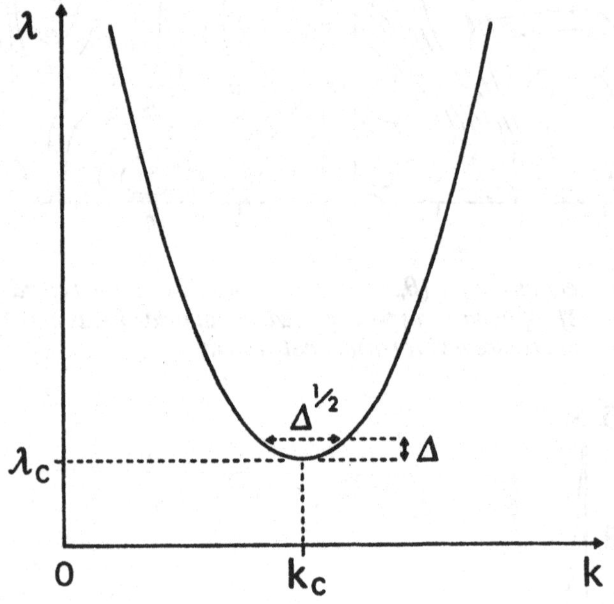

Figure 12: Schematic stability diagram, showing the assumed (quadratic) form of the critical curve in the vicinity of λ_c.

mathematics analytically tractable. The basic assumption of this approach is that the flow in which the wave grows is only weakly supercritical, and so only a small range of wavenumbers is unstable and grows relatively slowly. More detailed discussions can be found in [14], [15] and [16].

Consider a zonal flow under conditions just inside the stability threshold, with weak supercriticality Δ. Because the stability boundary is then asymptotically quadratic in k, a small range of $k \sim \Delta^{1/2}$ is destabilised - in a periodic x domain where k is discretised, this may permit only one unstable wavenumber. We introduce a 'long' timescale τ defined by

$$\tau = \Delta^{1/2} t; \quad \Delta << 1 \tag{20}$$

and solve for normal modes of the form

$$\phi = Re\{A(\tau)\, F(y,z)\, e^{ik(x-ct)}\} \tag{21}$$

together with the zonal flow of the form

$$U = \mathcal{U}(y,z) + V(\tau)G(y,z) \tag{22}$$

where A and V are respectively the slowly-varying amplitudes of the wave and correction to the zonal flow due to the self-interaction of the wave. The resulting evolution equations for A and V depend upon the relative magnitude of viscous dissipation.

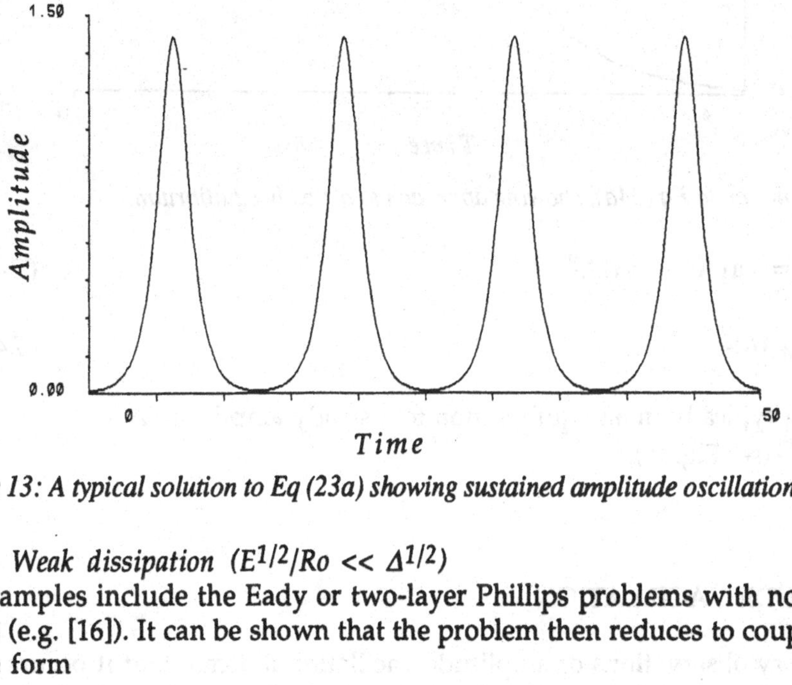

Figure 13: A typical solution to Eq (23a) showing sustained amplitude oscillations.

(i) Weak dissipation ($E^{1/2}/Ro << \Delta^{1/2}$)

Examples include the Eady or two-layer Phillips problems with no Ekman layers (e.g. [16]). It can be shown that the problem then reduces to coupled ODEs of the form

$$d^2A/d\tau^2 = a_1 A - |a_3| A^3 \tag{23a}$$

$$dV/d\tau = a_4 \, dA/d\tau \tag{23b}$$

typically resulting in a sustained amplitude modulation or 'vacillation' associated with the exchange of potential energy between wave and zonal flow (e.g. see Fig. 14).

(ii) Strong dissipation ($E^{1/2}/Ro = O(1)$)
Examples include the Eady problem with strong Ekman damping. It can be shown that the amplitude equations reduce to the well known Landau equation (e.g. see [14], [15])

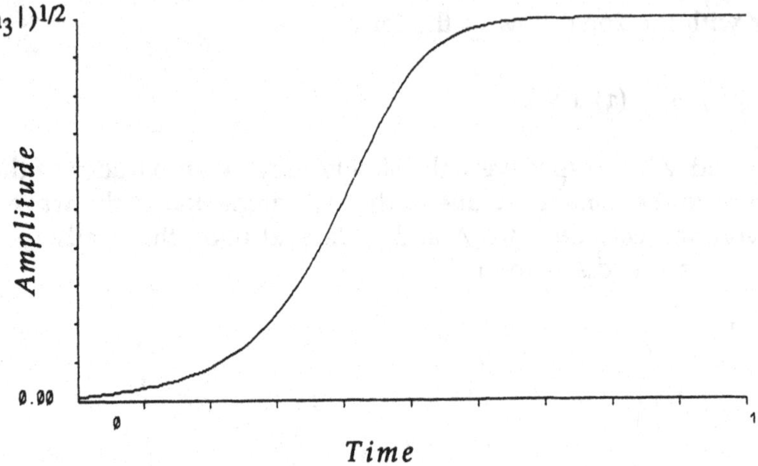

Figure 14: Solution to Eq (24a), showing approach to a steady equilibrium.

$$dA/d\tau = a_1 A - |a_3| A^3 \tag{24a}$$

$$V = a_4 |A|^2 \tag{24b}$$

resulting typically in an equilibration to a steady amplitude $A = (a_1/|a_3|)^{1/2}$ (see Fig. 15).

8. AMPLITUDE VACILLATION

Laboratory observations of amplitude vacillation indicate that it occurs close to the upper stability threshold of its wavenumber m, around where a transition from m to m-1 is observed, at moderate-high Taylor number. The 'vacillating' state comprises the periodic modulation of both the amplitude and drift frequency of the wave on a timescale ~ 10-100 'days'. More detailed diagnostics show periodic variations in total heat transport and in potential energy exchanges between the wave and zonal flow (see [17], [18]).

All these observations would seem to support an analogy with the weakly nonlinear wave-zonal flow interaction mentioned above. In practice, however, it may not be easy to distinguish this behaviour from interference arising from a quasi-linear superposition of two wave components with the same azimuthal wavenumber and differing vertical structure and drift frequencies ω_1 and ω_2. Apparent 'vacillation' then takes place at the difference frequency of the two components $|\omega_1 - \omega_2|$. If the two components cross-interact with the zonal flow, effects such as phase-locking and zonal flow modulation may occur, reproducing several aspects of the observed flows. Some observers claim to have identified this mechanism in measurements in the laboratory [20]), though the actual relevance of this mechanism in general remains controversial. Other recent work (in preparation, based on [18]) is more in keeping with the the wave-zonal flow interaction mentioned above. Fig. 16 shows cross-sections of mean azimuthal velocity and temperature at the two extremes of a numerically-simulated amplitude vacillation cycle. The zonal flow structure is seen to oscillate between a single jet pair at minimum wave amplitude and two double-jets at maximum amplitude. The isotherm

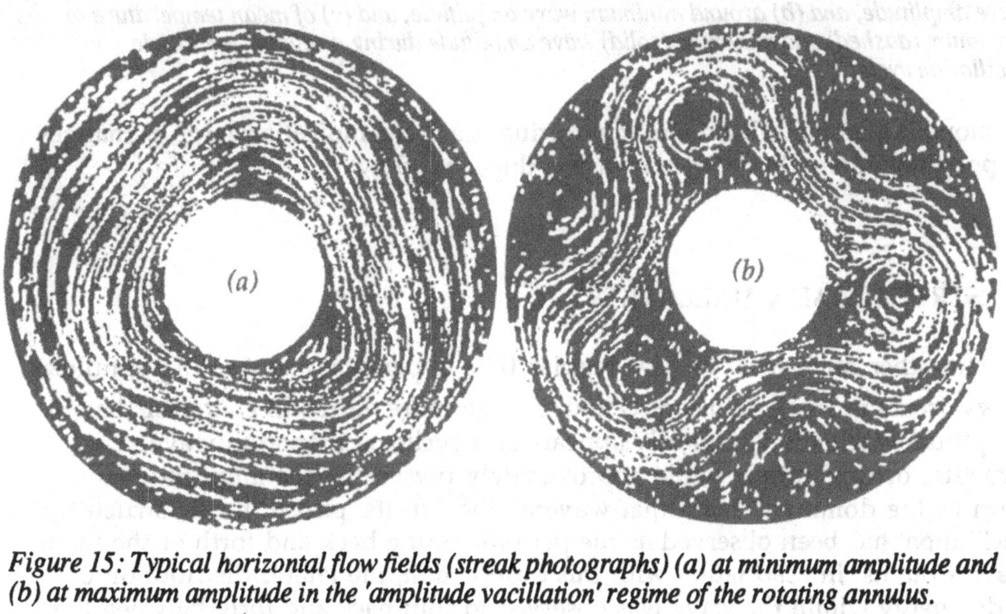

Figure 15: Typical horizontal flow fields (streak photographs) (a) at minimum amplitude and (b) at maximum amplitude in the 'amplitude vacillation' regime of the rotating annulus.

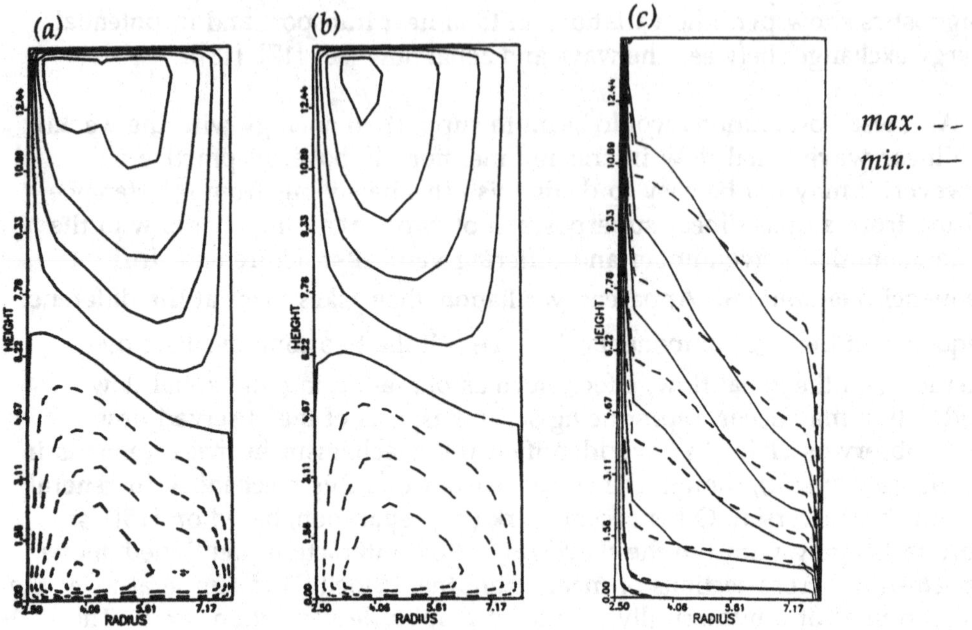

Figure 16: Cross-sections in the (r,z) plane of mean azimuthal velocity (a) around maximum wave amplitude, and (b) around minimum wave amplitude, and (c) of mean temperature at maximum (dashed) and minimum (solid) wave amplitude during a regular amplitude vacillation cycle.

slope is also modulated strongly during the cycle, indicating large exchanges of potential energy between waves and the zonal flow.

9. 'STRUCTURAL VACILLATION'

The other main form of near-regular time-dependence in rotating annulus flows occurs at small values of Θ and, though more varied in character than 'amplitude vacillation', typically appears as a cyclic oscillation in wave structure or orientation, while approximately preserving the amplitude and form of the dominant azimuthal wavenumber. In its 'purest' form, 'structural vacillation' has been observed as the periodic tilting back and forth of the main wave troughs. In association with this observation, the lateral distribution of eddy energy within the wave was observed to shift back and forth between the inner and outer sides of the channel. This observation led some workers to

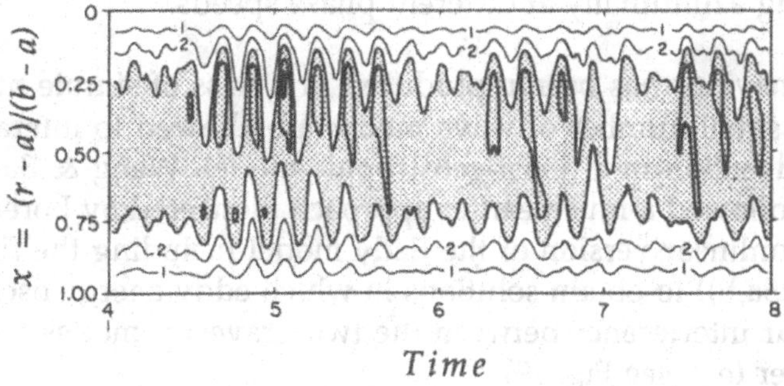

Figure 17: Behaviour with time of the radial distribution of temperature variance, measured in a rotating annulus during a structural vacillation (after [17]).

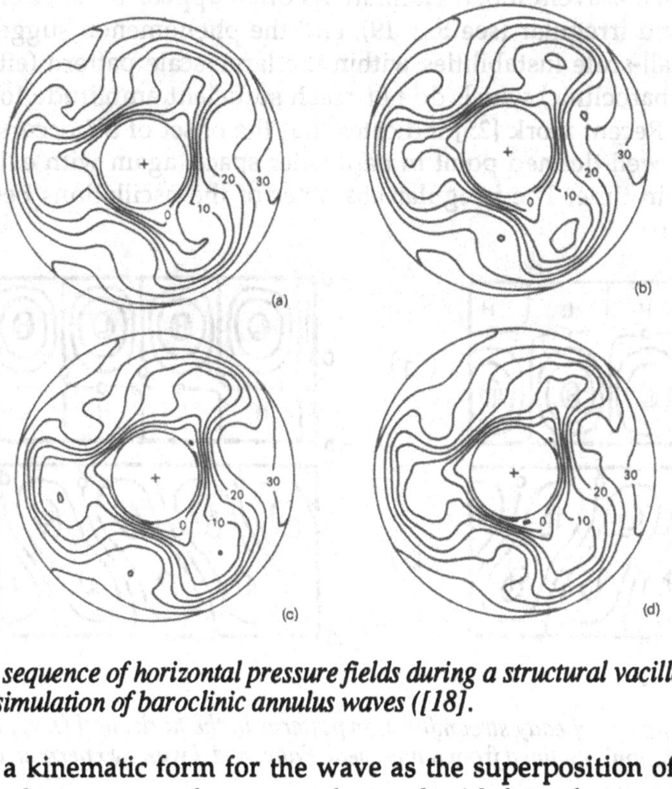

Figure 18: A sequence of horizontal pressure fields during a structural vacillation cycle, from a numerical simulation of baroclinic annulus waves ([18].

suggest a kinematic form for the wave as the superposition of dominant waves with the same zonal wavenumber and with lateral structures

$$\phi_1(y) = A_1 \sin \pi y \quad (25a)$$

$$\phi_2(y) = A_2 \sin 2\pi y, \quad (25b)$$

both propagating azimuthally at different phase speeds.

This type of behaviour has been reproduced in a class of simple numerical model, in which a small number of wave modes are allowed to interact through mutual advection in a quasi-geostrophic model. Weng & Barcilon [19], for example, followed a much earlier approach pioneered by Lorenz [21] and applied to a nonlinear version of the Eady model including the first two lateral modes (cf (25a,b)) to obtain solutions in which eddy energy oscillated in y through nonlinear interference between the two gravest y modes with the same x-wavenumber (e.g. see Fig. 19).

In practice, however, observed 'structural vacillations' are often more complicated than this picture would suggest, with transient small-scale features growing and decaying within a large-scale pattern dominated by a single azimuthal wavenumber. Oscillations often appear to be strongly intermittent and irregular (see Fig. 19), and the phenomenon suggests the growth of small-scale instabilities within the large-scale pattern (either barotropic or baroclinic) which do not reach sufficient amplitude to disrupt the main pattern. Recent work [25] indicates that the onset of SV occurs quite suddenly at a well-defined point in parameter space, again with evidence of intermittency in time. The irregular character of the oscillations becomes

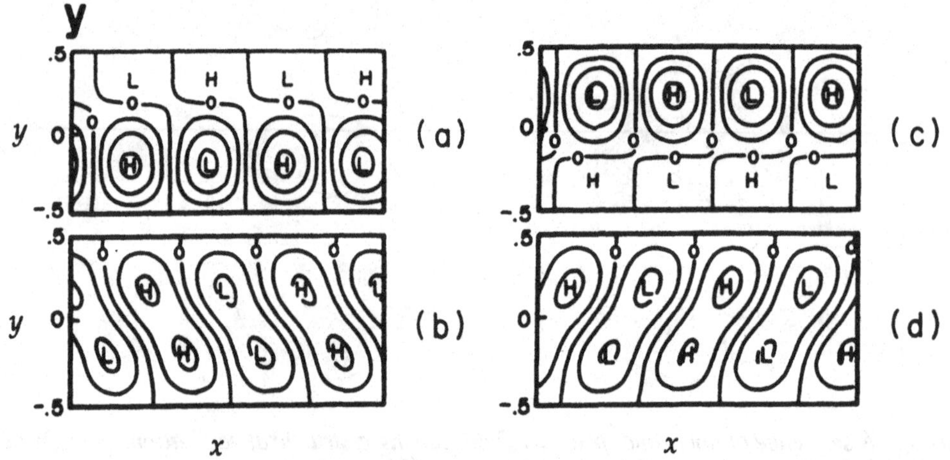

Figure 19: A sequence of eddy streamfunction patterns in the horizontal (x,y) plane during a 'structural vacillation' obtained from a nonlinear Eady model with two permitted lateral modes ([19]).

steadily more apparent as Ω is increased, and the large-scale pattern becomes gradually more distorted until it begins to break up into irregular flow. Hence, SV is frequently regarded as an intermediate state prior to the full onset of irregular wave flow or 'geostrophic turbulence'.

10. IRREGULAR FLOW AND 'BAROCLINIC CHAOS'

The onset of disordered or irregular flow is of interest as a simple 'model' of relevance to atmosphere/ocean flow and considerations of predictability. Laboratory measurements of the transition to irregular flow with increasing Ω (e.g. [22]) show the gradual broadening of the wavenumber spectrum and increasing significance of non-harmonically related azimuthal components. In extreme cases, the time-averaged spectrum does not display strong peaks at any particular wavenumber, but appears as a broad continuum with a characteristic quasi-power law decay towards the highest wavenumbers (simple theory predicts energies to decay as k^{-3}).

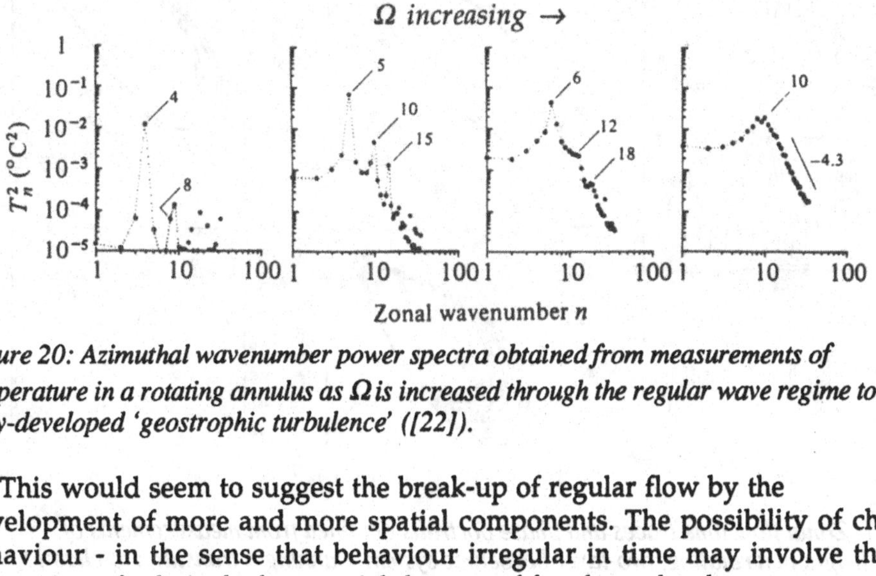

Figure 20: Azimuthal wavenumber power spectra obtained from measurements of temperature in a rotating annulus as Ω is increased through the regular wave regime towards fully-developed 'geostrophic turbulence' ([22]).

This would seem to suggest the break-up of regular flow by the development of more and more spatial components. The possibility of chaotic behaviour - in the sense that behaviour irregular in time may involve the interaction of relatively few spatial degrees of freedom - has been suggested as a first stage in the development of irregular flow ('weak geostrophic turbulence'). For example, even in weakly nonlinear theory, by taking a limit intermediate between limits (i) and (ii) of § 7 above (i.e. $E^{1/2}/Ro \sim \Delta^{1/2}$), several workers have shown that the single wave/zonal flow equilibration problem may reduce to a set of three coupled ODEs

$$dX/d\tau = \sigma(Y - X) \qquad (26a)$$

$$dY/d\tau = -XZ + R_aX - Y \tag{26b}$$

$$dZ/d\tau = XY - bZ \tag{26c}$$

(where σ, R_a and b are constants, X is related to A(t), Y is related to V(t) and Z ~ F(A,V) - e.g. see [23], [24]) which is the famous set of equations which can result in the *Lorenz Attractor* - a well known model of chaos. Thus, even a single wavenumber flow ought to be capable of behaving chaotically. Also, the onset of chaos in this model as parameters are smoothly varied would be characterised by a particular sequence of transitions typically involving either a sudden 'snap-through' bifurcation from the steady wave of § 7*(ii)* as

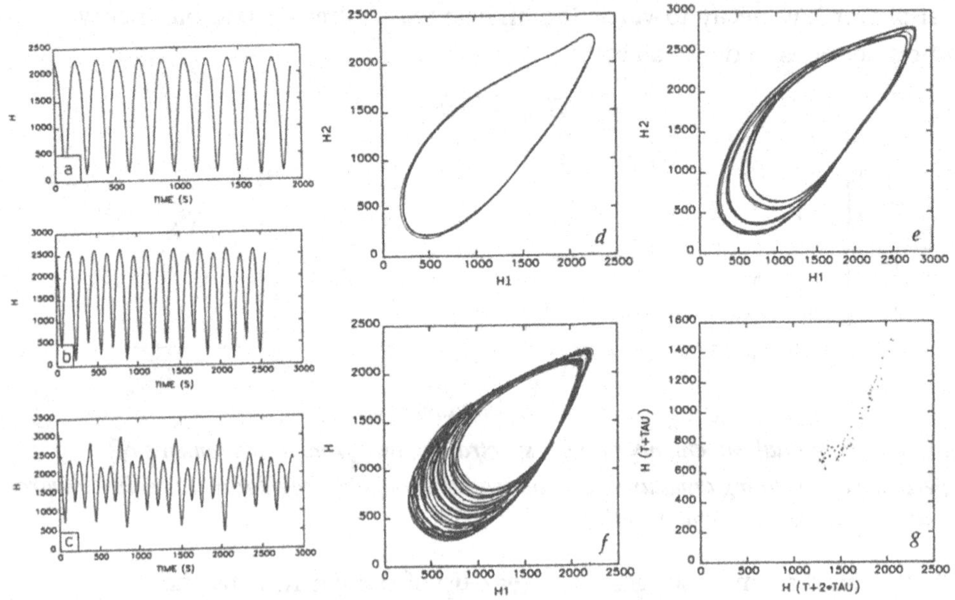

Figure 21: Zonal flow time traces and phase portraits obtained from measurements of interface height in a rotating, two-layer baroclinic system spanning a transition to chaotic flow ([25]).

dissipation is reduced, or a period-doubling cascade (in which a periodic 'limit cycle' develops an increasing number of kinks causing it to take 2^n (n integer) periods of the original oscillation to repeat its behaviour) from the 'amplitude vacillation' state of § 7*(i)* as dissipation is *increased* [24].

Some evidence for a period-doubling cascade has been detected in the work of Hart [25], who found a period-doubling route to chaos from an *amplitude vacillation* in his experiments on baroclinic waves in a two-layer mechanically-driven flow in an open cylinder from an m=2 state. The final state was not much like the classical 'Lorenz attractor', however, being more like a single chaotic band (see [24] for further discussion).

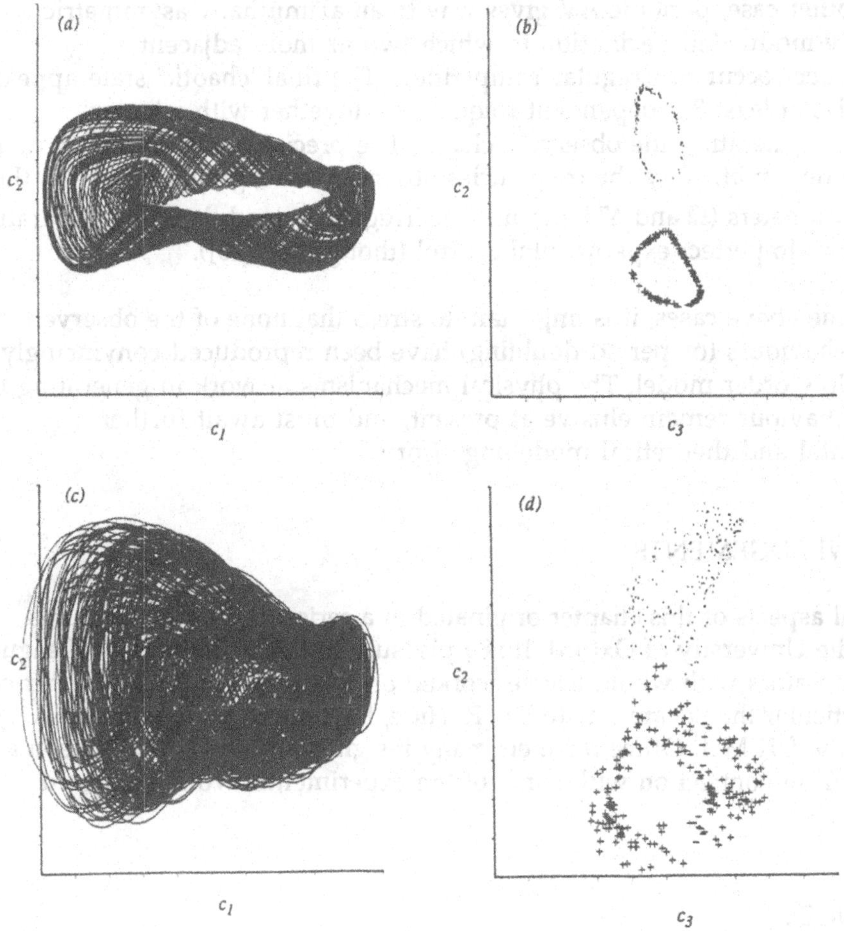

Figure 22: Phase portraits and Poincare sections obtained from measurements of temperature in a rotating annulus showing a transition from amplitude vacillation to a chaotic 'double vacillation' ([26]).

In the thermal annulus, the situation seems more complicated, with the possibility of at least two distinct routes to chaotic behaviour. We have already seen one route involving the onset of SV, which appears to occur suddenly and is immediately chaotic (no intermediate period-doubled periodic states?)

following the appearance of intermittent bursts of irregular activity. This does not seem to be readily consistent with notions of chaos in the formal sense - and its precise nature is still not fully understood (see [26] for further discussion).

In another case, periodic AV gives way to an azimuthally asymmetric chaotically-modulated vacillation in which two or more adjacent wavenumbers occur in irregular competition. The final 'chaotic' state appears to comprise at least 3 independent frequencies, together with a 'noisy' component generating the observed 'chaos'. The precise nature of this flow is also in some doubt, since the frequencies observed are highly sensitive to the control parameters (Ω and ΔT) and may be irregularly modulated by apparatus drifts due to imperfect experimental control (though see [26]).

In all the above cases, it is important to stress that none of the observed chaotic behaviours (or period-doubling) have been reproduced convincingly in a simple low-order model. The physical mechanisms at work in generating the chaotic behaviour remain elusive at present, and must await further experimental and theoretical modelling efforts.

ACKNOWLEDGEMENTS

Several aspects of this chapter originated in a series of graduate lectures given at the University of Oxford. It is a pleasure to thank the many colleagues and collaborators with whom I have worked on this problem for a number of years. Particular thanks are due to Drs R. Hide, P. Hignett, M. J. Bell and A. A. White of the UK Met. Office for their many insights, and to D. W. Johnson and R. M. Small in connection with some of the experimental work discussed herein.

REFERENCES

1. Read, P. L.: The 'philosophy' of laboratory experiments and studies of the atmospheric general circulation, *Met. Mag.*, **117** (1988), 35-45.

2. Vettin, F.: Experimentale darstellung von luftbewegungen unter dem einflusse von temperatur-unterschieden und rotations-impulsen, *Meteorol. Z.*, **1** (1884), 227-230 and 271-276.

3. Exner, F. M.: Über die bildung von windhosen und zyklonen, *Sitzungsber der Akad. der Wiss. Wien*, Abt. IIa, **132** (1923), 1-16.

4. Fultz, D.: Experimental analogies to atmospheric motions, in: *Compendium of Meteorology* (Ed. Malone, T. F.), American Meteorological Society 1951.

5. Hide, R.: An experimental study of thermal convection in a rotating fluid, *Phil. Trans. R. Soc. Lond.*, **A250** (1958), 441-478.

6. Pippard, A. B.: *Response and Stability*, Cambridge University Press 1985.

7. Hide, R. & Mason, P. J.: Sloping convection in a rotating fluid, *Adv. in Phys.*, **24** (1975), 47-100.

8. White, A. A.: The dynamics of rotating fluids: numerical modelling of annulus flows, *Met. Mag.*, **117** (1988), 54-63.

9. Hignett, P., Ibbetson, A. & Killworth, P. D.: On rotating thermal convection driven by non-uniform heating from below', *J. Fluid Mech.*, **109** (1981), 161-187.

10. Read, P. L.: Regimes of axisymmetric flow in an internally heated rotating fluid, *J. Fluid Mech.*, **168** (1986), 255-289.

11. Hignett, P.: A note on the heat transfer by the axisymmetric thermal convection in a rotating fluid annulus, *Geophys. Astrophys. Fluid Dyn.*, **19** (1982), 293-299.

12. Williams, G. P. (1969) 'Numerical integration of the three-dimensional Navier-Stokes equations for incomressible flow', *J. Fluid Mech.*, **37** (1969), 727-750; Baroclinic annulus waves, *J. Fluid Mech.*, **49** (1971), 417-

13. Bell, M. J. & White, A. A.: The stability of internal baroclinic jets: some analytical results, *J. Atmos. Sci.*, **45** (1988), 2571-2590.

14. Drazin, P. G.: Variations on a theme of Eady, in: *Rotating Fluids in Geophysics (ed. P. H. Roberts & A. M. Soward)*, Academic Press 1978, 139-169.

15. Hocking, L. M.: Theory of hydrodynamic stability, in: *Rotating Fluids in Geophysics (ed. P. H. Roberts & A. M. Soward)*, Academic Press. 437-469.

16. Pedlosky, J.: *Geophysical Fluid Dynamics (2nd Edition)*, Springer-Verlag 1987.

17. Pfeffer, R. L., Buzyna, G. & Kung, R.: Time dependent modes of behavior of thermally-driven rotating fluid', *J. Atmos. Sci.*, **37** (1980), 2129-2149.

18. Hignett, P., White, A. A., Carter, R. D., Jackson, W. D. N. & Small, R. M.: A comparison of laboratory measurements and numerical simulations of baroclinic wave flows in a rotating cylindrical annulus, *Quart. J. R. Met. Soc.*, **111** (1985), 131-154.

19. Weng, H.-Y. & Barcilon, A.: Wave structure and evolution in baroclinic flow regimes, *Quart. J. R. Met. Soc.*, **113** (1987), 1271-1294.

20. Lindzen, R. S., Farrell, B. & Jacqmin, D.: Vacillation due to wave interference: applications to the atmosphere and to annulus experiments, *J. Atmos. Sci.*, **39** (1982), 14-23.

21. Lorenz, E. N.: The mechanics of vacillation', *J. Atmos. Sci.*, **20** (1963), 448-464.

22. Buzyna, G., Pfeffer, R. L. & Kung, R.: Transition to geostrophic turbulence in a rotating differentially heated annulus of fluid, *J. Fluid Mech.*, **145** (1984), 377-403.

23. Brindley, J. & Moroz, I. M.: Lorenz attractor behaviour in a continuously stratified baroclinic fluid, *Phys. Lett.*, **77A** (1980), 441-444; Gibbon, J. D. & McGuiness, M. J.: A derivation of the Lorenz equations for some unstable dispersive physical systems, *Phys. Lett.*, **77A** (1980), 295-299; Pedlosky, J. & Frenzen, C.: Chaotic and periodic behavior of finite-amplitude baroclinic waves, *J. Atmos. Sci.*, **37** (1980), 1177-1196.

24. Klein, P.: Transition to chaos in unstable baroclinic systems: a review, *Fluid Dyn. Res.*, **5** (1990), 235-254.

25. Hart, J. E.: A laboratory study of baroclinic chaos on the f-plane, *Tellus*, **37A** (1985), 286-296.

26. Read, P. L., Bell, M. J., Johnson, D. W. & Small, R. M.: Quasi-periodic and chaotic flow regimes in a thermally-driven, rotating fluid annulus, *J. Fluid Mech.*, (1992 in press).

27. Read, P. L.: Dynamics and Instabilities of Ekman and Stewartson Layers, this volume 1992.

PART V

VORTEX DYNAMICS

Chapter V.1

DYNAMICS OF VORTEX FILAMENTS

D.W. Moore
Imperial College, London, UK

PREFACE

In the flow generated by a solid body started from rest, fluid particles have zero vorticity initially. If the fluid is homogeneous, only those fluid particles which are at some stage of their history close to the body acquire vorticity by diffusion from the boundary. Were it not for the occurence of flow separation, the flow away from the boundary of the body would remain irrotational.

Flow separation will always occur when the body has sharp edges not aligned with the flow (unless the body is oscillating sufficiently rapidly). This separation can be beneficial. For example, the leading-edge separation on slender wing aircraft like Concorde enhances the lift. An authoritative account of the aerodynamics of such flows has been given by Kuchemann (1).

When the Reynolds number of the flow is large, the shed vorticity lies in a thin shear layer, which adopts a characteristic spiral form. This process of shedding and spiral formation is clear in the experiments of Pullin and Perry (2), in which fluid initially at rest is forced past a projecting wedge by a piston. Figure 1 displays the evolution of the separated flow. When the piston stops, the separated shear layer breaks away from the tip of the wedge. We can anticipate that viscous action will diffuse away the peaks of the vorticity distribution, to produce an isolated smooth distribution of vorticity, of roughly circular shape. An estimate for the time scale for this smoothing process can be found in (3). Since vorticity needs to diffuse only between neighbouring turns of the spiral, this time scale is much shorter than the diffusion time based on the overall length scale.

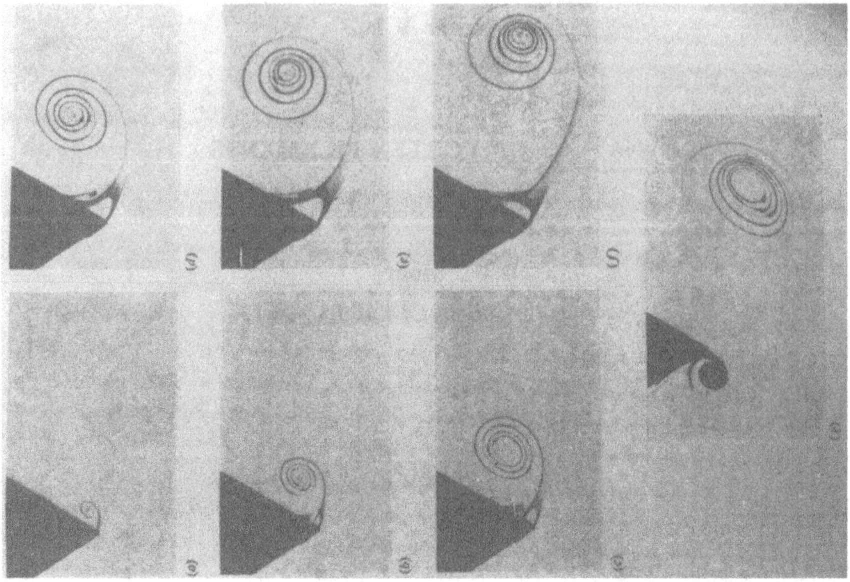

Figure 1. Vortex sheet separating at a sharp corner visualised by Pullin and Perry (2)

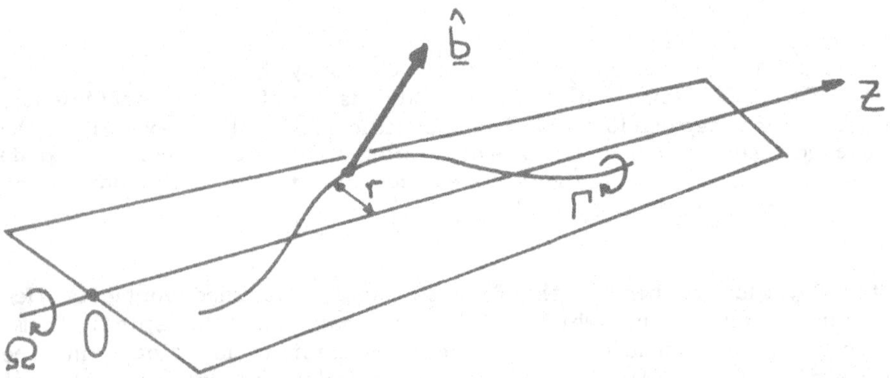

Figure 2. Geometry of a rotating plane-wave. The velocity is along the binormal and is consistent with rigid rotation if equation (2.23) holds.

A similar sequence of events is involved in the formation of a vortex ring at an orifice. Furthermore, if we replace the downstream evolution of the vorticity shed from the leading edge of a slender delta wing by the temporal evolution which would be seen in a plane, moving past the wing with the main stream - the Treffitz plane approximation - we can carry over these ideas to the aeronautical case.

To study the evolution of the shed vorticity mathematically, we have to make some approximations. Viscous effects are small away from the body, so we base our analysis on the Euler equations (1.1) and (1.2). However, these equations are intractable and progress has been possible only by making geometrical simplifications. These simplifications concern the support of the vorticity field, which is taken to be either a surface - giving a vortex sheet - or a space curve - giving a vortex filament. We might thus seek to model the roll-up of the shear layer by replacing it by a vortex sheet. The remarkable evolution of an initially elliptical vortex ring was studied (4) using a vortex filament model.

It is a simple matter to express the velocity field as an integral over the support of the vorticity field - equation (1.5) -, so these geometrical approximations reduce the dimension of the original 3D problem to 2D for the sheet and 1D for the filament. However there is a price to pay for this simplification. In both cases ordinary convergence of the integral is lost at points of the support itself. This presents no difficulty if we merely want to calculate the velocity field at distant points, but to follow the evolution of the sheet or filament we need, as we shall see, the velocity at every point of the support.

In the case of the vortex sheet the difficulty can be bypassed by working with the potential flow away from the sheet and matching pressures and normal velocities across it. It was not until the work of Birkhoff and Fisher (5) and Rott (6) that it was realised that the divergence could be removed in the case of two space dimensions, by treating the integral as a Cauchy principal value (although this was known for straight sheets in the context of thin-wing theory). The divergence cannot be dealt with in this way for the case of the vortex filament. Such an intractable divergence is a warning that our simplifications have been carried too far. We cannot completely ignore the detailed structure of the vortex. But if the support of the vorticity is a thin tube, Crow (7) suggested a method of modifying the integral so that it had a finite value. The 'cut off' theory is easy to describe and easy to employ. I describe it in Section 2 of these lectures and apply it to the helical filament in section 3.

The theory can be justified by a formal asymptotic analysis based on a thinness parameter. However, this is a rather intricate matter and I prefer to try to explain why the cut-off theory works in an intuitive manner. The key to understanding the problem is, I suggest in section 4, the elucidation of wave propagation on the filament. Thus I describe the relevant results from Lord Kelvin's celebrated paper of 1880 (8).

In section 5 I show how local scale and large scale dynamics interact and how the method of matched asymptotic expansions can be used to study their interaction.

Finally, I consider in section 6, two types of instability to which vortex filaments are prone. Of course, a vortex filament can display shear flow instability provided the flow along the filament axis is sufficiently strong, but the instabilities I describe are due to the action of weak irrotational strain. Such a strain is experienced by every point of a circular vortex filament and it was by invoking this type of instability that Widnall and Tsai (9) were able to suggest a mechanism for the observed break up.

§1 BASIC IDEAS

I shall consider only inviscid, incompressible flow in these lectures. Then the equations of motion are

$$\frac{\partial \mathbf{u}}{\partial t} + (\mathbf{u}.\nabla)\mathbf{u} = -\frac{1}{\rho_0} \nabla p \tag{1.1}$$

and

$$\nabla.\mathbf{u} = 0 \tag{1.2}$$

Here $\mathbf{u}(\mathbf{x},t)$ is the fluid velocity at position $\mathbf{x} = x\hat{\mathbf{i}} + y\hat{\mathbf{j}} + z\hat{\mathbf{k}}$ relative to fixed axes Oxyz. The density ρ_0 is a constant and $p(\mathbf{x},t)$ is the pressure.

Equations (1.1) and (1.2) are awkward to solve numerically, because there is no equation for the evolution of the pressure and the constraint (1.2) has to be satisfied.

We are led to the vorticity vector field $\omega(\mathbf{x},t)$ when we try to deal with this difficulty by eliminating the pressure. If we apply the operator $\nabla_\wedge$ to equation (1.1) we find that

$$\frac{\partial \omega}{\partial t} + (\mathbf{u}.\nabla)\omega = (\omega.\nabla)\mathbf{u} \quad , \tag{1.3}$$

where $\omega(\mathbf{x},t)$ is the vorticity field, defined by

$$\omega = \nabla_\wedge \mathbf{u}. \tag{1.4}$$

Suppose the fluid is unbounded and is at rest at infinity. Suppose, also, that ω is piecewise continuous and vanishes outside some sphere $|\mathbf{x}| = R_0$. Then it can be shown (a good treatment of the equivalent Poisson problem is given by Jeffreys and Jeffreys (10), Chapter 6) that

$$\mathbf{u}(\mathbf{x},t) = \frac{1}{4\pi} \int \frac{\omega(\mathbf{y}) \wedge (\mathbf{x}-\mathbf{y})}{|\mathbf{x}-\mathbf{y}|^3} d^3\mathbf{y} \tag{1.5}$$

is the solution of (1.2) and (1.4) which vanishes at infinity.

The region of integration is just the support of the vorticity field, which makes the representation of the velocity field efficient. We note that the integrand is infinite when $\mathbf{y} = \mathbf{x}$. The integral can be shown to converge when $\omega(\mathbf{y})$ is piecewise continuous, but difficulties arise when we try to go further and let ω be a generalised function, as in the case of a vortex sheet or vortex filament.

We can see that (1.3) and (1.5) can be combined to form an evolution equation

$$\frac{\partial \omega}{\partial t} = \mathbf{N}(\omega) \tag{1.6}$$

where $\mathbf{N}$ is a non-linear function of ω and its spatial derivatives, and our objective is gained.

The situation is at its simplest if we consider a "plane" flow in which $\mathbf{u}$ is independent of z. Then

$$\omega(x, y, t) = \hat{\mathbf{k}}\, \zeta(x, y, t) \tag{1.7}$$

and the vorticity equation (1.3) reduces to

$$\frac{\partial \zeta}{\partial t} + u \frac{\partial \zeta}{\partial x} + v \frac{\partial \zeta}{\partial y} = 0 \tag{1.8}$$

This equation, when viewed from a fluid-particle or "Lagrangian" standpoint, has a simple meaning. Let $(x_p(t), y_p(t))$ be the coordinates of a fluid particle p; $x_p(t)$ and $y_p(t)$ satisfy the ordinary differential equations

$$\left.\begin{aligned} \frac{dx_p}{dt} &= u(x_p(t), y_p(t), t) \\ \frac{dy_p}{dt} &= v(x_p(t), y_p(t), t) \end{aligned}\right\} . \tag{1.9}$$

Define

$$\zeta_p(t) = \zeta(x_p(t), y_p(t), t) \tag{1.10}$$

to be the vorticity carried by the fluid particle p. Then

$$\frac{d\zeta_p}{dt} = \frac{\partial \zeta}{\partial t} + \frac{\partial \zeta}{\partial x}\frac{dx_p}{dt} + \frac{\partial \zeta}{\partial y}\frac{dy_p}{dt} = 0 , \tag{1.11}$$

using (1.8) and (1.9). Thus in plane motion each fluid particle carries a constant vorticity. We don't have to solve the vorticity equation because we can find the vorticity distribution by tracking fluid particles.

In a practical calculation a finite set of fluid particles is followed using (1.9), where the velocity of each fluid particle is obtained from the two-dimensional form of (1.5)

$$u(s,t) - i\, v(s,t) = -\frac{i}{2\pi} \iint \frac{\zeta(s',t)\ dx'dy'}{s - s'} , \tag{1.12}$$

discretised over the chosen set of fluid particles: here $s = x + iy$. This is the point-vortex method. It is more often used for vortex sheets, where high accuracy can be achieved if care is taken with the evaluation of the integral in (1.12), see (11) and (12).

In three-dimensional flow, the fluid particle description is less simple. Equation (1.13) shows that

$$\frac{d\omega_p}{dt} = (\omega.\nabla)\mathbf{u}\,\Big|_{\mathbf{x} = \mathbf{x}_p} , \tag{1.13}$$

so that the vorticity carried by p is modified by the strain field at p. To rescue the fluid particle formulation we must find a representation of the strain field in terms of fluid particle motion. We introduce a second fluid particle q which is at an infinitesimal displacement Δ from p. It is easily seen that

$$\frac{d\Delta}{dt} = (\Delta.\nabla)\mathbf{u}\,\Big|_{\mathbf{x} = \mathbf{x}_p} . \tag{1.14}$$

The strain field enters (1.13) and (1.14) in the same way, so that if μ is any constant, (1.13) and (1.14) can be combined to give

$$\frac{d}{dt}(\omega_p - \mu\Delta) = ((\omega_p - \mu\Delta).\nabla)\mathbf{u}\,\Big|_{\mathbf{x} = \mathbf{x}_p} . \tag{1.15}$$

We now decide that t = 0 q lies on the vortex line through p, which which implies choosing μ so that

$$\omega_p(0) = \mu\Delta(0) . \tag{1.16}$$

Then, provided $\mathbf{u}$ is sufficiently well-behaved - the precise requirements are embodied in Picard's theorem (13) - the system of ordinary differential equations (1.15) has the unique solution

$$\omega_p(t) = \mu\Delta(t) , \tag{1.17}$$

which means that q continues to lie on the vortex line through p, and

$$|\omega_p(t)| \,/\, |\omega_p(0)| \quad = \quad |\Delta(t)| \,/\, |\Delta(0)| \; .^{\dagger} \qquad (1.18)$$

Just as in two dimensions, we need not solve the vorticity equation. Instead we represent each vortex line by a set of fluid particles $p_1, p_2, p_3, \ldots$ whose vorticity at time t is computed from $\omega(t)$ using (1.5). The vorticity of each "vortex stick" $p_s \, p_{s+1}$ is updated using (1.18). Leonard (15) has given an account of attempts to use Lagrangian ideas for three dimensional flow calculations.

† For an interesting history of this method of proof see (14).

§2 VORTEX FILAMENTS

In many flows of interest the vorticity is organised into discrete structures embedded in otherwise irrotational flow. The trailing vortices behind an aircraft are an example: many wing-spans downstream the flow is well represented by a pair of parallel vortex filaments. Since this vortex trail can persist 10 kilometres behind the aircraft, nothing is lost by ignoring the process of formation of the vortices and treating the flow as that of a pair of infinite vortex filaments, as in the pioneering work of Crow (7).

By a vortex filament we mean a flow field in which the vorticity zero outside a tube whose radius of curvature is everwhere large compared to its diameter. The cross-sectional shape is assumed to vary only on the same length scale - we rule out sudden changes in cross-sectional area or shape. I shall try to show why these properties hold in actual flow later.

I can be more specific and define a straight, circular vortex filament to be a flow in which

$$\mathbf{u} \;=\; V(r)\,\hat{\boldsymbol{\theta}} \;+\; W(r)\,\hat{\mathbf{k}} \qquad (2.1)$$

in cylindrical polars (r,θ,z) where, additionally,

$$\left.\begin{aligned} V(r) &= V(a)\,\frac{a}{r} \qquad (r > a) \\ W(r) &= 0 \qquad\qquad (r > a) \end{aligned}\right\} \qquad (2.2)$$

Thus we have potential flow for $r > a$ and the vorticity is given by

$$\left.\begin{aligned} \omega(r) &= -\hat{\theta}\,\frac{dW}{dr} + \hat{k}\,\frac{1}{r}\,\frac{d}{dr}\,(rV) && (r < a) \\ \omega(r) &= 0 && (r > a) \end{aligned}\right\} \quad (2.3)$$

The vorticity is not parallel to the filament axis, but the θ component proves to have no net effect on the flow induced by the filament. The net axial vorticity is $\Gamma = 2\pi a V(a)$.

I can now pose the problem I wish to consider in these lectures. Suppose we switch on an irrotational flow $\mathbf{V}_E\,(\mathbf{x},t)$ at time $t = 0$. This flow varies on a length scale L and on a time scale T which are large compared to the vortex core radius a and circulation time a^2/Γ; thus $a/L \ll 1$ and $a^2/\Gamma T \ll 1$. We will, in fact, demand that $T = O(L^2/\Gamma)$. We also insist that the strain rate $\|\nabla\,\mathbf{V}_E\|$ be much smaller than the core vorticity, which implies that $|\mathbf{V}_E| \ll \Gamma/a$. These restrictions will enable us to develop an approximate theory for the filaments response.

The assumption of small strain means that the filament cross sections are nearly circular, so we regard the filament as being the volume swept out when a circular disc of radius a(s) is swept along a centre-line curve

$$\mathbf{X} = \mathbf{R}(s,t)\ , \qquad (2.4)$$

the disc being always normal to this curve. Now, by assumption, using the notation of equation (1.5),

$$\mathbf{x} - \mathbf{y} \;=\; (\mathbf{x} - \mathbf{R})(1 + O(a/L)) \qquad (2.5)$$

so that (1.5) becomes

$$\mathbf{u}(\mathbf{x},t) \;=\; -\,\frac{1}{4\pi}\int\left[\frac{\mathbf{x} - \mathbf{R}}{|\mathbf{x} - \mathbf{R}|^3} \wedge \int \omega\, dA\right] ds \qquad (2.6)$$

To evaluate the inner integral we take local rectangular axes $O'x'y'z'$ with O' at the centre of the cross-section. Then

$$\omega \;=\; (\hat{\mathbf{x}}'\sin\theta' - \hat{\mathbf{y}}'\cos\theta')\,\frac{dW}{dr'} + \frac{\hat{\mathbf{z}}'}{r'}\,\frac{d}{dr}\,(r'V) \qquad (2.7)$$

where (r',θ') are polar coordinates in the plane of the cross section. The integral is easily evaluated to give

$$\int \omega \, dA = \hat{z}' \, 2\pi a V(a) = \Gamma \, \hat{t}(s,t) \qquad (2.8)$$

where $\hat{t}$ is the tangent to the centre line and Γ is the circulation around the boundary $r' = a$ of the cross-section. Since the flow outside the filament is irrotational, Γ is in fact the circulation around <u>any</u> circuit enclosing the filament. Furthermore, Kelvin's circulation theorem shows that Γ is independent of time.

Thus we find the Biot-Savart line integral

$$\mathbf{u}(\mathbf{x},t) = \frac{\Gamma}{4\pi} \int_{-\infty}^{\infty} \frac{\hat{\mathbf{t}} \wedge (\mathbf{x} - \mathbf{R})}{|\mathbf{x} - \mathbf{R}|^3} \, ds \quad . \qquad (2.9)$$

We need not confine ourselves to initially straight filaments and we can consider filaments which are closed curves $0 \le s \le \ell(t)$ with $\mathbf{x}(0,t) = \mathbf{x}(\ell(t),t)$, where $\ell(t)$ is the length of the filament. In this case $\ell(t)$ has to be determined as part of the problem.

It is worth stressing that Γ is <u>not</u> affected by stretching the filament: vortex stretching is not - in this approximation - a significant process.

As we have seen, each constituent vortex line in the filament is moving with the fluid. The dominant motion is, of course, a a rapid rotation. We might suppose, however, that if we could somehow subtract this motion what would be left would be the motion of the filament as a whole. As a first shot, we might evaluate (2.9) on the centre-line itself, even though (2.9) is strictly valid only if $|\mathbf{x} - \mathbf{R}| \gg a$. Here we arrive at the fundamental problem of filament dynamics - the Biot-Savart integral defined in (2.9) diverges on the curve $\mathbf{x} = \mathbf{R}(s,t)$.

To see this, let us try to evaluate $\mathbf{u}(\mathbf{R}_0)$ where $\mathbf{R}_0 = \mathbf{R}(s_0,t)$ is a point of the centre line. The expansion

$$\mathbf{R}(s) = \mathbf{R}(s_0) + (s - s_0)\mathbf{R}'(s_0) + \tfrac{1}{2}(s - s_0)^2 \, \mathbf{R}''(s_0) + \ldots \qquad (2.10)$$

enables us to obtain the dominant behaviour of the integral in (2.9) for s close to s_0; this behaviour is

$$\frac{\hat{t} \wedge (R_0 - R)}{|R_0 - R|^3} \sim \frac{1}{2} \frac{(s - s_0)^2}{|s - s_0|^3} \frac{\hat{b}(s_0)}{\rho(s_0)} , \quad (2.11)$$

where $\hat{b}$ is the binormal and ρ the radius of curvature of the filament.

The integral thus diverges logarithmically, but is not of Cauchy principal-value type because the factor involving $s - s_0$ is positive. Unlike the vortex sheet, there is no local cancellation of the large induced velocities.

A method of dealing with this difficulty was proposed by Crow (7) This consists of removing the small interval $s_0 - \delta \le s \le s_0 + \delta$ from the range of integration in equation (2.9). The cut-off length is to be chosen by comparison with the known solution for a vortex ring. Suppose the vortex ring is parametrically

$$R(s) = \hat{i} R \cos\left(\frac{s}{R}\right) + \hat{j} R \sin\left(\frac{s}{R}\right) \quad (2.12)$$

so that $\ell = 2\pi R$. We can take $s_0 = 0$ to find that on substituting (2.12) into (2.9) we have to evaluate the integral

$$u(\hat{i} R) = \frac{\Gamma \hat{k}}{4\pi R} \int_{\delta/R}^{2\pi-\delta/R} \frac{(1 - \cos\theta)}{(2 - 2\cos\theta)^{3/2}} d\theta , \quad (2.13)$$

after an obvious change of variable. This gives

$$u(\hat{i} R) = - \frac{\Gamma \hat{k}}{4\pi R} \ln \left[\tan\left(\frac{\delta}{4R}\right)\right] \quad (2.14)$$

We can see that, as asserted, the integral diverges logarithmically as $\delta/R \to 0$.

Now the velocity of propagation a vortex ring has been calculated for a general distribution of core vorticity and the propagation speed U is given by

$$U = \frac{\Gamma}{4\pi R}\left[\ln\frac{8R}{a} - \frac{1}{2} + \frac{8\pi^2}{\Gamma^2}\int_0^a r\left[\frac{1}{2}V^2(r) - W^2(r)\right]dr\right]; \qquad (2.15)$$

see (16) and (17), where $V(r)$ and $W(r)$ are as defined in (2.1) and (2.2).

Thus if we compare (2.14) and (2.15) we find that

$$\ln(2\delta/a) = \frac{1}{2} + \frac{8\pi^2}{\Gamma^2}\int_0^a r\,(W^2(r) - \frac{1}{2}V^2(r))\,dr \quad .(2.16)$$

We see that δ/a is of order unity and that δ is independent of R. It depends only on the local core properties. Thus to apply the cut-off idea to a non-circular filament, we merely use the core properties $a(s_0)$, $W(r,s_0)$, $V(r,s_0)$ <u>at</u> the cut-off point itself. But now a difficulty arises. How are these functions of s_0 to be determined? Saffman and I (18) tackled this problem, using the method of matched asymptotic expansions to link the large (L) and small core scale (a) dynamics. We found that the cut-off theory emerged when leading order terms only were retained in an asymptotic expansion in the small parameter a/ρ; here a and ρ are typical values of $a(s,t)$ and $\rho(s,t)$. We gave arguments to show that $a(s,t)$ was independent of s, while $a(t)$ was determined from the condition

$$a^2(t)\ell(t) = a^2(0)\ell(0) \quad , \qquad (2.17)$$

demanded by conservation of filament volume: recall that the filament surface consists of vortex lines and is thus material. An equation of motion more accurate than the cut-off theory was derived by retaining the next term in the asymptotic expansion. Equations for the evolution of V and W were derived, but I will not go into details here. The higher-order theory has been re-derived recently by Fukumoto and Miyazaki (19), using a different approach.

We have thus arrived at a first-order theory for filament motion. Let us choose a material coordinate ξ to replace arc distance, so that the filament has the equation

$$\mathbf{x} = \mathbf{R}(\xi,t) \quad (0 \le \xi \le 1) \ , \qquad (2.18)$$

where we insist that

$$\frac{\partial \mathbf{R}}{\partial t}(\xi_0,t) = \mathbf{u}(\xi_0,t) + \mathbf{V}_E\,(\mathbf{R}(\xi_0,t),t) \ . \qquad (2.19)$$

In equation (2.20), **u** is obtained from the Biot-Savart integral (2.9) with the cut-off applied. The arc distance function is obtained from the equation

$$\frac{\partial s}{\partial \xi} = \left|\frac{\partial \mathbf{R}}{\partial \xi}\right| \qquad (2.20)$$

and $\ell(t)$ needed, for the computation of a(t), follows by integration. The evolution of an elliptical vortex filament has been computed in this fashion by Dhanak and de Bernardnis (4).

The form of the integral near the singularity which is displayed in equation (2.11) shows that the cut-off theory can be written

$$\frac{\partial \mathbf{R}}{\partial t} = \frac{\Gamma \hat{\mathbf{b}}}{4\pi\rho} \ln (L^{*}/\delta_0) \qquad (2.21)$$

where L* is an unknown constant of the order of the overall length scale L of the flow; notice that V_E has been absorbed into L*. If we arbitrarily neglect the variation of L*, we can choose a re-scaled time τ so that

$$\frac{\partial \mathbf{R}}{\partial \tau} = \frac{\partial \mathbf{R}}{\partial s} \wedge \frac{\partial^2 \mathbf{R}}{\partial s^2} . \qquad (2.22)$$

This is the "local-induction approximation" due to Arms and Hama (20). It is practically useless because the values of a/L* encountered in experiment are much too large for the logarithmic term to dominate. For example, in the trailing vortex problem, L* is comparable to the wing-span while a is roughly one tenth of this. However (2.22) has stimulated interesting mathematical analysis by Hasimoto (21) and Kida (22). In particular, it can be shown (21) that (2.22) can be reduced to the cubic Schroedinger equation, for which a general solution is available. It follows that soliton solutions and multi-soliton solutions can be found.

If we restrict our attention to vortex filaments which lie in a plane, an elementary treatment of (2.22) is possible. The self-induced velocity is at every point of the filament along the binormal, hence normal to the plane instantaneously containing the filament. This will result in pure rotation about a fixed axis OZ so that the filament will continue to be planar, provided that

$$\Omega r = \frac{1}{\rho(z)} , \qquad (2.23)$$

where Ω is the angular velocity and r is distance from the z axis; see Figure 2. This equation can be shown to govern the equilibria of a thin elastic rod (23), where some interesting solutions are described.

If, in addition to the assumption of planar form, we insist that the filament is nearly straight, then $\rho(z)$ can be approximated to give

$$\Omega r = \frac{d^2 r}{dz^2} . \tag{2.24}$$

For sinusoidal solutions $\Omega < 0$ and we note the sense of rotation is opposite to the rotation around the vortex.

I shall not consider this local induction approximation further, but look instead at wave propagation on a vortex filament when the full cut-off equations are used. I choose to examine the case of the helical wave because this is the simplest form of propagating disturbance and has been subject to experimental investigation (24).

§3 THE PROPAGATION OF HELICAL WAVES

We consider a helical vortex filament of core radius a which is independent of position along the filament. The equation of the helix is given parametrically by

$$R(\theta, t) = \hat{i} D\cos\theta + \hat{j} D\sin\theta + \hat{k} \frac{(\theta + nt)}{\gamma} , \tag{3.1}$$

where n is to be determined.

The constant D is the radius of the circular cylinder on which the helix is inscribed and $2\pi/\gamma$ is the distance between successive turns of the helix measured along any generator of the cylinder; γ is thus the wave number of the disturbance. If we differentiate equation (3.1) with respect to θ we find that

$$\hat{t} = (-\hat{i} D\sin\theta + \hat{j} D\cos\theta + \hat{k}\, \gamma^{-1}) \frac{d\theta}{ds} \tag{3.2}$$

where s is arc distance measured along the helix. If we choose the sense of measurement so that s increases with z, we have

$$\frac{ds}{d\theta} = \sqrt{1+\gamma^2 D^2}/\gamma \tag{3.3}$$

and

$$\hat{t} = \frac{\gamma D}{\sqrt{1+\gamma^2 D^2}} \left[-\hat{i}\sin\theta, + \hat{j}\cos\theta + \hat{k}\, (\gamma D)^{-1}\right]. \tag{3.4}$$

Thus the angle between the helix and the generators is $\tan^{-1}(\gamma D)$. If we differentiate (3.4) again with respect to θ we find

$$\frac{\hat{n}}{\rho} = -\frac{\gamma^2 D}{1+\gamma^2 D^2}(\hat{i}\cos\theta + \hat{j}\sin\theta) \tag{3.5}$$

so that the radius of curvature of the helical filament is given by

$$\rho = (1 + \gamma^2 D^2)/(\gamma^2 D) \tag{3.6}$$

We expect the cut-off theory to hold if the core size is much less than both the distance between successive turns and the radius of curvature, so that we must insist that

$$a \ll \min\left(\frac{2\pi}{\gamma}, \frac{1+\gamma^2 D^2}{\gamma^2 D}\right). \tag{3.7}$$

The trace of the helix in a fixed plane normal to the axis rotates with angular velocity $-n$ while the trace of the helix in a fixed axial plane translates with velocity $-n/\gamma$, which is thus the phase speed of the helical wave.

We are now in a position to apply the cut-off theory explained in the previous section. We find that the velocity at the point $\hat{i}D$ is given at time $t=0$ by

$$u_0 = \frac{\Gamma}{4\pi} \mathop{\int\!\!\!\!\times}_{-\infty}^{\infty} \frac{[-\hat{i}\sin\theta + \hat{j}\cos\theta + \hat{k}(\gamma D)^{-1}] \wedge [\hat{i}D(1-\cos\theta) - \hat{j}D\sin\theta - \hat{k}\frac{\theta}{\gamma}]d\theta}{[\theta^2/\gamma^2 + 2D^2(1-\cos\theta)]^{3/2}} \tag{3.8}$$

where the cross on the integral sign is to remind us to cut out the interval $|s - s_0| \le \delta$; here, in view of equation (3.3), we remove the range

$$|\theta| \le \theta_0 = \gamma\,\delta/\sqrt{1+\gamma^2 D^2} \tag{3.9}$$

to find, invoking the symmetries of the integral about $\theta = 0$,

$$u_0 = \frac{\Gamma\gamma^2 D}{2\pi}\int_{\theta_0}^{\infty} \frac{[\hat{j}(1-\cos\theta - \theta\sin\theta) + \hat{k}\gamma D(1-\cos\theta)]\,d\theta}{[\theta^2 + 2\gamma^2 D^2(1-\cos\theta)]^{3/2}} \tag{3.10}$$

These integrals have to be evaluated numerically, but this presents no difficulty. Instead we consider long waves for which $\gamma D \ll 1$ and neglect terms of order $\gamma^2 D^2$. The second term in the denominator is uniformly smaller than the first, leading to

$$\mathbf{u}_0 = -\frac{\Gamma\gamma^2 D}{2\pi}\int_{\gamma\delta}^{\infty}\left[\hat{\mathbf{j}}\left(\frac{1-\cos\theta}{\theta^3} - \frac{\sin\theta}{\theta^2}\right) + \hat{\mathbf{k}}\gamma D\,\frac{(1-\cos\theta)}{\theta^3}\right]d\theta \quad (3.11)$$

where we have also consistently approximated in the definition, equation (3.9) of θ_0. These integrals can be expressed in terms of Sine and Cosine integrals. However, the inequality (3.7) shows that $\gamma a \ll 1$ and since we saw in the previous section that $\delta/a = O(1)$, it follows that $\gamma\delta \ll 1$. We can thus replace the Sine and Cosine integrals by their asymptotics to get

$$\mathbf{u}_0 = \frac{\Gamma\gamma^2 D}{4\pi}\left[\hat{\mathbf{j}}\left(-\frac{1}{2} + C + \ln(\gamma\delta)\right) + \hat{\mathbf{k}}\gamma D\left(\frac{3}{2} - C - \ln(\gamma\delta)\right)\right] \quad (3.12)$$

where $C = \cdot5772\ldots$ is Euler's constant.

Before proceeding to the calculation of the propagation velocity n/γ it is worth comparing (3.12) with the local induction result (2.21). The binormal at any point of the filament is $\mathbf{t}_\wedge\mathbf{n}$, so that, using (3.4), (3.5) and (3.6) and again neglecting $O(\gamma^2 D^2)$, we find that the self-induction velocity $\mathbf{u}_0'$ at Di is given by

$$\mathbf{u}_0' = +\frac{\Gamma\gamma^2 D}{4\pi}(-\hat{\mathbf{j}} + \gamma D\hat{\mathbf{k}})\ \ln(L^*/\delta) \quad (3.13)$$

By comparison with (3.12) we see that the terms of $O(|\ln\delta|)$ agree, but that no choice of L^* can make the $O(1)$ terms agree. This is because the velocity given by the cut-off theory is <u>not</u> along the binormal: local induction is wrong in magnitude and direction.

Finally, we must equate the self-induced velocity $\mathbf{u}_0$ at the point $\hat{\mathbf{i}}D$, corresponding to $\theta = 0$ and $t = 0$, to the velocity of the helix at the same point. However, the velocity of a space curve along its tangent is undefined, so that all we can insist on is that

$$\hat{\mathbf{n}}\cdot\left(\mathbf{u}_0 - \frac{\partial\mathbf{R}}{\partial t}(0,0)\right) = 0 \quad (3.14)$$

and

$$\hat{b}\cdot\left[u_0 - \frac{\partial R}{\partial t}(0,0)\right] = 0 \qquad (3.15)$$

Now from the defining equation (3.1)

$$\frac{\partial R}{\partial t}(0,t) = \gamma^{-1}\, n\, \hat{k} \ , \qquad (3.16)$$

so that (3.14) is trivially satisfied, while (3.15) leads to

$$n = \frac{\Gamma\gamma^2}{4\pi}\left[\frac{1}{2} - C - \ln(\gamma\delta)\right]. \qquad (3.17)$$

The final step is to use equation (2.16) to obtain δ in terms of core properties. Then we find (8) that

$$n = \frac{\Gamma\gamma^2}{4\pi}\left[\ln\frac{2}{\gamma a} - C + \frac{8\pi^2}{\Gamma^2}\int_0^a\left[r\left(\frac{1}{2}V^2(r)\right) - W^2(r)\right]dr \right] \qquad (3.18)$$

This formula agrees fairly well with experiment (24); a comparison is shown in Figure 3.

One case is of special interest, because we can find the result in another way. This is the Rankine vortex which has rigid rotation and no axial flow in the unperturbed core, so that

$$V(r) = \frac{\Gamma r}{2\pi a^2} \qquad (0 \le r \le a) \qquad (3.19)$$

and

$$W(r) = 0 \qquad (0 \le r \le a) \qquad (3.20)$$

which gives, after evaluating the integrals in (3.19),

$$n = \frac{\Gamma\gamma^2}{4\pi}\left[\ln\frac{2}{\gamma a} - C + \frac{1}{4}\right], \qquad (3.21)$$

a result obtained in a completely different way by Lord Kelvin (8) in his paper of 1880. Kelvin looked at waves of amplitude D small compared to the core radius a. This is a very informative calculation, which we take up in the next section.

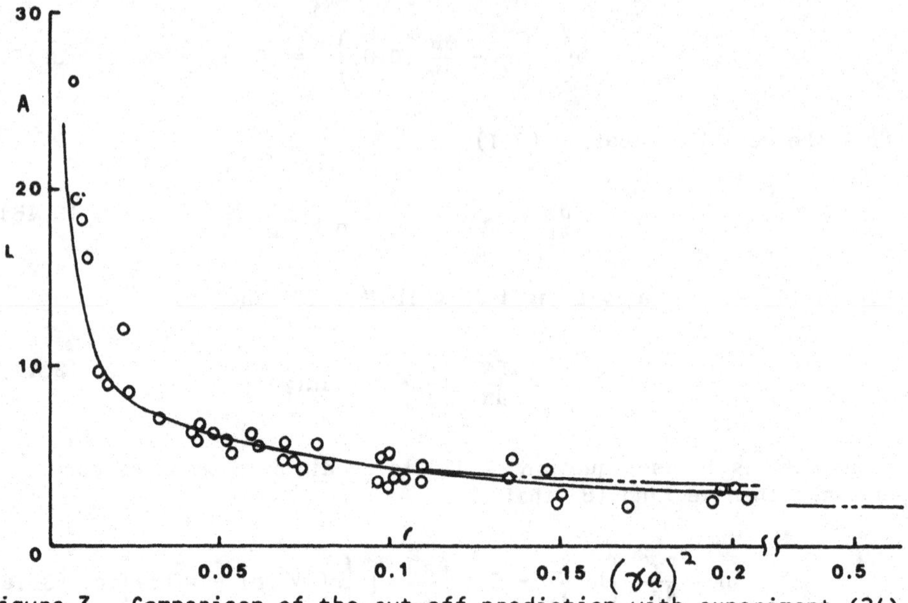

Figure 3. Comparison of the cut-off prediction with experiment (24). Maxworthy et al write

$$\omega = \frac{\Gamma\gamma^2}{4\pi} \ln A$$

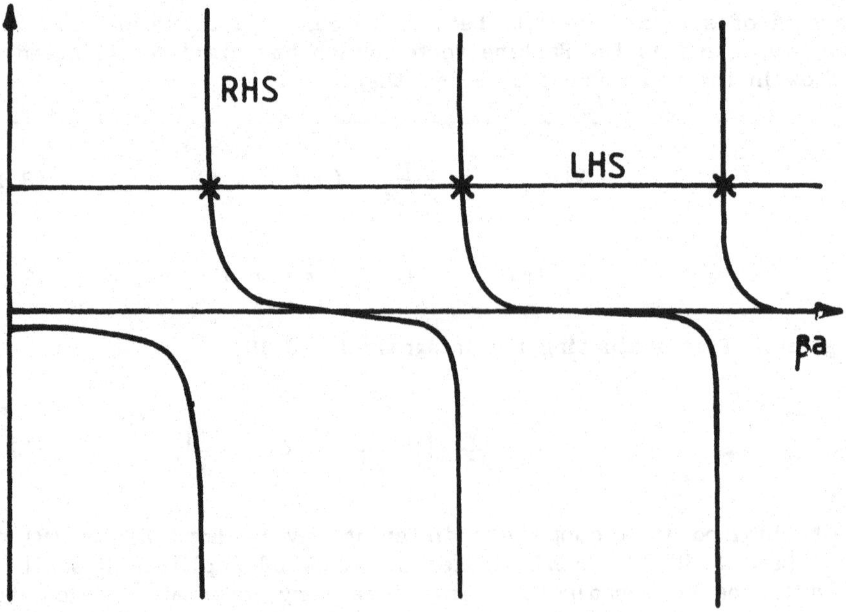

Figure 4. The dispersion equation (4.19) for m = 0.

§4 WAVES OF SMALL AMPLITUDE IN A STRAIGHT VORTEX FILAMENT - KELVIN'S ANALYSIS

We consider small perturbations to the Rankine vortex whose velocity field $\hat{\theta}\, V(r)$ is given by

$$\left.\begin{aligned} V(r) &= \Omega r & (0 \le r \le a) \\ &= \Omega a^2/r & (r \ge a) \end{aligned}\right\} \tag{4.1}$$

This corresponds to an infinite straight filament with uniform vorticity 2Ω in the core, $0 \le r \le a$. The perturbations deform the boundary of the filament so that it becomes

$$r = a + D \exp(i(m\theta + \gamma z + nt)) , \tag{4.2}$$

where the disturbance amplitude D is much less than a. We seek the dispersion function $n(m,\gamma)$ for $\gamma \ge 0$ and for $m = 0, 1, 2, \ldots$

We assume that the velocity field $\mathbf{u}$ in the perturbed motion is

$$\mathbf{u}(r,\theta,z,t) = \hat{\theta}V(r) + u\hat{r} + v\hat{\theta} + w\hat{z} \tag{4.3}$$

and that the perturbed pressure is $\rho_0\, P(r,\theta,z,t)$. If we substitute into the equations of motion and linearise we find

$$\frac{\partial u}{\partial t} + \frac{V}{r}\frac{\partial u}{\partial \theta} - 2\frac{Vv}{r} = -\frac{\partial P}{\partial r} , \tag{4.4}$$

$$\frac{\partial v}{\partial t} + \frac{V}{r}\frac{\partial v}{\partial \theta} + \left(\frac{dV}{dr} + \frac{V}{r}\right)u = -\frac{1}{r}\frac{\partial P}{\partial \theta} , \tag{4.5}$$

$$\frac{\partial w}{\partial t} + \frac{V}{r}\frac{\partial w}{\partial \theta} = -\frac{\partial P}{\partial z} , \tag{4.6}$$

$$\frac{\partial u}{\partial r} + \frac{u}{r} + \frac{1}{r}\frac{\partial v}{\partial \theta} + \frac{\partial w}{\partial z} = 0 . \tag{4.7}$$

On noting that the coefficients in this linear system of partial differential equations are functions of r only, we realise that the same exponential factor as in (4.2) can be assumed for the perturbed velocity field. The linearised equations (4.4) - (4.7) then reduce to

$$i f u - \frac{2V}{r} v = - \frac{dP}{dr} , \tag{4.8}$$

$$i f v + \left(\frac{dV}{dr} + \frac{V}{r}\right) u = - \frac{imP}{r} , \tag{4.9}$$

$$i f w = -i\gamma P \tag{4.10}$$

and

$$\frac{du}{dr} + \frac{u}{r} + \frac{imv}{r} + i\gamma w = 0 , \tag{4.11}$$

where

$$f(r) = n + \frac{mV}{r} . \tag{4.12}$$

We can express u v and w in terms of P without recourse to differentiation, so that substitution into the equation of continuity (4.11) will lead to a second order ordinary differential equation for P. In our case, with V(r) as defined in equation (4.1), this equation is

$$\frac{d^2P}{dr^2} + \frac{1}{r}\frac{dP}{dr} + \left[\left(\frac{4\Omega^2}{f^2} - 1\right)\gamma^2 - \frac{m^2}{r^2}\right]P = 0 , \tag{4.13}$$

<u>inside</u> the perturbed filament where now

$$f = n + m\Omega \tag{4.14}$$

so that f is independent of r.

<u>Outside</u> the perturbed filament the flow is irrotational and it follows from the linearised Bernoulli's theorem that

$$P = -i\left(n + \frac{m\Omega a^2}{r^2}\right)\phi(r) , \tag{4.15}$$

where the perturbed velocity potential $\phi(r)$ satisfies

$$\frac{d^2\phi}{dr^2} + \frac{1}{r}\frac{d\phi}{dr} - \left(\gamma^2 + \frac{m^2}{r^2}\right)\phi = 0 \ . \tag{4.16}$$

We must insist that the perturbed boundary (4.2) is a material surface and the fluid pressure is continuous across it. The first requirement turns out to be satisfied if the perturbed radial velocity u is continuous at r = a; hence

$$\left[\frac{i\, f\, \frac{dP}{dr} + \frac{2im\Omega P}{a}}{f^2 - 4\Omega^2}\right]_{r=a-0} = \frac{d\phi}{dr}(a + 0) \tag{4.17}$$

Since the basic pressure and its first derivative with respect to r are continuous across the boundary of the boundary of the unperturbed filament, the second requirement is met provided that the perturbation pressure P is continuous across r = a, leading to

$$P(a - 0) = P(a + 0) = -i\, f\, \phi(a+0). \tag{4.18}$$

Both (4.13) and (4.16) are forms of Bessel's equation, so that on choosing the appropriate solutions to ensure regularity at r=0 and r=∞ we obtain the dispersion equation

$$-\frac{K_m'(\gamma a)}{\gamma a\, K_m(\gamma a)} = \frac{1}{\beta a}\frac{J_m'(\beta a)}{J_m(\beta a)} + \frac{2\Omega_m}{\beta^2 a^2 f} \tag{4.19}$$

where

$$\beta^2 = \gamma^2\left[\frac{4\Omega^2}{f^2} - 1\right] \tag{4.20}$$

This result was obtained by Lord Kelvin (8).

One or two properties of equation (4.19) are worth noting before we proceed to discuss how it is to be solved for the frequency n. First we note that n occurs only in the definition (4.14) of f. We can see the physical significance if we use (4.14) to write the disturbance in the form

$$r = a + D\exp(i(ft + \gamma z + m(\theta - \Omega t))) \ . \tag{4.21}$$

This shows that f is the frequency of the disturbance viewed by an observer moving with the core boundary. Next we note some symmetries. The transformation $m \to -m$ and $n \to -n$ leaves (4.18) unchanged, so that modes with $m < 0$ merely propagate in the opposite direction. This reflects the physical symmetry of the vorticity field. Thus we have not sacrificed generality by assuming that $m \geq 0$. We can see also that $\gamma \to -\gamma$ and $n \to n$ leaves the dispersion equation unchanged. Thus we can construct a disturbance

$$r = a + D\exp(i(nt + m\theta + \gamma z)) + D\exp(i(nt + m\theta - \gamma z)) \qquad (4.22)$$

or

$$r = a + 2D\cos\gamma z\exp(i(nt + m\theta))$$

representing a rotating plane wave, the angular velocity of the rotation being $-n/m$.

It is the existence of a closed-form solution for the perturbation pressure P in the vortex core which enables a general discussion of wave propagation relatively simple. This is not an accident but is due to the fact that P is harmonic in stretched spatial variables, the stretching being frequency dependent. The recognition of this fact is basic in the theory of rapidly-rotating fluids, and is related to the Proudman-Taylor theorem.

The dispersion equation (4.19) - (4.20) can be discussed graphically. If γa is chosen, that is we fix the axial wavelength, the left-hand side of (4.18) is merely a prescribed positive constant. I choose to examine first the case $m = 0$, which are axi-symmetric "sausaging" modes. The functions $J_0(\beta a)$ and $J_0'(\beta a)$ have non-trivial zeros $j_{0,s}$ and $j'_{0,s}$ respectively and, by a standard theorem about second-order ordinary differential equations, these are interlaced. A sketch of $J_0(u)$ shows that $J_0 > 0$ and $J_0' < 0$ in $0 < u < j_{0,1}$ so that the right-hand side of (4.19) is negative in $0 < \beta a < j_{0,1}$. The right-hand side, which is qualitatively like $-\tan\beta a$, has simple poles at $\beta a = j_{0,s}$ and we can now see from Figure 4 that (4.19) has a root $\beta_s(\gamma a)$ satisfying $j_{0,s} < a\beta_s(\gamma a) < j'_{0,s}$ (s = 1,2,3 ...(4.22)), each value of the integer s corresponding to a particular radial structure. This allows us to see that the dispersion equation is

$$n_s(\gamma a) = \pm \frac{2\Omega\gamma a}{\sqrt{\gamma^2 a^2 + a^2\beta_s^2(\gamma a)}}, \qquad (s = 1,2,\ \ldots) \quad .(4.23)$$

Thus waves can propagate in either direction with phase speed.

$$C_s(\gamma a) = \frac{2\Omega}{\sqrt{\gamma^2 a^2 + a^2\beta_s^2(\gamma a)}}, \quad (s = 1,2,\ldots) \quad .(4.24)$$

Now the left-hand side of (4.19) is a monotonically decreasing function of γa, so that $a\beta_s$ is a monotonically increasing function of γa, with $a\beta_s \to j'_{0's}$ as $\gamma a \to \infty$ and $a\beta_s \to j_{0,s}$ as $\gamma a \to 0$. It follows that $C_s(\gamma a)$ is a monotonically decreasing function of γa. Thus each radial mode s = 1,2,3,... achieves its maximum phase speed $2\Omega a/j_{0,s}$ as $\gamma a \to 0$, so that for each mode the long waves travel fastest. The overall maximum is $2\Omega a/j_{0,1} = \cdot 8317\cdots\Omega a$. Since $\frac{dC_s}{d\gamma} \leq 0$ and $C_s = n_s/\gamma$ it follows that

$$\frac{dn_s}{d\gamma} \leq C_s \qquad (s = 1,2\ldots) \qquad (4.25)$$

so that the group velocity has the same bound, achieved as $\gamma a \to 0$.

It is worth returning briefly to look at the core-pressure equation (4.13). We have just seen that as $\gamma a \to 0$, then $n_s \sim 2\Omega a/a\beta_s$. This means that the last term in (4.13) is not negligible for long waves: in other words, axial derivatives must be retained in the long-wave limit.

The graphical treatment is rather more complicated for $m \geq 1$. We retain $a\beta$ as the key variable to make the argument of the Bessel function simple. Then we must solve

$$\frac{-K'_m(\gamma a)}{\gamma a\, K_m(\gamma a)} - \frac{m\,\mathrm{sgn}(f)\,(\beta^2+\gamma^2)^{1/2}}{a^2\beta^2\gamma} = \frac{J'_m(\beta a)}{\beta a\, J_m(\beta a)}$$

where

$$f = \frac{\mathrm{sgn}\,(f)\,2a\gamma}{\sqrt{a^2\gamma^2 + a^2\beta^2}} \qquad (4.27)$$

It now transpires (see Fig. 5) that there are two roots in each interval $j_{m,s} < a\beta < j_{m,s+1}$ for $s = 1,2,3,\ldots$, corresponding to modes which are either contra-rotating ($f > 0$) or co-rotating ($f < 0$) when viewed from a frame rotating with the core circumference. Long waves, however, have a frequency f given by

$$f = \frac{2a\gamma\ \mathrm{sgn}(f)}{j_{m,s}}, \tag{4.28}$$

so that the contra- and co-rotating modes then have the same speed.

The modes with $m \geq 3$ resemble helical grooves cut in the vortex core. The mode $m = 2$ is a twisted elliptical cylinder, while $m = 1$ represents distortion of the core into a helix of small radius D, the core itself remaining circular. The analysis of this section is valid if $D \ll a$, while that of §3, based on the cut-off, required $a \ll \gamma^{-1}$ when $\gamma D \ll 1$. Thus if

$$\gamma D \ll \gamma a \ll 1 \tag{4.29}$$

both theories should hold and we should be able to recover (3.21) from Kelvin's analysis. Clearly, putting $m = 1$ in the long-wave dispersion equation (4.28) does <u>not</u> give agreement. We are rescued by the existence of a mode with $0 < \beta a < j_{m,1}$, which has no analogue in the axi-symmetric case $m = 0$. If we use the power-series expansions of the Bessel function $J_m(\beta a)$ we find that

$$\frac{1}{\beta a}\frac{J'_m(\beta a)}{J_m(\beta a)} = \frac{m}{\beta^2 a^2} - \frac{1}{2(m+1)} + O(\beta^2 a^2). \tag{4.30}$$

The first term on the left-hand side of equation (4.26) is positive. It follows that if $f < 0$ - the co-rotating case - the left-hand side of (4.26) is greater than $m/\beta^2 a^2$, for $\beta a \ll 1$, while for $\beta a \ll 1$, the right-hand side is less than this value. Thus there is no co-rotational mode in this limit. There is, however, a contra-rotational mode. Provided that f is $O(\Omega)$ and $f \neq 2\Omega$, which we must check afterwards, we find that $\gamma a = O(\beta a 0$ and so $\gamma a \ll 1$. It is therefore consistent to expand the modified Bessel functions. This process gives us an explicit formula for f;

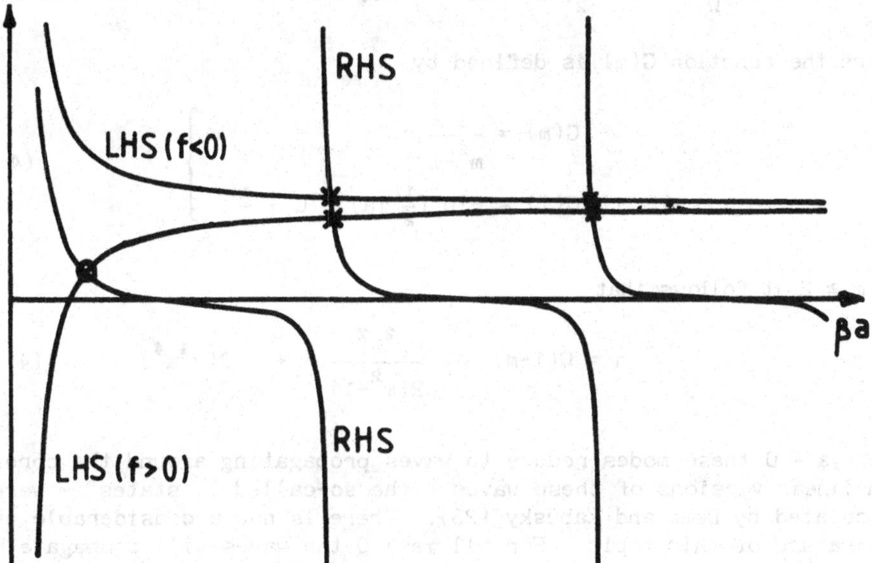

Figure 5. The dispersion equation (4.19) for m ≠ 0.

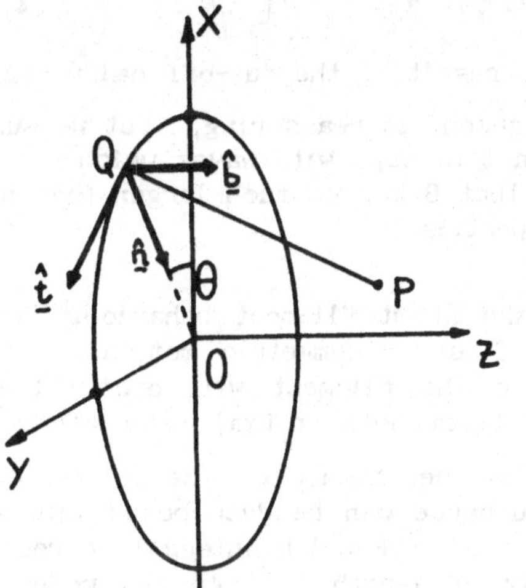

Figure 6. The geometry of the vortex ring.

$$\frac{f}{\Omega} = 1 + \frac{1}{2}\gamma^2 a^2 G(m) + O(\gamma^4 a^4) , \qquad (4.31)$$

where the function G(m) is defined by

$$\left.\begin{aligned} G(m) &= \frac{1}{m^2-1} \\ G(1) &= -\ln\left(\frac{1}{2}\gamma a\right) - C + \frac{1}{4} \end{aligned}\right\} . \qquad (4.32)$$

If $m \geq 2$ it follows that

$$n = \Omega(1-m) + \frac{\Omega\gamma^2 a^2}{2(m^2-1)} + O(\gamma^4 a^4) . \qquad (4.33)$$

When $\gamma a = 0$ these modes reduce to waves propagating around the core. Non-linear versions of these waves - the so-called "V states" - were calculated by Deem and Zabusky (25). There is now a considerable literature of this topic. For all $\gamma a > 0$ the waves will propagate but their group velocity is $O(\gamma a)$.

If $m = 1$

$$n = \frac{1}{2}\Omega\,\gamma^2 a^2 \left[\ln(2/\gamma a) - C + \frac{1}{4}\right] , \qquad (4.34)$$

which agrees with the result of the cut-off calculation (3.17), since $\Omega = \Gamma/2\pi a^2$. This argument is reassuring, but we must recall that the cut-off theory is meant to cope with waves which are strongly non-linear, in the sense that D can be much larger than a; also, it allows for general core properties.

What can we learn about filament behaviour from this complicated analysis? Consider first axisymmetric motion. An axially-periodic initial disturbance to the filament will excite the axi-symmetric modes of oscillation frequencies $n_s(\gamma a)$, the weighting of each mode depending on the radial dependence of the initial condition. A non-periodic initial disturbance can be decomposed into a sum of such disturbances by means of a Fourier integral representation. If this initial disturbance is of length L, then the modes excited will be dominated by modes with $\gamma L \sim 1$. Our results now show that, provided L is not much smaller than a, energy will be carried away from the initial location of the disturbance at a velocity of order Ωa. Thus the imposed variations of cross-sectional area will be smoothed out in a time of order $L/\Omega a$ or La/Γ. This time is much shorter than the time scale L^2/Γ inherent in the cut-off model. If we believe that the same orders of magnitude will characterise propagation on a curved filament we are forced to the conclusion that the cross-sectional area of the core will not vary along the vortex.

The situation in the non-axisymmetric case is similar, only now there are the slow modes with $\beta a \ll 1$ to consider. As we have seen there is one such mode for each azimuthal wave number $m = 1,2,\ldots$. These modes have group velocity of order Γ/L and are left behind b the remaining modes with group velocities of order Γ/a. The modes which are left behind have a simple radial structure, because for small βr

$$J_m(\beta r) \sim Ar^m \quad . \tag{4.35}$$

In particular, when $m=1$, we have merely a lateral displacement of the core without any change of shape.

Let us reconsider the switch on problem in §2. We impose an initial disturbance of length scale $L \gg a$, whose amplitude is small. The imposed area variations and shape variations with complex radial structures will propagate away from the disturbed region with group velocities of order Ωa. We are left with modes of simple radial structure. However, the excitation of modes with $m \geq 2$ will be small, so that we are left with helical modes in which the core is displaced without distortion. It is with these modes that the cut-off theory is concerned.

§5. THE VELOCITY OF PROPAGATION OF A VORTEX RING

The calculation of the velocity of propagation of a vortex ring is not a simple matter. If we assume steady axi-symmetric inviscid motion with Stokes stream function ψ, and no swirl component of the velocity field, the vorticity field $\zeta\hat{\underline{\theta}}$ must satisfy

$$\frac{\zeta}{r} = F(\psi) \text{ in } A$$

and

$$\zeta = 0 \qquad \text{outside } A,$$

where cylindrical polars (r,θ,z) are employed. The function F is given, but A, the intersection of the core with an axial plane, is unknown, as is U, the speed of propagation. A and U have to be found by requiring continuity of the velocity field across ∂A and by specifying the area of A and its distance from the axis of symmetry. This is a formidable free-boundary problem and no exact solution is known for any choice of F. The method of successive approximations, or numerical methods, must be employed.

However, an approximate calculation of the propagation velocity of vortex ring is instructive. This is because the calculation must deal systematically with the interaction between the core scale and filament scale dynamics. The most efficient approximate method of calculating the propagation speed is to use an identity given by Lamb (26, p.240). I have chosen instead the momentum flux balance method employed by Saffman and me (18) in our justification and extension of the cut-off theory. This is based on the fact that in steady flow and for an arbitrary closed surface S,

$$\int_S P_{ij}\, n_j \, ds = 0 , \tag{5.1}$$

where P_{ij} is the momentum flux tensor defined by

$$P_{ij} = \rho_0 u_i u_j + p\,\delta_{ij} , \tag{5.2}$$

and $\hat{\mathbf{n}}$ is the inward normal to S. I have chosen this method because I believe it best brings out the physics of the problem.

I shall take the centre-line of the vortex ring to be the circle $x^2 + y^2 = R^2$ lying in the plane $z = 0$. In axes thus fixed in the vortex ring the flow is steady and equal to $-U\hat{\mathbf{k}}$, where U is the unknown propagation speed of the ring. The core vorticity is in the positive sense with respect to Oz. I now introduce orthogonal coordinates as follows. Let P be any point not on the z-axis. Then there is a unique point Q on the circle such that PQ is a minimum. Suppose this point Q has coordinates $(X_Q, Y_Q, 0)$ given by

$$\left.\begin{aligned} X_Q &= R\cos\theta \\ Y_Q &= R\sin\theta \end{aligned}\right\} \tag{5.3}$$

Then I choose θ to be my first coordinate. Now we consider the plane which passes through Q and the z-axis. This plane is normal to the tangent $\hat{\mathbf{t}}(\theta)$ to the circle at Q. But $\overrightarrow{QP}$ is perpendicular to $\hat{\mathbf{t}}(\theta)$, so that P must lie in this plane. Suppose $\hat{\mathbf{n}}(\theta)$ and $\hat{\mathbf{b}}$ are the unit normal and binormal to the circle at Q. The normal $\hat{\mathbf{n}}$ points towards the axis of symmetry while $\hat{\mathbf{b}}$ completes the orthogonal triad; see Figure 6. Explicitly

$$\left.\begin{aligned} \hat{t}(\theta) &= -\hat{i}\sin\theta + \hat{j}\cos\theta \\ \hat{n}(\theta) &= -\hat{i}\cos\theta - \hat{j}\sin\theta \\ \hat{b} &= \hat{k} \end{aligned}\right\} \tag{5.4}$$

If we express $\overrightarrow{QP}$ in the form

$$\overrightarrow{QP} = \hat{n}\, u_2 + \hat{b}\, u_3 \tag{5.5}$$

then $(\theta,\ u_2,\ u_3)$ are my coordinates. The position vector $\overrightarrow{OP} = \mathbf{x}$ in the frame Oxyz is given by

$$\mathbf{x} = \hat{i}\, R\cos\theta + \hat{j}\, R\sin\theta + \hat{n}\, u_2 + \hat{b}\, u_3. \tag{5.6}$$

We can easily verify that this system is orthogonal with

$$d\mathbf{x}^2 = (R - u_2)^2\, d\theta^2 + du_2^2 + du_3^2 . \tag{5.7}$$

In this coordinate system the core is taken to be the interior of the torus

$$u_2^2 + u_3^2 = a^2 , \qquad 0 \le \theta \le 2\pi . \tag{5.8}$$

This is of course an approximation, because the core boundary is not exactly circular. The precise shape must be determined as part of the solution. However, the deviation from a circle is a higher order effect and we can consistently ignore it.

For the control surface S in the momentum flux balance we choose the surface of a thin slice of the core with plane ends E_1 at θ and E_2 at $\theta + d\theta$. To calculate the contribution from the plane fact at θ it is convenient to introduce local polar coordinates (s, η) such that

$$\left.\begin{aligned} u_2 &= s\cos\eta \\ u_3 &= s\sin\eta \end{aligned}\right\}$$

Then if the velocity in the core is as defined as in Section 2, so that the core velocity field is given by

$$\mathbf{u} = \hat{\eta}\, V(s) + \hat{t}\, W(s) , \tag{5.10}$$

the contribution to the flux integral from E_1 is $+\hat{t}(\theta)I$ where

$$I = \int_0^a 2\pi s \left[p(s) + \rho_0 W^2(s)\right] ds \quad ; \qquad (5.11)$$

the contributions from the off-diagonal elements of P_{ij} vanish by symmetry.

The pressure p(s) is determined from the equation

$$-\frac{V^2(s)}{s} = -\frac{1}{\rho_0}\frac{\partial p}{\partial s} \qquad (5.12)$$

Now

$$\int_0^a s\, p(s)\, ds = \left[\frac{1}{2} s^2\, p(s)\right]_0^a - \frac{1}{2}\int_0^a s^2 \frac{\partial p}{\partial s}\, ds \qquad (5.13)$$

and if we choose to define the pressure so that p(a) = 0, it follows that

$$I = 2\pi\rho_0 \int_0^a s\left[W^2(s) - \frac{1}{2} V^2(s)\right] ds \quad . \qquad (5.14)$$

The contribution from the plane face E_2, at $\theta + d\theta$, is $-\hat{t}(\theta + d\theta)I$,

so the net contribution from the two faces E_1 and E_2 is $-\hat{n}I\, d\theta$,

using the fact that

$$\frac{\partial \hat{t}}{\partial \theta} = \hat{n} \quad . \qquad (5.15)$$

I must stress that, so far, we have carried out our dynamical calculations as if the filament were straight. The assumptions that the core is circular and that the velocity fields can be functions of s only are true to leading order only.
The curvature enters <u>only</u> through (5.15) which tells us that the two ends of the control volume together contribute a force **F** per unit length, where

$$\mathbf{F} = -\frac{\mathbf{n}\, I}{R} \qquad (5.16)$$

If I < 0, this force acts inwards.

The origin of this force is the lowering of the core pressure by the swirl velocity V(s). The lowered pressure tries to expand our control volume, so that the core behaves as if it were in tension. This inward force is like that which causes a stretched elastic band to grip a circular cylinder. The centrifugal force associated with W(s) tries to expand the core, so its contribution comes in with opposite sign.

This inward force must be balanced by an outward force acting on the curved surface of our control volume V. Now there is a circulation Γ around the vortex ring, so the curved surface of our control volume will experience a lift force. If the filament were straight, the Kutta formula would give a lift L per unit length, where

$$\mathbf{L} = \hat{n}\,\rho_0\,U\,\Gamma \quad . \tag{5.17}$$

If we insist that L + F = 0, we get an equation for U: however, it gives only the integrated term in formula (2.15) for the propagation speed. The obvious notion of balancing the contracting force due to the tension against the lift will not work and we must embark on a detailed calculation of the flow outside the core.

The flow outside the core is irrotational, which is a great simplification. We start by expanding its velocity potential $\Gamma\phi/2\pi$ in in terms of $\varepsilon = a/R$, so that

$$\bar{\phi} = \bar{\phi}_0 + \varepsilon\,\bar{\phi}_1 + \ldots . \tag{5.18}$$

In terms of dimensionless coordinates $(\theta, a\bar{s}, \eta)$ Laplace's equation is

$$\frac{\partial}{\partial\theta}\left[\frac{\bar{s}}{\bar{h}}\frac{\partial\bar{\phi}}{\partial\theta}\right] + \frac{\partial}{\partial\bar{s}}\left[\bar{h}\,\bar{s}\,\frac{\partial\bar{\phi}}{\partial\bar{s}}\right] + \frac{\partial}{\partial\eta}\left[\frac{\bar{h}}{\bar{s}}\,\frac{\partial\bar{\phi}}{\partial\eta}\right] = 0 \tag{5.19}$$

where the line element is given by

$$\bar{h} = 1 - \varepsilon\,\bar{s}\cos\eta \tag{5.20}$$

The leading term in the expansion (5.17) must represent the swirling flow around a straight cylinder, so we choose

$$\bar{\phi}_0 = \eta \tag{5.21}$$

Then $\bar{\phi}$ must satisfy

$$\frac{\partial^2\bar{\phi}_1}{\partial\bar{s}^{\,2}} + \frac{1}{\bar{s}}\,\frac{\partial\bar{\phi}_1}{\partial\bar{s}} + \frac{1}{\bar{s}^2}\,\frac{\partial^2\bar{\phi}_1}{\partial y^2} = -\,\frac{\sin\eta}{\bar{s}} \tag{5.22}$$

subject to the boundary condition

$$\frac{\partial \bar{\phi}_1}{\partial \bar{s}} = 0 \qquad \text{on } \bar{s} = 1 . \tag{5.23}$$

The general solution of (5.22) which satisfies (5.23) is, invoking the symmetry of the flow,

$$\bar{\phi}_1 = \left[A_1 \bar{s} + (A_1 - \tfrac{1}{2})\bar{s}^{-1}\right]\sin\eta + \tfrac{1}{2}\bar{s}\ln\bar{s}\,\sin\eta + \sum_{2}^{\infty} A_n\left[\bar{s}^{\,n} + \bar{s}^{\,-n}\right]\sin n\eta \tag{5.24}$$

We expect this solution to be valid when $1 \le \bar{s} << \varepsilon^{-1}$, but it will fail when $\bar{s} = O(\varepsilon^{-1})$. This is because the core-scale dynamics is based on regarding the filament as nearly straight.

The flow field at distances O(R) is the uniform stream $-U\hat{k}$ plus a contribution from the Biot-Savart integral. In practice, the equivalent Stokes stream function $\frac{\Gamma R}{2\pi}\bar{\psi}$ is easier to work with and a straightforward calculation (26, p.237) gives

$$\psi = -\frac{\pi U r^2}{R\Gamma} + k^{-1}\left[\frac{r}{R}\right]^{1/2}\left[(2-k^2)\,K(k) - 2E(k)\right] + O(\varepsilon^2) \quad , \tag{5.25}$$

where

$$k^2 = 4Rr\left[z^2 + (R+r)^2\right]^{-1} . \tag{5.26}$$

The functions E(k) and K(k) are complete elliptic integrals. The last term in (5.25) represents the effect of the detailed core structure on the velocity field outside the core: this is subsumed under Γ only for a straight filament.

We expect (5.25) to hold when the distance from the vortex ring is much greater than the core radius a. If we express r and z in terms of core variables we find that

$$\left.\begin{aligned} r &= R - as\cos\eta \\ z &= a\bar{s}\,\sin\eta \end{aligned}\right\} \tag{5.27}$$

so that

$$(1-k^2)^{1/2} = \frac{1}{2}\varepsilon\bar{s} + O(\varepsilon^2) \tag{5.28}$$

It is now straightforward to substitute into (5.26) and invoke standard results to show that, for $\varepsilon \to 0$ with fixed $\bar{s}$,

$$\bar{\phi} = \eta + \varepsilon\left[-\frac{1}{2}\bar{s}\ln\bar{s} + \frac{1}{2}\bar{s}\ln\frac{8}{\varepsilon} - \frac{1}{2}\bar{s} - \frac{2\pi a}{\Gamma}U\bar{s}\right]\sin\eta + O(\varepsilon^2). \tag{5.29}$$

We expect this to be valid when $1 \ll \bar{s} \ll \varepsilon^{-1}$. This must therefore match with the approximate form of $\bar{\phi}_0 + \varepsilon\bar{\phi}_1$ when $1 \ll \bar{s} \ll \varepsilon^{-1}$. This matching requirement shows that

$$A_1 = \frac{1}{2}\ln\frac{8}{\varepsilon} - \frac{1}{2} - \frac{2\pi aU}{\Gamma} \tag{5.30}$$

and $A_n =)$ $(n \geq 2)$.

It is now a simple matter to calculate the force **G** on the curved surface of the control volume as

$$\mathbf{G} = \rho_0\,\hat{n}\left[-\Gamma U + \frac{\Gamma^2}{4\pi R}\left[\ln\left(\frac{8R}{a}\right) - \frac{1}{2}\right]\right] . \tag{5.31}$$

The final step is to require that **F** + **G** = 0, which leads to the result (2.15).

How accurate is the cut-off theory? Formally, the relative error is $O(\varepsilon^2)$, but a better test is to compare with the numerical results given by Norbury (27). Norbury considered the family of vortex rings in which

$$\zeta/r = \text{constant} \tag{5.32}$$

where r is distance from the axis of symmetry. This variation of the vorticity is a consequence of the stretching law (1.18). This effect has been neglected in our treatment, because it contributes only at higher order. If we set $V(r) = \Omega r$ and $W(r) = 0$ we find that (2.15) reduces to

$$U = \frac{\Gamma}{4\pi R}\left[\ln\left(\frac{8R}{a}\right) - \frac{1}{4}\right] . \tag{5.33}$$

To compare with the results presented in (27) we define R to be the mean of the smallest and largest distances of the core boundary from the symmetry axis, and define a by πa^2 = core cross-sectional area. Then we find the results shown in the Table where UR/Γ is displayed

a/R	(21)	eqn(5.33)
•2	•27	•273
•4	•21	•218
•6	•17	•186

Evidently equation (5.33) does remarkably well, although we should be cautious about assuming equal success for the cut-off theory, which - in the absence of axial flow - has formally equivalent acccuracy (18).

§6 INSTABILITY

A single filament with no axial flow and whose core vorticity is uniform - a Rankine vortex - is stable, as we saw in Section 4. Thus any instability must be due to modification of the filament by its environment.

Let us consider a filament lying initially along the z axis and and subject to a small irrotational plane strain field $\mathbf{U}_s$ given by

$$\mathbf{U}_s = -e\,y\,\hat{\mathbf{i}} - e\,x\,\hat{\mathbf{j}} \tag{6.1}$$

where e (> 0) is the strain rate. By small we mean that $e \ll \zeta_0$, where the uniform core vorticity $\zeta_0 = \Gamma/\pi a^2$. As a matter of fact, this problem can be solved exactly (28) and it transpires that the core becomes elliptical, provided $e/\zeta_0 < \cdot 15\ldots$. If e/ζ_0 exceeds this critical value, no steady solution with elliptical form is possible.

However, we will assume that $e/\zeta_0 \ll 1$, and consider what happens when the filament is distorted into a long plane wave

$$\mathbf{x}_1 = (x_1\,\hat{\mathbf{i}} + y_1\,\hat{\mathbf{j}})\cos\gamma s + s\,\hat{\mathbf{k}} \ , \tag{6.2}$$

where $\gamma a \gg 1$ and s is arc distance. We have seen that a long plane wave will rotate with angular velocity n given by

$$n = \frac{\Gamma\gamma^2}{4\pi}\left[\ln\left(\frac{2}{\gamma a}\right) - C + \frac{1}{4}\right] . \tag{6.3}$$

Thus the external strain (6.1) and self-induced contra-rotation of the filament will combine to give , for all values of s,

$$\frac{dx_1}{dt} = y_1(-e + n) \tag{6.4}$$

and

$$\frac{dy_1}{dt} = x_1(-e - n) \; ; \tag{6.5}$$

the $\cos\gamma s$ factor cancels. Thus we have instability if $e > |n|$; the self-induced rotation is stabilising and removes the instability if $|n| > e$.

If $\gamma = 0$, so that (6.2) represents a transition of the filament without distortion, $n = 0$ and the displacement grows exponentially. The motion is plane and the instability is a consequence of the hyperbolic pattern of streamlines in the plane flow (6.1). As γa increases, $|n|$ grows and the instability disappears when

$$\frac{1}{4}\gamma^2 a^2\left[\ln\left(\frac{2}{\gamma a}\right) - C + \frac{1}{4}\right] = \frac{e}{\zeta_0} . \tag{6.6}$$

Since we have assumed that $e/\zeta_0 \ll 1$, the critical value of γa will be small, so the cut-off theory which yields n is valid and our treatment is consistent.

Let us try to apply these ideas to a flow of practical interest, the flow due to the trailing vortices behind an aircraft. We follow Crow (7) model and so this flow by a pair of straight filaments, one with circulation Γ at $x = y = 0$ and one with circulation $-\Gamma$ at $x = 0$, $y = b$. In this frame the mutually induced translational motion appears as a uniform stream $i\Gamma/2\pi b$. We want to apply the cut-off theory, so we assume $a/b \ll 1$. This is reasonable, since an approximate theory of the formation of the trailing vortex system (30, p. 186) gives $a/b = \cdot 11$.

Each vortex filament exerts a strain on the other, where $|e| = \Gamma/2\pi b^2$. If we could apply the simple theory I have just described, we would predict instability, the most unstable mode being two-dimensional. However, we cannot restrict perturbations of shape to the vortex at $x = y = 0$. We must allow the vortex at $x = 0$, $y = b$ to undergo a plane-wave distortion as well, so that its locus is

$$\mathbf{x}_2 = + b\hat{j} + (x_2\hat{i} + y_2\hat{j}) \cos\gamma s' + s'\hat{k} . \tag{6.7}$$

To allow for the effect of this filament on the original filament we have to calculate the Biot-Savart integral

$$\mathbf{u}_2(\mathbf{x}_1) = -\frac{\Gamma}{4\pi}\int_{-\infty}^{\infty}\frac{\frac{\partial\mathbf{x}_2}{\partial s'}\wedge\left[\mathbf{x}_1-\mathbf{x}_2\right]}{|\mathbf{x}_2-\mathbf{x}_1|^3}\,ds' \quad , \qquad (6.8)$$

where $\mathbf{x}_2$ is defined by (6.7), and where we take s = 0 in (6.2) to give give $\mathbf{x}_1 = x_1\mathbf{i} + y_1\mathbf{j}$. It is legitimate to do this, because the $\cos\gamma$s will simply cancel if we take $s \neq 0$. We find that, on linearising in x_1, x_2, y_1 and y_2,

$$\mathbf{u}_2(\mathbf{x}_1) = \frac{\Gamma}{2\pi b^2}\left\{\hat{\mathbf{i}}(-b - y_1 + y_2\,G(\gamma b)) + \hat{\mathbf{j}}\,(-x_1 + x_2 H(\gamma b))\right\} \qquad (6.9)$$

where

$$G(u) = -u^2\int_0^{\infty}\frac{\cos x + x\sin x\,dx}{(x^2+u^2)^{3/2}} + 3\,u^4\int_0^{\infty}\frac{\cos x\,dx}{(x^2+u^2)^{5/2}} \qquad (6.10)$$

and

$$H(u) = u^2\int_0^{\infty}\frac{\cos x + x\sin x\,dx}{(x^2+u^2)^{3/2}} \qquad (6.11)$$

These integrals are easily evaluated (31, p. 185) and

$$G(u) = u\,K_1(u) \qquad (6.12)$$

and

$$H(u) = -u^2 K_1'(u) \quad . \qquad (6.13)$$

The appearance of the modified Bessel function K_1 should not surprise us, because the far field of the distorted filament is derivable from from the velocity potential ϕ_2 given by

$$\phi_2 = -\frac{\Gamma\theta_2}{2\pi} + \cos\gamma z\,(A\cos\theta_2 + B\sin\theta_2)K_1(\gamma r_2) \quad , \qquad (6.14)$$

where (r_2, θ_2, z) are cylindrical polars centred on the line x=0, y=b. We can now construct the equations of motion for the vortex filament. The first term in (6.9) is cancelled by the uniform stream seen by the axes fixed in the pair, so we get

$$\frac{dx_1}{dt} = \frac{\Gamma}{2\pi b^2}\left\{-y_1(1-\bar{\omega}) + y_2\,G(\gamma b)\right\} \qquad (6.15)$$

and

$$\frac{dy_1}{dt} = \frac{\Gamma}{2\pi b^2}\left\{-x_1(1+\bar{\omega}) + x_2\,H(\gamma b)\right\} , \qquad (6.16)$$

where

$$\bar{\omega} = \frac{1}{2}\gamma^2 b^2\left[\ln\frac{2}{\gamma a} - C + \frac{1}{4}\right] = \frac{2nb^2}{\pi\Gamma} \qquad (6.17)$$

Next, we ought to obtain a second equation from the motion of the filament perturbed from x=0, y=b. Instead, we appeal to symmetry and put $x_1 = x_2$ and $y_1 = y_2$, so that the two filaments are mirror images in the symmetry plane $y = \frac{b}{2}$ (Figure 7).

We then find, after numerical evaluation of the coefficients in (6.16) and (6.17), that - taking a/b = ·1 - the most unstable mode has wave length λ_c, where λ_c/b = 8·5, and that the planes containing the vortices are inclined at an angle close to 45° to the x-axis; that is to say that the two plane waves are closest together below the descending pair. Non-linear calculations using a cut-off method (32)show that the two troughs meets and it is then plausible to assume that re-connection takes place, so that an array of vortex rings is formed, as is actually observed in aircraft vortex trails, (33). However, there-connection process takes place on a scale comparable to the core size and necessarily involves the viscosity. It is thus beyond the scope of these lectures.

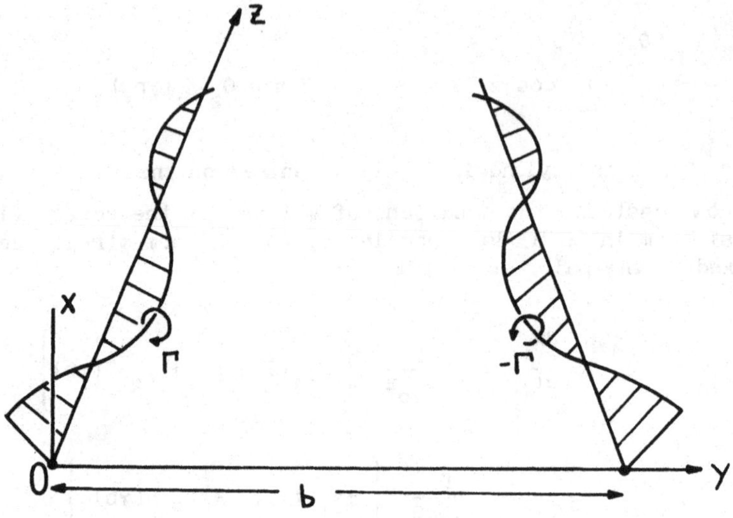

Figure 7. The geometry of the Crow (7) instability.

Let us return to the case of a single filament in a fixed strain, whose perturbed motion is governed by (6.4) and (6.5). We have seen that instability will arise if $e > |n|$. Now if $e/\zeta_0 \ll 1$ instability is certain for $\gamma \to 0$ and persists until $e = |n|$, at some critical value of γa which is small, we might imagine that the filament is stable for all larger values of γa. However, when γa satisfies the condition

$$\gamma a = 2\exp(\tfrac{1}{4} - C) = 1{\cdot}442\ldots \tag{6.18}$$

$n = 0$ and we anticipate that a narrow band of unstable wavelengths will appear, centred on the critical value given by (6.18). Unfortunately, the critical value of γa is outside the range of validity of the cutt-off theory, so the calculation is inconsistent. In fact, if we pursue $n(\gamma a)$ numerically to values of γa of order unity, we find it never vanishes. However, Widnall, Bliss and Tsai (28) showed that the co-rotating modes with $m = 1$ and $s \gg 1$ had frequencies which vanished at critical values of γa. These higher "bending" modes have radial nodal circles: when $s = 1$ the inner and outer parts of the filament move in opposite directions. If we label these frequencies n_{1s}, where, as before, $s = 0$ is the mode yielded by the cut-off theory, so $n = n_{10}$ in this notation, numerical solution of (4.19) shows that n_{11} vanishes at $\gamma a = 2{\cdot}5$ and n_{12} vanishes at $\gamma a = 4{\cdot}35$, etc. In (34) it was suggested that this would cause a narrow band of instability to appear, as in my previous argument.

However, even though the rotation frequency is now correct, the simple argument which treats the filament as a line is not valid, and a systematic stability calculation is called for. Saffman and I (35) showed that a slightly strained filament of arbitrary vorticity would be subject to this instability, provided the rotation frequency vanished, and Widnall and Tsai (9) showed that the same mechanism rendered the vortex ring unstable, thereby removing a long-standing disagreement between theory and experiment.

It is clear that the mechanism gives a sharply defined wavelength and Saffman (36) calculated this for the distribution of core vorticity to be expected if the ring is formed at a sharp edge, obtaining good agreement with experiment. Thus existence of the Widnall short wave instability is established theoretically and the detailed predictions agree with experiment. Since filaments are almost always subject to strain in real flow situations, it is arguably the most important type of instability.

REFERENCES

1. Küchemann, D. (1978) The Aerodynamic Design of Aircraft. Pergamon Press.

2. Pullin, D.I. and Perry, A.E. (1980) JFM 97, 239.

3. Moore, D.W. and Saffman, P.G. (1973) Proc. Roy. Soc. A 333, 491.

4. Dhanak, M.R. and de Bernardnis, B. (1981) JFM 109, 189.

5. Birkhoff, G. and Fisher, J. (1959) Rend Circ. Mat. Palermo, Sec. 2, 8, 77.

6. Rott, N. (1956) JFM 1, 111.

7. Crow, S.C. (1970) AIAAJ 8, 2172.

8. Kelvin, Lord (1880) Phil. Mag. 10, 155.

9. Widnall, S.E. and Tsai, C-Y. (1977) Phil.Trans.Roy.Soc. 287, 273.

10. Jeffreys, H. and Jeffreys, B. (1972) Methods of Mathematical Physics. Cambridge University Press.

11. Van der Vooren, A.I. (1980) Proc. Roy. Soc. A373, 67.

12. Moore, D.W. (1981) SIAM J.Sci. Stat. Comp. 2, 65.

13. Ince, E.L. (1926) Ordinary Differential Equations. Dover Publications.

14. Whitham, G.B. (1963) in Laminar Boundary Layers ed. Rosenhead, L. Oxford University Press.

15. Leonard, A. (1985) Ann. Rev. Fluid Mech. 17, 523.

16. Saffman, P.G. (1970) Stud. Appl. Math. 49, 371.

17. Widnall, S.E., Bliss, D. and Zalay, A. (1971) in Aircraft Wake Turbulence And Its Detection. Plenum Press.

18. Moore, D.W. and Saffman, P.G. (1972) Phil.Trans.Roy.Soc. A 272, 38.

19. Fukumoto, Y. and Miyazuki, T. (1991) JFM 222, 369.

20. Arms, R.J. and Hama, J.L. (1965) Phys. Fluid. 8, 553.

21. Hasimoto, H. (1972) JFM 51, 577.

22. Kida, S. (1981 JFM 112, 397.

23. Lamb, H. (1928) Statics. Cambridge University Press.

24. Maxworthy, T., Hopfinger, E.J. and Redekopp, K.G. JFM 151, 141.

25. Deem, G.S. and Zabusky, N.J. (1978) Phys. Rev. Lett. 40, 859.

26. Lamb, H. 1931) Hydrodynamics. Cambridge University Press.

27. Norbury, J. (1973) JFM 57, 417.

28. Moore, D.W. and Saffman, P.G. (1972) in Aircraft Wake Turbulence And Its Detection. ed. Olsen, J.H. and Goldberg, A. Plenum Press.

29. Kida, S. (1981). J. Phys. Soc. Japan 50, 3517.

30. Milne-Thomason, L.M. (1971) Theoretical Hydrodynamics. Macmillan.

31. Watson, G.N. (1941) Theory of Bessel Functions. Cambridge University Press.

32. Moore, D.W. (1972) Aeronautical Quarterly, 23, 307.

33. Bisgood, P.L., Maltby, R.L. and Dee, F.W. (1972) in Aircraft Wake Turbulence and its Detection. ed. Olsen, J.H. and Goldberg, A. Plenum Press.

34. Widnall, S.E., Bliss, D. and Tsai, C-Y (1974) JFM 66, 35.

35. Moore, D.W. and Saffman, P.G. (1975) Proc. Roy. Soc. A 346, 413.

36. Saffman, P.G. (1978) JFM 84, 625.

Chapter V.2

TWO-DIMENSIONAL BAROTROPIC AND BAROCLINIC VORTICES

E. J. Hopfinger
University J.F. and CNRS, Grenoble Cedex, France

1. INTRODUCTION

In the atmosphere and the oceans and on other planets, coherent structures or vortices are easily identified. For this reason, the study of the stability of isolated vortices and the dynamics of the interaction of pairs, triades or more is of fundamental interest for refined models of the general circulation and of geostrophic turbulence. Processes of heat transfer and the dispersion of biochemical components are closely connected with coherent structures. Two dimensional vortex dynamics in barotropic fluid is also a key problem in free shear flows with or without rotation. In this case, the main question is connected with the stability of these vortices to three-dimensional disturbances.

The merging of 2D vortex pairs in inviscid barotropic fluid has been studied in some detail numerically [1], [2]. A clear stability boundary has been shown to exist. The first physical experiments on merging of barotropic isolated pairs of vortices were conducted by Capéran and Maxworthy [3].

Considerable progress has been made on the instability conditions of barotropic circular vortices. The emergence of dipoles and tripolar vortices was demonstrated theoretically [4], numerically [5] and recently by experiments [6]. In a strain field the vortices have an elliptic shape. A model describing closely the properties of elliptic vortices and their stability was developed by Dritschel and Legras [7].

Vortices in rotating, stratified fluid are subject to subtle physical processes. If we have a fairly good idea about the stability conditions (boroclinic instability) of such vortices, the interaction properties raise many questions. Experiments by Griffiths and Hopfinger [8] indicate that merger of baroclinic vortices in a two-layer stratified fluid depends strongly on the radius of deformation compared with the vortex size. Contour dynamics calculations of the same problem by Polvani et al [9], representing the vortices by uniform potential vorticity patches, show, however, that merging is fairly insensitive to stratification. When the vortices are represented as relative vorticity patches a quite different result emerges [10]. A related problem is the vertical alignment of baroclinic vortices and finite area 'heton' interactions [11], [12] . In what follows some of these problems will be discussed.

2. STABILITY OF BAROTROPIC VORTICES

The stability of barotropic vortices depends on the radial distribution of the vorticity. A general idea about the stability is obtained from Rayleigh's circulation theorem (three-dimensional flows) and the inflexion point theorem (two-dimensional flows).These provide necessary but not sufficient conditions for instability and are simple ways of telling whether or not a vortex might be unstable but do not tell anything about the mode of instability or the growth rate.

The linear stability of a simplified vortex, having piecewise constant vorticity, was first considered by Flierl [4], [13]. In these studies and the following numerical simulations by Gent and McWilliams [14] and Carton et al [5], the two-dimensional instability mode 2 was the fastest growing mode giving rise to dipole formation or a tripolar structure.

In recent laboratory experiments by Kloosterziel and van Heijst (KvH) [6] it was found that the stability behaviour depends on the sign^{+} of the vortex indicating that three-dimensional effects cannot be excluded in real situations. The background rotation of the system causes symmetry breaking. In order to prevent confusion it must, however, be mentioned that when the Rossby number is small anticyclones and cyclones have a similar behaviour and the flow is purely 2D.

To see the effect of the background rotation it is instructive to extend Rayleigh's circulation theorem to absolute vorticity $\zeta + f$. Rayleigh's theorem states that an inviscid stationary swirling flow is stable to axisymmetric disturbances when

------

+ By convention, a positive or cyclonic vortex has counter-clockwise rotation.

$$\frac{d(vr)^2}{dr} \geq 0, \tag{1}$$

where v(r) is the azimuthal velocity. For circular flow the v momentum equation reads:

$$\frac{D(v\,r + \frac{1}{2}f\,r^2)}{Dt} = 0 \tag{2}$$

where D/Dt signifies the material derivative and u = Dr/Dt is the radial velocity component. The equation for u is

$$\frac{Du}{Dt} - \frac{v^2}{r} - f\,v = -\frac{1}{r}\frac{\partial p}{\partial r}. \tag{3}$$

In the initial state the pressure gradient for the vortex with azimuthal velocity $v_0(r)$ is therefore

$$\frac{1}{r}\frac{\partial p_0}{\partial r} = \frac{v_0^2}{r} + f\,v_0\,. \tag{4}$$

KvH [6] then introduce an axisymmetric perturbation, letting a fluid element be displaced from r_0 to $r' = r_0 + \delta r$ and, using angular momentum conservation and (4), it has been established that a fluid element displaced outward by $\delta r = u\delta t$ will continue to move outward when

$$\frac{d(v_0\,r + \frac{1}{2}f\,r^2)^2}{dr} < 0 \tag{5}$$

In terms of non-dimensional variables $v^* = v_0/v_m$, $r^* = r/r_m$, $\zeta^* = \zeta\, r_m/v_m$, equation (5) takes the form:

$$(2\,Ro\,v^* + r^*)(Ro\,\zeta^* + 1) < 0 \tag{6}$$

for instability, where $Ro = v_m/ f\,r_m$. KvH considered two types of vortex structures, a Gaussian vortex and a Lamb vortex withazimuthal velocity identical to the Burgers vortex. The Gaussian vortex, called also isolated vortex is given by

$$v^* = \frac{1}{2} r^* \exp(-\frac{1}{2} r^{*2}) , \qquad (7)$$

with corresponding vorticity

$$\zeta^* = (1 - \frac{1}{2} r^{*2}) \exp(-\frac{1}{2} r^{*2}) . \qquad (8)$$

The Lamb vortex has the form:

$$v^* = \frac{1}{r^*} (1 - \exp(-\frac{1}{2} r^{*2})) , \qquad (9)$$

with corresponding vorticity:

$$\zeta^* = \exp(-\frac{1}{2} r^{*2}) . \qquad (10)$$

In the experiments of KvH a Gaussian vortex was approximately obtained by the stirring technique, called stirred vortices. The other type of vortex is more closely approximated by the source or sink technique. In Figs. 1 and 2 the product of the righthand side of equation (6) is reproduced from KvH as a function of r^* for different values of $\varepsilon = 2$ Ro.

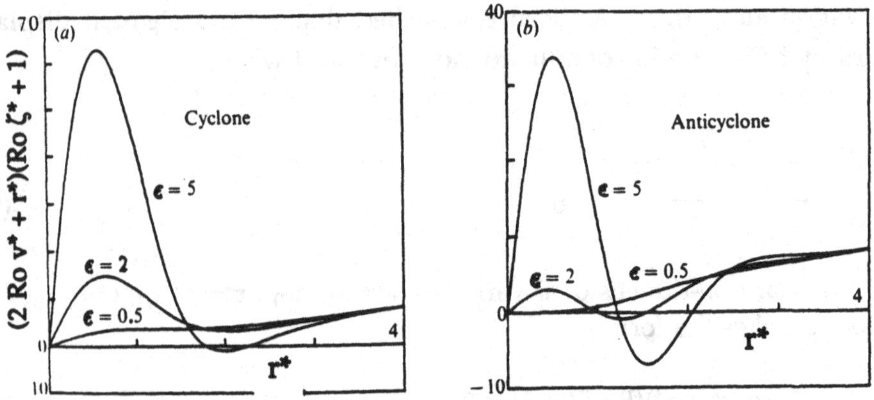

Fig.1 Product of non-dimensional vorticity and velocity, equation (6), as a function of r^* for a Gaussian vortex, equation (7), having three different Rossby numbers Ro= $\varepsilon/2$. (a), cylonic vortex, (b) anticyclones. From [6] with permission.

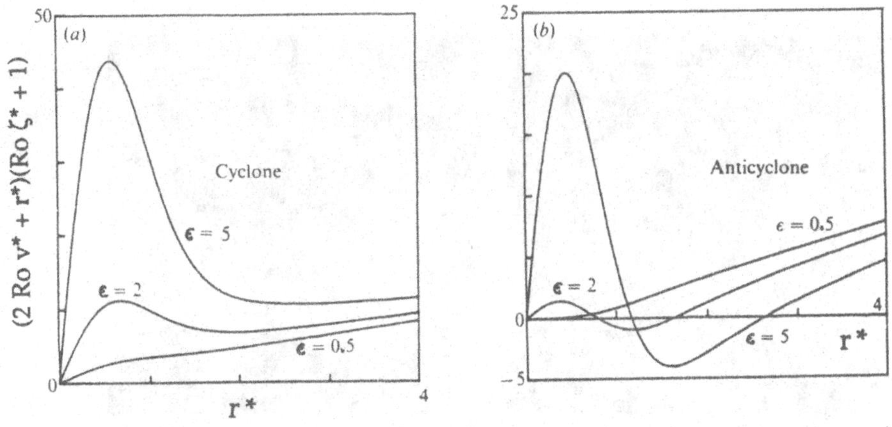

Fig. 2 Product of non-dimensional vorticity and velocity, equation (6), as unction of r* for a Lamb vortex, equation (9), and three values of the Rossby number $\varepsilon/2$. (a), cyclones, (b), anticyclones. From [6] with permission.

It is seen from these figures that cyclones are much more stable to axisymmetric disturbances than anticyclones. A cyclonic Lamb vortex is always stable, whereas a stirred cyclone can be unstable when Ro ≥ 2.3. Anticyclones are unstable when Ro ≥ 0.3 in both cases.

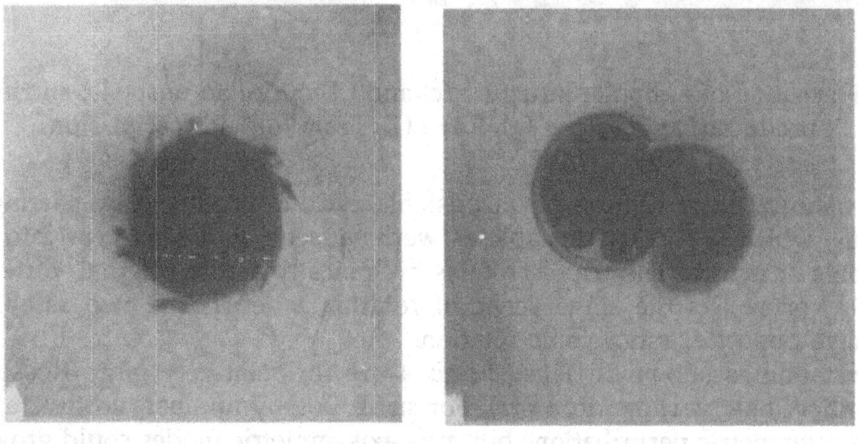

Fig. 3 Evolution of an unstable stirred cyclonic vortex into a tripole structure. From [6] with permission.

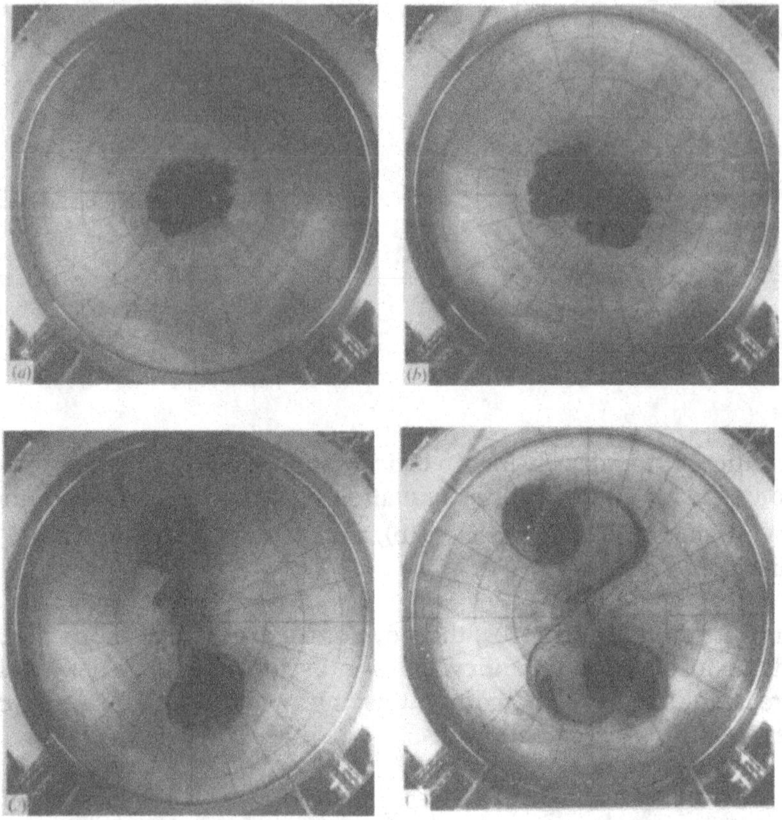

Fig 4 Formation of a dipolar structure (c) and (d) out of an unstable, stirring produced, anticyclone (a), Ro= o(1). From [6] with permission.

Fig 3 , shows the evolution of an unstable cyclone produced by stirring. This vortex, which is weakly unstable as we have seen above, evolves into a stable tripole structure. The tripole rotates cyclonically as the original vortex. The center vortex has the some sense of rotation whereas the two satelite vortices have opposite, anticyclonic rotation.

The instabilities shown in Figs. 3 and 4 are for relatively large Rossby number anticyclonic vortices, Ro= o(1). For small Rossby number, vortices are stable to axisymmetric perturbations but non-axisymmetric modes could grow. To show this would require a stability analysis as was considered by Flierl [4]. A necessary condition for instability is given by Rayleigh's inflexion point theorem, i.e. the gradient of vorticity of the basic state must change sign somewhere in the interval $r \in [0, \infty]$. For a Lamb vortex no inflexion point exists. For a Gaussian vortex the gradient of vorticity is

$$\frac{d\zeta_0}{dr^*} = (\frac{1}{2} r^{*3} - 2 r^*) \exp(-\frac{1}{2} r^{*2}) \quad (11)$$

Equation (11) shows that the vorticity changes sign at $r^* = 2$. The experiments with the low Rossby number, stirred vortices of KvH showed the existence of mode 2 and mode 3 instabilities. The former led to dipole splitting and the latter gave rise first to a tripolar structure and then to further instability and dipole formation.

In numerical studies by Carton et al [5] a family of vorticity profiles of the form

$$\zeta_0 = (1 - \frac{\alpha'}{2} r^* \alpha') \exp(- r^* \alpha')$$

were considered. the parameter α' is a steepness parameter and is $\alpha' \geq 2$. The Gaussian vortex corresponds to $\alpha' = 2$. The numerical simulations by Carton et al show that mode 2 is the fastest growing mode with tripole formation for moderate α' and dipole splitting for larger values of α' ($\alpha' > 3$).

3. MERGER OF BAROTROPIC VORTICES.

3.1 Theory and Contour Dynamics Calculations:

The interaction of barotropic corotating vortices has been extensively studied by contour dynamics calculations solving the two-dimensional Euler equation. The conditions of the existence of co-rotating equilibria states (V-state) were first examined by Saffman and Szeto [15] and extended to more than two vortices by Dritschel [2]. Saffman and Szeto examined also the stability of a V-state and predicted instability when the ratio of the distance between the vortex centers d to the radius R, $d/R < 3.2$. This critical distance for instability was unambiguously documented by the contour dynamics calculations of Overman and Zabusky [1]. For two initially circular vortices, the critical merging distance is very similar, $d/R = 3.3$.

Merging or coalescence occurs normally in a time t_1 short compared with the orbital time $\Gamma/2\pi^2 d^2$, where Γ is the circulation ($t_1 / (\Gamma/2\pi^2 d^2) \approx 0.1$). When the distance is just below critical the two vortices move rapidly together by mutual straining and then overlap to form an elliptical vortex. At the extremities two spiral arms form which carry away circulation. This process of pealing off of vorticity is responsible for axisymmetrisation [16].

3.2 Experiments:

Merging of strained elliptical 2D vortices in shear flow is a well known process responsible for the growth of a shear layer for example [17]. Experiments with an isolated pair of 2D vortices were made for the first time by Capéran and Maxworthy [3]. The vortices were created at the edge of a vertical plate displaced in shallow water. No critical merging distance was found in their experiments, contrary to the contour dynamics calculation results. The vortices were attracted even when $d/R \approx 5$ (note that R, defined below, is about twice the mean radius defined by Capéran and Maxworthy). The corresponding non-dimensional time was $t_1(2\pi^2 d^2/\Gamma) \approx 3$ to 4 . This long merging time is indicative of weak interaction when R is large. The reason for the existence of such long range interactions is that the vortices created by Capéran and Maxworthy had vorticity tails which do not exist in the contour dynamics vortex models.

Griffiths and Hopfinger[8] produced source (anticyclones) and sink (cyclones) 2D vortices in a rotating fluid. The Vortex structure was of the Lamb type and was closely approximated by a Rankine model. A critical merging distance was found for anticyclones but not for cyclonic vortices. When defining a vortex radius by $R' = \Gamma/2\pi V_m$, where V_m is the maximum velocity, the observed ratio of critical merging distance to radius, d/R', was for the anticyclones $d/R' = 3.3 \pm 0.2$. Cyclonic vortices behaved in a way similar to what was observed by Capéran and Maxworthy. The reason here might be of different origin. The surface deformation acts as a β plane and cyclonic vortices move north-west, hence toward smaller depth by a dipolar perturbation [18].

4. STRUCTURE OF TWO-LAYER STRATIFIED VORTICES

4.1 Discrete Vortices:

Solutions to the potential vorticity equation for discrete vortices in a two-layer stratified fluid were obtained by Gryanik [19] and in a more compact form by Hogg and Stommel [20] who considered also the interaction of four discrete vortices including 'heton' pair interactions. Discrete or point vortices in stratified fluid illustrate nicely the stratification effect and it is, therefore, useful to reproduce the results of Hogg and Stommel. The streamfunctions Ψ_i satisfy the potential vorticity conservation equations [21]:

$$D\Pi_i / D_i t = 0 ,$$

where $D / D_i t = \partial / \partial t + J(\Psi_i , \ )$ and the potential vorticity is:

$$\Pi_i = \nabla^2 \Psi_i + (-1)^i \frac{f^2}{g' H_i} (\Psi_1 - \Psi_2) , \qquad (12)$$

with i = 1, 2 for a two-layer fluid, $\nabla = \nabla_H$ and g' the reduced gravity $g' = g\,\Delta\rho$, where $\Delta\rho = \rho_2 - \rho_1$. For simplicity the layers are taken to be of equal rest depth $H_1 = H_2 = H$, so that $f^2/g'H = \Lambda^{-2}$; the horizontal scale Λ is the internal Rossby radius of deformation. The left-hand side of equation (12) is $\Pi_i = \Pi'_i H - f$. If it is assumed that both layers have the same uniform potential vorticity $\Pi' = f/H$, except for a singular vortex in layer 1, then $\Pi_1 = s\,\delta(r)$ and $\Pi_2 = 0$, where $\delta(r)$ is the delta function for a vortex centred on r=0 and s is the vortex intensity, positive for a cyclone and negative for an anticyclone. The solution to equation (12) is then:

$$\Psi_1 = \frac{1}{2} s \left(\ln r - K_0\left(\frac{r}{\lambda}\right)\right) \quad , \Psi_2 = \frac{1}{2} s \left(\ln r + K_0\left(\frac{r}{\lambda}\right)\right)$$

or

$$v_1 = \frac{1}{2} s \left(\frac{1}{r} + K_1\left(\frac{r}{\lambda}\right)\right) \quad , v_2 = \frac{1}{2} s \left(\frac{1}{r} - K_1\left(\frac{r}{\lambda}\right)\right), \qquad (13)$$

where λ, also refered to as Rossby deformation radius is $\lambda = \Lambda/\sqrt{2}$ and K_0 and K_1 are the modified Bessel functions. Fig. 5 shows the azimuthal velocity as a function of r/λ in the upper and lower layers. The velocity field is modified by the interface displacement and what is most important to note is that the far field velocity is the same in both layers. This implies that two vortices which are in different layers interact as if they were in the same layer, provided the distance between them is large.

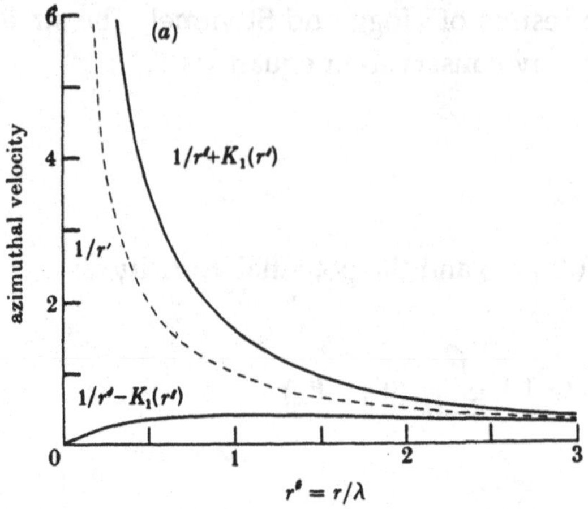

Fig. 5 Azimuthal velocities for a singular vortex created in one layer of a two-layer stratified fluid (from Hogg and Stommel [20])

4.2 Two-layer Finite Core Size Vortex Model:

Baroclinic, finite core vortices are characterized by two horizontal length scales that are generally independent: the internal Rossby radius of deformation and a core radius. The simple model of a two-layer geostrophic vortex considered by Griffiths and Hopfinger [8] is sketched in Fig. 6. Both layers were assumed to have equal 'rest' depths H. The fluid having anomalous potential vorticity is confined to the vortex core of radius R in the upper layer. The model assumes that the core has uniform potential vorticity Π_0, different from the potential vorticity in the rest of the layer away from the vortex core which is $\Pi = f/H$. In general, $\Pi = (f + \zeta)/\eta$, where ζ is the relative vorticity, so that when $\eta = H$, the relative vorticity $\zeta = 0$.

The model considered here is more simple than that expected for oceanic or atmospheric vortices, where the potential vorticity may not be entirely uniform within the core, the density varies continuously with depth and fractional variations in layer depth (in oceanic frontal eddies) can be large. However, the model does capture the essential features of real flows. The quasi-geostrophic assumption neglects local accelerations ($\partial \mathbf{u}/\partial t$) relative to Coriolis acceleration ($2\Omega \times \mathbf{u}$), and takes the relative vorticity to be comparable with vorticity associated with stretching of fluid columns by variations in layer depth, but small compared with the background vorticity f. While departures from geostrophic balance are allowed, the model does not include centrifugal forces v^2/r that are associated with finite Rossby numbers ζ/f and which are significant in frontal ocean eddies.

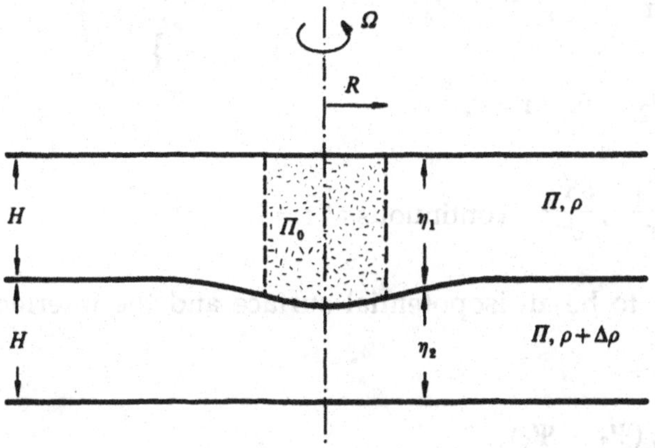

Fig. 6. A sketch of the two-layer geostrophic vortex with piecewise uniform potential vorticity; from [8].

In the model the streamfunctions Ψ_i are thus assumed to satisfy the inviscid quasi-geostrophic potential vorticity equations in the form:

$$\left.\begin{aligned} &\nabla^2\Psi_1 + \frac{1}{2}\lambda^{-2}(\Psi_2 - \Psi_1) = H\Pi_0 - f\,, && r < R,\\ &\nabla^2\Psi_1 + \frac{1}{2}\lambda^{-2}(\Psi_2 - \Psi_1) = 0\,, && r > R,\\ &\nabla^2\Psi_2 + \frac{1}{2}\lambda^{-2}(\Psi_1 - \Psi_2) = 0, && \end{aligned}\right\} \qquad (14)$$

where subscripts 1 and 2 refer again respectively to the top and pottom layer and $\lambda = \frac{1}{\sqrt{2}}(g'\,H)^{1/2}/f$. The right-hand side of (14) is simply the relative vorticity $\zeta_0 = H\,\Pi_0 - f$, which vanishes everywhere outside the core.

In polar coordinates centered at $r = 0$, an axisymmetric vortex has azimuthal velocities $v_1 = \partial\psi_1/\partial r$ and $v_2 = \partial\psi_2/\partial r$. Suitable boundary conditions on the solution to (13) are:

$$\left.\begin{array}{l} \dfrac{\partial \Psi_1}{\partial r} = \dfrac{\partial \Psi_2}{\partial r} = 0, \quad r = 0, \\[2ex] \dfrac{\partial \Psi_1}{\partial r} \to \dfrac{\partial \Psi_2}{\partial r} \to 0, \; r \to \infty, \\[2ex] \Psi_2 = 0, \quad r = 0, \\[2ex] \Psi_1, \Psi_2, \dfrac{\partial \Psi_1}{\partial r}, \dfrac{\partial \Psi_2}{\partial r} \quad \text{continuous at } r = R. \end{array}\right\} \tag{15}$$

The bottom is taken to be an isopotential surface and the interface height is given by:

$$\eta_2 = H + \frac{f}{g'}(\Psi_1 - \Psi_2), \tag{16}$$

which is continuous by virtue of the continuity of the stream function.

Solutions to (14) are found by addition and subtraction of equations for the top and bottom layers in each of the regions $r < R$ and $r > R$. The resulting solutions, for azimuthal velocities $v_i = \partial \Psi_i / \partial r$ in each layer and region, of equations (14), subject to the boundary conditions (15), are [8], [22], [23]:

$$\left.\begin{array}{ll} \left.\begin{array}{l} \dfrac{v_1}{R\zeta_0} = \dfrac{1}{4}\dfrac{r}{R} + \dfrac{1}{2} K_1\!\left(\dfrac{R}{\lambda}\right) I_1\!\left(\dfrac{r}{\lambda}\right) \\[2ex] \dfrac{v_2}{R\zeta_0} = \dfrac{1}{4}\dfrac{r}{R} - \dfrac{1}{2} K_1\!\left(\dfrac{R}{\lambda}\right) I_1\!\left(\dfrac{r}{\lambda}\right) \end{array}\right\} & r < R \\[5ex] \left.\begin{array}{l} \dfrac{v_1}{R\zeta_0} = \dfrac{1}{4}\dfrac{R}{r} + \dfrac{1}{2} I_1\!\left(\dfrac{R}{\lambda}\right) K_1\!\left(\dfrac{r}{\lambda}\right) \\[2ex] \dfrac{v_2}{R\zeta_0} = \dfrac{1}{4}\dfrac{R}{r} - \dfrac{1}{2} I_1\!\left(\dfrac{R}{\lambda}\right) K_1\!\left(\dfrac{r}{\lambda}\right), \end{array}\right\} & r > R \end{array}\right\} \tag{17}$$

where I_1 and K_1 are the modified Bessel functions. A vortex in the top layer displaces the density interface so that

$$\frac{\eta_2}{H} = \left\{ \frac{1 + \frac{\zeta_0}{f}\left[1 - \frac{R}{\lambda} K_1\left(\frac{R}{\lambda}\right) I_0\left(\frac{r}{\lambda}\right)\right]}{1 + \frac{\zeta_0}{f}\frac{R}{\lambda} I_1\left(\frac{R}{\lambda}\right) K_1\left(\frac{r}{\lambda}\right)} \right\} \tag{18}$$

In equation (17) the velocities are normalized by the velocity $R\zeta_0$ (rather than the alternative scale $\lambda\,\zeta_0$) as this is the correct scale for the maximum azimuthal velocity in the vortex. It is noteworthy that the shapes of the velocity profiles are not dependent on the sign of the vorticity in the core. Thus cyclones and anticyclones with equal and opposite relative vorticities ζ_0 have equal and opposite velocities.

The velocity profiles (17) are plotted in Fig. 7 for three values of the ratio λ/R.

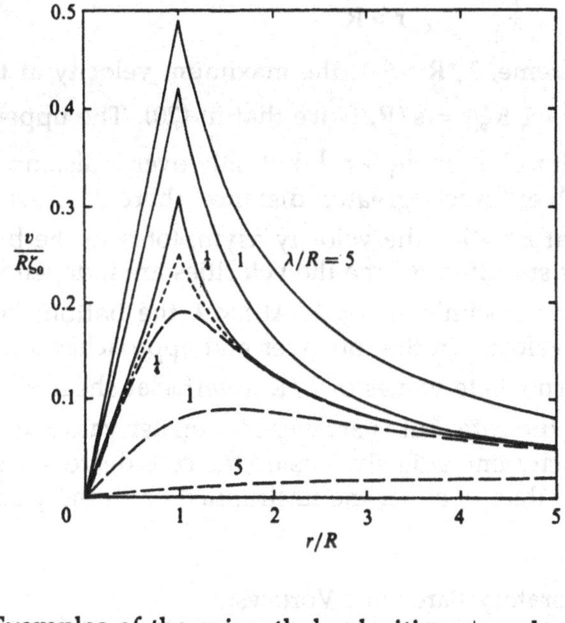

Fig. 7 Examples of the azimuthal velocities given by equation (17). ——, top layer velocities; ------, bottom layer velocities; ········, barotropic limit; from [8].

In the limit $R/\lambda \to 0$, we have $I_1(R/\lambda) \to \frac{1}{2}R/\lambda$. and we recover, for $r > R$, the result for a point vortex [20]:

$$v = \frac{1}{2}\frac{s}{r}\left[1 \pm \frac{r}{\lambda}K1\left(\frac{r}{\lambda}\right)\right], \tag{19}$$

where the positive sign is for the top layer and the negative sign for the bottom layer. The constant $s = \frac{1}{2}R^2\zeta_0$ is the vortex intensity and the vortex strength is $2\pi s$. The intensity serves as a convenient multiplying constant in (17). The circulation Γ outside the vortex core is a function of r except in the barotropic limit $\lambda = 0$, where $\Gamma = \pi s$. In the limit $\lambda/R \to 0$, where the influence of density gradients vanishes and the flow becomes independent of depth, the structure approaches the Rankine vortex:

$$v_1 = v_2 = \begin{cases} \frac{1}{2}\frac{s}{R^2}\, r, & r < R, \\ \frac{1}{2}\frac{s}{r}, & r > R. \end{cases} \tag{20}$$

At the opposite extreme, $\lambda/R >> 1$, the maximum velocity at the edge of the core approaches $v_1 = \frac{1}{2}R\zeta_0 = s/R$, twice that in (20). The upper layer velocity again decreases with radius as $v_1 \sim r^{-1}$, but only over a distance comparable to the core radius. Over much greater distance there is now an additional decrease such that at $r >> \lambda$, the velocity asymptotes to the barotropic value (20). In this strongly stratified regime the velocities are independent of depth at $r >> \lambda$ and strongly baroclinic at $r < \lambda$. At $r < \lambda$ the bottom layer velocity is much less than the velocity in the top layer and approaches a maximum value at $r \approx 1.1\,\lambda$. For intermediate values of λ/R, interfacial shear is always greatest at the outer edge of the core, but is smaller for smaller values of λ/R. At $\lambda/R \approx 1$, the top layer velocity immediately outside the core decreases with increasing radius more rapidly than in either the unstratified or strongly stratified cases.

4.3 Structure of Laboratory Baroclinic Vortices:

Griffiths and Hopfinger [8] performed experiments in a circular tank 100 cm in diameter, 45 cm deep and rotating about a vertical axis through the center. The rotation rate was $\Omega = 1.0$ rad s^{-1} in the (positive) anticlockwise direction. One rotation period was thus $T_\Omega = 2\pi\Omega^{-1} = 6.28$ s and the background vorticity was f = 2. 0 s $^{-1}$. The fluid layers were of equal depth H = 20 cm. The density difference $\Delta\rho = \rho_2 - \rho_1$ between the layers was varied so that the Rossby radii $\Lambda = (g' H)^{1/2}/ f$ were 0, 1.5, 5, 10, and 15 cm. The vortices were produced by sources and sinks and visualized by dye. Velocities were obtained from streak photographs but only for Λ = 5 cm. The other values of Λ were used in the study of vortex merging.

In Fig. 8 the velocity profiles in the top and bottom layer are shown for an anticyclonic vortex. The vortex was initiated in the top layer of the two-layer stratified fluid with Λ = 5 cm.

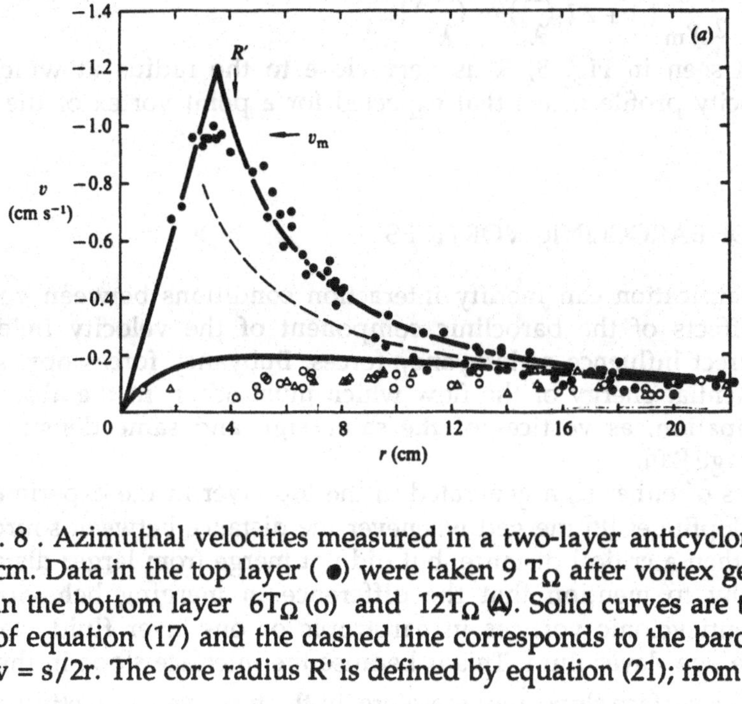

Fig. 8 . Azimuthal velocities measured in a two-layer anticyclone with Λ = 5 cm. Data in the top layer (•) were taken 9 T_Ω after vortex generation and in the bottom layer 6T_Ω (o) and 12T_Ω (▲). Solid curves are the profiles of equation (17) and the dashed line corresponds to the barotropic vortex v = s/2r. The core radius R' is defined by equation (21); from [8].

Fig. 8 shows a clear deviation from $v \sim r^{-1}$ beyond the core radius. Furthermore, the motion in the bottom layer is much less than that in the top layer for $r < 2\Lambda$., while for $r > 3\Lambda$ the flow is practically independent of depth (nearly the same velocity in top and bottom layers). The comparison between the measured and model velocities was made by fitting the predicted form (17) for $v_1(r > R)$ to the data from the top layer by using an estimate for the ratio

Λ/R, evaluating a multiplying constant s, and then using that constant to compute the bottom layer velocity v_2. The measured bottom layer velocities are modelled satisfactorily although they are somewhat smaller than those predicted. The discrepancy might be due to bottom friction and to finite interface thickness. For the cyclonic vortices the comparison between the model and observations is similar.

The definition of the relevant core radius is somewhat arbitrary. It was assumed to be the radius at which the potential vorticity deviates significantly from its constant value outside the core. Hence the core radius R' was defined by the intersection of the predicted velocity v_1 with the maximum velocity observed v_m. For unstratified vortices this gives $R' = \Gamma/2\pi v_m$. For baroclinic vortices the expression is:

$$R' = \frac{1}{2}\frac{s}{v_m}\left[1 + 2 I_1\left(\frac{R}{\lambda}\right) K_1\left(\frac{R'}{\lambda}\right)\right]. \tag{21}$$

In practice, as seen in Fig. 8, R' is very close to the radius at which the measured velocity profile meets that expected for a point vortex of the same strength.

5. MERGING OF BAROCLINIC VORTICES

Density stratification can modify interaction conditions between vortices through the effects of the baroclinic component of the velocity field and through the direct influence of buoyancy forces. Buoyancy forces oppose the increase in potential energy of the flow which must occur, in the absence of sufficient dissipation, as vortices of the same sign and same density level approach or merge [20].

Two vortices of either sign generated in the top layer in the experiment of Griffiths and Hopfinger [8] merged whenever the distance between sources or sinks was less than a critical distance, but did not merge from larger distances. It is noteworthy to mention that the difference in merging behaviour of cyclonic and anticyclonic vortices in barotropic or one-layer fluid was not observed in the two -layer fluid. This is because the interface slope in the tank is the same as the surface slope and, therefore, in the top layer no β effect exists.

Fig. 9 shows photographs of an experiment [8] with two anticyclones of $\Lambda/R' = 2.5$ and $d/R' = 4.5$, where d is the initial distance between vortices.

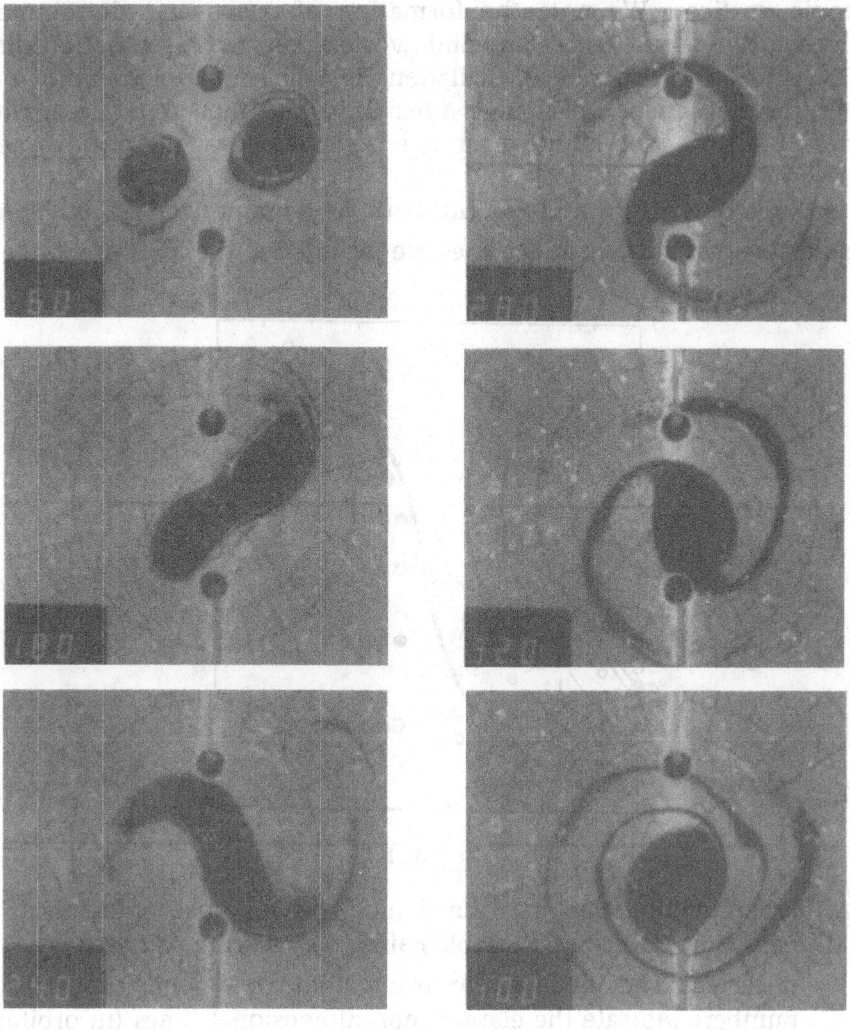

Fig.9. Photographs of merging of two anticyclones in the top layer : d = 18 cm, Λ = 10 cm, R' = 4.0 cm s = 7.4 cm^2s^{-1}. The numbers indicate the elapsed orbital time in rotation periods. The orbital period is 44 T_{Ω}. The two vortices have merged at about the second frame; from [8].

The two vortices are observed to merge after $t_* \approx 15T_\Omega$ (second frame in Fig. 9) or $t_*s/2\pi d^2 \approx 0.34$. Qualitatively the appearance of the interacting baroclinic vortices and the merging process are similar to that observed for barotropic vortices. We note the formation of cusps and detrainment of vorticity via two spiral arms. Baroclinic vortex pairs with separation distance close to critical also experience oscillations in which the vortices move closer together, develop cusps , exchange dyed fluid and then separate again. The important point to note is, however, the dependance of the critical merging distance on Rossby radius.

The results of Griffiths and Hopfinger [8] are summarized in Fig. 10, where the results for unstratified anticyclones are included at the limit $\Lambda = 0$.

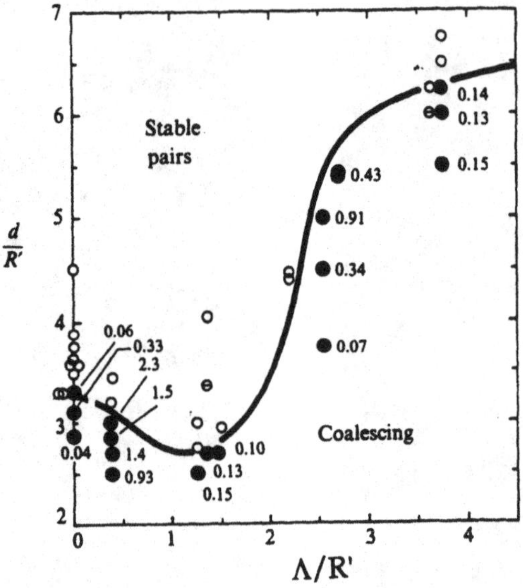

Fig. 10. Merging boundary as a function of Λ/R' for a pair of identical baroclinic vortices: (o), stable pairs; (•), merging pairs; (ө), uncertain. The one layer barotropic limit corresponds to $\Lambda = 0$. The numbers indicate the elapsed non-dimensional times (in orbital periods) for merging occurance; from [8].

The solid curve in Fig. 10 indicates the boundary between stable pairs (open circles) and rapidly merging pairs (solid circles). Beginning at $\Lambda = 0$, where the critical distance is $d_*/R' \approx 3.3$, the stability boundary falls to slightly smaller values of the dimensionless separation with increasing Rossby radius of deformation.A minimum of $d_*/R' \approx 2.7$ is reached at $\Lambda/R' \approx 1$. For Rossby radii greater than the core radius, the critical distance increases rapidly with Λ/R' and reaches $d_*/R' \approx 6.3 \pm 0.3$, corresponding to $d_*/\Lambda \approx 1.6$, at $\Lambda/R' \approx 4$ Thus, merging of identical baroclinic vortex pairs can occur over much larger distances than of barotropic vortex pairs. There are observations of tropical cyclones which have merged over distances of about 10 times their size. Merging occurs in a time t_* short compared with the orbital time $2\pi d^2/s$, except when $\Lambda/R' < 1$. The numbers in Fig. 10 are dimensionless times $t_*/(2\pi d^2/s)$ before merging. In terms of T_Ω, the time t_* ranged from $2T_\Omega$ to $50T_\Omega$. The larger merging times were observed when Λ/R' was small, with t^* extending up to 47 T_Ω or 2.3 orbital periods. Complete amalgamation generally required a further $10T_\Omega$.

At small values of Λ/R' the curve in Fig. 10 is discontinuous which means that probably no merging boundary can be defined. Baroclinic instability is likely to occurs on an eddy scale in this narrow range of Λ/R' [23].

6. NUMERICAL SIMULATIONS OF MERGING OF BAROCLINIC VORTICES

Numerical simulations use the quasi geostrophic formulation. The model for the two-layer stratified system on an f plane is then given by the potential vorticity equation (12)

$$\frac{D\Pi_i}{Dt} = \frac{D}{Dt}\{\nabla^2\Psi_i + (-1)^i \frac{f^2}{g'H_i}(\Psi_1 - \Psi_2)\} = -V_i \quad , i = 1, 2 \qquad (22)$$

where V_i represents lateral friction. The case which corresponds to the experiments discussed in § 5 is $H_1 = H_2 = H$ for which

$$\Lambda = \frac{\sqrt{g'H}}{f} = \sqrt{2}\,\lambda .$$

In the contour dynamics calculations $V_i = 0$ and in the finite difference method V_i can be made negligibly small so that potential vorticity is pratically conserved in each layer and is equal to the initial values Π_{10} and Π_{20} :

$$\nabla^2 \Psi_1 - \frac{1}{2}\lambda^{-2}(\Psi_1 - \Psi_2) = \Pi_1 = \Pi_{10}$$

$$\nabla^2 \Psi_2 + \frac{1}{2}\lambda^{-2}(\Psi_1 - \Psi_2) = \Pi_2 = \Pi_{20}$$

It turns out that the initialization is crucial in the stability behaviour of baroclinic vortex pairs [10]. Three initial states were considered: i) potential vorticity initialization, ii) relative vorticity initialization and iii) intermediate initialization.

i) Potential vorticity initialization (**PVI**) . This corresponds to the models developed in sections 4.1 and 4.2 and was used by Polvani et al [9]. If the upper layer vortices are of pure potential vorticity of the Rankine type (denoted by function Ra) we can write

$$\Pi_{10} = \Sigma_{k=1,2}\, Ra^k$$

$$\Pi_{20} = 0,$$

where the summation k =1,2 on Ra^k indicates here a pair of vortices in the same layer.

ii) Relative vorticity initialization (**RVI**). In this case it is assumed that the initiated upper layer vortices have relative vorticity Ra, the lower layer being at rest. The potential vorticity is then

$$\Pi_{10} = \Sigma_{k=1,2}\, Ra^k - \frac{1}{2}\lambda^{-2}\Psi_{10}$$

with Ψ_{10} calculated from $\nabla^2 \Psi_{10} = \Sigma_{k=1,2}\, Ra^k$. The potential vorticity in the lower layer is consequently

$$\Pi_{20} = \frac{1}{2}\lambda^{-2}\Psi_1$$

iii) Intermediate initialization (**II**). In this case it is assumed that the upper layer vortices are of relative vorticity but the lower layer has zero potential vorticity.

$$\Pi_{10} = \Sigma_{k=1,2} R_a^k - \frac{1}{2}\lambda^{-2}(\Psi_{10} - \Psi_{20})$$

$$\Pi_{20} = 0$$

The initial stream functions Ψ_{10} and Ψ_{20} are then solutions of the system

$$\nabla^2 \Psi_{10} = \Sigma_{k=1,2} R_a^k$$

$$\nabla^2 \Psi_{20} - \frac{1}{2}\lambda^{-2}\Psi_{2}) = -\frac{1}{2}\lambda^{-2}\Psi_{10}$$

The results obtained for these three initial states are given in [24] and are summarized in Fig. 11

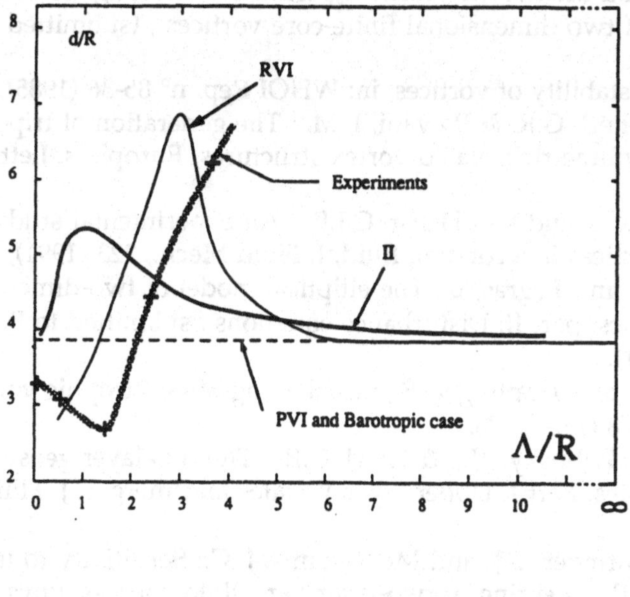

Fig 11 Critical merger boundaries d*/R' as a function of Λ/R' for PVI, RVI and II initializations and comparison with experiments; from[24].

Fig 11 indicates that the RVI initialization reproduces closest the experimental results. The initial state in the experiments is, however, more of the PVI type. The extreme sensitivity of merging to the initial state is indicated by the (II) model. These are only simulations and satisfactory explanations are in need.

Acknowledgements:
I am particularly indebted to Ross W. Griffiths, whose influence and collaboration is found all along these notes. Many discussions with T. Maxworthy, J. McWilliams and my close colleagues Ph. Capéran and J. Verron have also greatly contributed to my understanding of the problems discussed here.

REFERENCES

1. Overman, E.A. and Zabusky, N.J.: Evolution and merger of isolated vortex structures, Phys. Fluids, 25 (1982), 1297.
2. Dritschel, D.G.: The stability and energetics of co rotating uniform vortices, J. Fluid Mech., 157 (1985), 95.
3. Capéran, P. and Maxworthy,T.:An experimental investigation of the coalescence of two dimensional finite-core vortices , (submitted Phys. Fluids, 1986)
4. Flierl, G. R. Instability of vortices, in: WHOI Rep. n° 85-36 (1985), 119.
5. Carton,X. J., Flierl, G.R. & Polvani, L.M. : The generation of tripoles from unstable axisymmetric isolated vortex structures. Europhys. Lett. 9, (1989), 339.
6. Kloosterziel, R.C. and van Heijst, G.J.F.: An experimental study of unstable barotropic vortices in a rotating fluid, J. Fluid Mech., 223 (1991), 1.
7. Dritschel, D.G. and Legras, B.: The elliptical model of two-dimensional vortex dynamics: part II: Disturbance equations . submitted to Phys Fluids A (1991).
8. Griffiths, R.W. and Hopfinger, E.J.: Coalescing of geostrophic vortices. J. Fluid Mech. 178 (1987), 73.
9. Polvani, L. M., Zabusky, N.J. & Flierl, G.R.: The two-layer geostrophic vortex dynamics, Part I: Upper layer V-states and merger. J. Fluid Mech. 205 (1989), 215.
10. Verron,J., Hopfinger, E.J. and McWilliams, J. C.: Sensitivity to initial conditions in the merging of two-layer baroclinic vortices.,Phys Fluids A, 2 (6), (1990), 886.
11. Polvani, L. M.: Two-layer geostrophic vortex dynamics . Part II: Alignment and two-layer V-states, J. Fluid Mech., 225 (1991), 241.

12. Griffiths, R.W. and Hopfinger, E.J.: Experiments with baroclinic vortex pairs in a rotating fluid, J. Fluid Mech., 173 (1986), 501.
13. Flierl, G. R.: On the stability of geostrophic vortices.,J. Fluid Mech., 197 (1988), 349.
14. Gent, P. R. and Mc Williams, J. C.: The instability of circular vortices, Geophys. Astrophys. Fluid Dyn., 35 (1986), 209.
15. Saffman, P. G. & Szeto, R.: Equilibrium states of a pair of equal uniform vortices, Phys. Fluids, 23 (1980), 2339.
16. Melander, M.V.,Zabusky,N.J. and McWilliams, J. C.: Axisymmetrization and vorticity filamentation. J.Fluid Mech. 178 (1987), 137.
17. Winant C. D. and Browand F.K.: Vortex pairing: a mechanism of turbulent mixing layer growth at moderate Reynolds number. J. Fluid Mech. 63, (1974), 237.
18. Carnevale, G.F., Kloosterziel, R.C. and van Heijst, G.J.F.: Propagation of barotropic vortices over topographyin a rotating tank. J. Fluid Mech.(1991).
19. Gryanik, V. M.: Dynamics of singular geostrophic vortices in a two-level model of the atmosphere or ocean, Izv. Akad. Nauk. USSR Atmos. Oceanic Phys. 19 (1983), 171.
20. Hogg, N.G. & Stommel, H.M.: The heton, an elementary interaction between discrete baroclinic geostrophic vortices and its implications concerning eddy heat flow, Proc. Roy. Soc. Lond., A 397 (1985), 1.
21. Pedlosky, J.: Geophysical Fluid Dynamics, Springer Verlag, 1979
22. Pedlosky, J. : The instability of continuous heton clouds, J. Atmos. Sci. 42 (1985), 1477.
23. Helfrich, K.R. and Send, U.: Finite amplitude evolution of two-layer geostrophic vortices, J. Fluid Mech. 197 (1988), 331.
24. Verron, J. and Hopfinger, E.J.: The enigmatic merger conditions of two-layer baroclinic vortices. CRAS , t 313 S.II (1991), 737.

Chapter V.3

LONG-LIVED EDDIES IN THE ATMOSPHERES OF THE MAJOR PLANETS

P.L. Read
University of Oxford, Oxford, UK

ABSTRACT

We review current knowledge of the giant, long-lived eddies observed in the atmospheres of the Major Planets. Particular emphasis is placed on determining the dynamical processes which may be at work, and we discuss a range of both theoretical and laboratory models which have been suggested as analogues of these features.

1. STRUCTURE OF THE MAJOR PLANETS

The Major Planets (Jupiter, Saturn, Uranus and Neptune) are essentially hydrogen-rich fluid bodies of great size, with only relatively small rocky cores [1]. Their visible surfaces are typically shrouded in clouds of ammonia and/or methane, and only the cloudy surface layers are amenable to direct observation. The observed atmospheric motions are therefore probably not influenced by any solid surface. All major planets are weak sources of energy through the slow release of gravitational potential energy in their interiors. They are also electrically conducting throughout most of their interiors, which may influence the balance of forces in their outer regions through magnetohydrodynamic (MHD) interactions with the planetary magnetic fields.

The static stability is highly variable with height, with a stable stratosphere, weakly stable troposphere and roughly adiabatic convective interior, but N^2 is affected by latent heat and compositional layering effects associated with cloud condensation [2], conversion between the ortho- and para- forms of molecular hydrogen [3] and other exotic thermodynamic effects at depth. Clouds and zonal winds are observed to be arranged in latitudinal bands ('belts' and 'zones') which do not appear to vary, but the zonal jets decay in strength with height above the visible clouds; the processes responsible for driving the bands and zonal jets, and the depth to which they penetrate, are largely unknown.

The major eddies are typically oval in shape, elongated to a greater or lesser extent in the zonal direction and of variable (though immense) scale, and great persistence (e.g. Jupiter's Great Red Spot (GRS) discovered in 17th century). Many similarities are found between eddies of a similar sense of circulation, particularly in the morphology and surface texture of their clouds. Their horizontal velocity structures as obtained from tracking discrete cloud features all appear to show strong jet streams at their periphery and relatively stagnant cores. The eddies also appear to exhibit a characteristic signature in the infrared, indicative of certain features relating to systematic variations in cloud height and vertical motion (which cannot be measured directly). Anticyclones are typically found to be cool at the centre (indicating elevated cloud tops and implied upwelling) with warm peripheral 'collars' (indicating clear regions and implied downwelling). Cyclones may have opposite signatures, though this is less clear owing to poor resolution of the observations.

All major eddies also take on a characteristic scale which varies with latitude and occur only at certain preferred latitudes, both of which seem to be connected with the pattern of mean zonal flow in their vicinity. Thus([4],[6]), the eddies take on a latitudinal scale comparable with the separation of adjacent zonal jets, and are located in ambient zonal shear of the same sign as the relative vorticity ζ of each eddy (near to where $u_{yy} - \beta = 0$). Scales are such that the local Rossby number is observed to be small (< 0.3) and the Richardson number is inferred to be relatively large. Significant uncertainty (and some controversy), however, remains concerning the size of particularly the largest features with respect to the Rossby deformation radius L_R (= ND/f), owing to major uncertainties in the appropriate values for N and vertical scale D. All that can be said conclusively to date [4],[5] is that most eddies are larger than or comparable with L_R.

Other observations to note include [6] the marked tendency on all major planets (except Uranus) to form anticyclones in preference to cyclones (though persistent cyclones are not unknown - e.g. 'barges' on Jupiter). A number of the smaller eddies have also been observed [6] to interact and on occasions to merge in strongly 'inelastic' collisions. A variety of structural oscillations have also been observed, often taking the form of roughly periodic 'tilting' of the major axes of eddies such as the GRS and the Great Dark Spot (GDS) on Neptune [7].

Any viable model of these features should therefore be able to address such questions as (a) the origin of the eddies, and how they are generated from plausible initial states; (b) what processes govern their structure and morphology; (c) to what do they owe their remarkable persistence and apparent stability (simple inertia or intrinsic stability); (d) what processes maintain their energy and vorticity against the inevitable effects of 'friction'; and (e) what is their role within the general circulation e.g. as a means of transport of heat, momentum, vorticity etc.?

2. MODELS OF LONG-LIVED EDDIES: FORMULATION

As Table 1 indicates, a great diversity of models have been put forward over many years to account for at least some of the above observations, including several laboratory studies of particular dynamical processes. One of the earliest models of Jupiter's Great Red Spot supposed it to be formed by the interaction of a zonal flow in Jupiter's atmosphere with a large solid obstacle beneath the visible clouds - i.e. a form of quasi-inviscid Taylor column (see [8]). This model has since been largely discounted since Jupiter cannot have a solid surface on compositional grounds.

Most (though not all) recent theoretical/numerical models consider solutions to an adiabatic, inviscid potential vorticity equation; either the QG form

$$[\partial/\partial t + \mathbf{u}.\nabla_{\mathbf{h}}]\,[\nabla^2\psi + f^2\partial/\partial z(1/N^2\ \partial\psi/\partial z) + \beta y] = 0 \qquad (1)$$

(where $\mathbf{u}$ is the geostrophic velocity, y the geostrophic streamfunction, N the buoyancy frequency and β the lateral gradient of Coriolis parameter f), or (especially recently) the fully nonlinear (reduced gravity) shallow water equations

$$\partial\mathbf{u}/\partial t + (f + \zeta)\mathbf{k}\times\mathbf{u} = -\nabla[g'(h + h_2) + K]\{-(u - u_J(y))/\tau\} \quad (2)$$

TABLE 1

Models of long-lived eddies in the atmospheres of the major planets.

MODEL	SOURCE OF ENERGY/VORTICITY	SCALE	ANALOGOUS SYSTEM	REMARKS
'Taylor Column' [8]	Topographic obstacle and zonal flow	$L >> L_R$	Mathematical idealisation (MI)/ Laboratory model	Needs isolated obstacle at depth
'Hurricane'	Latent heat of water (CISK)	$L < L_R$?	Loose analogy with terrestrial tropical storm.	Needs large H_2O abundance and a stress-bearing surface (for Ekman moisture flux)?
'Modon' [9]	None	$L >> L_R$	MI	Strong solitary eddy in shear flow. Barotropic (no vertical structure)
Weakly nonlinear 'Soliton' [25]	None (horizontal shear, 'wave CISK'....?)	$L \sim L_R$	MI	Weak solitary eddy in shear flow. Can include vertical structure.
'Coalescing Modon' [10],[11],[23]	Smaller eddies (& horizontal shear?)	$L > L_R$	Numerical model	Strong solitary eddy of large scale. Equiv. barotropic or baroclinic structure.
Shallow water ('intermediate geostrophic') eddy [15],[17]	Smaller eddies (& horizontal zonal shear?)	$L >> L_R$	Numerical model	Strong solitary eddy of very large scale. No vertical structure.
'Rossby soliton' [16],[12]	Horizontal zonal shear	$L > L_R$	MI/Laboratory model	Single or multiple trains of eddies in shallow fluid layer.
'Baroclinic eddy' [19],[4],[20],[22],[23]	PE associated with tilted isotherms/ differential heating	$L \sim L_R$	Numerical model & Laboratory expts.	Single or multiple trains of strong eddies. Vertical structure crucial
'Oscillating eddy' [7]	None? (cf 'Modon' models)	$L > L_R$	Numerical model	Compact isolated eddy which tilts periodically in orientation

$$\partial h/\partial t + \nabla.(h\mathbf{u}) = 0 \qquad (3)$$

where h is the depth of an upper 'weather layer', h_2 the depth of an assumed quiescent lower layer, K = KE/unit mass ($\mathbf{u}.\mathbf{u}/2$), $\zeta = \mathbf{k}.(\nabla\times\mathbf{u})$ and g' is the reduced gravity. The terms on the RHS of (2) represent forcing towards a given zonal flow state, $u_J(y)$, derived either arbitrarily or from observations, with relaxation timescale τ. The terms in h_2 are assumed to represent steady zonally-symmetric flow in the deep interior, such that

$$\partial(gh_2)/\partial y = -fu_2 \qquad (4)$$

Both these sets of equations yield conservation equations for potential vorticity, which can be integrated numerically with appropriate initial and boundary conditions. Various authors have used differing assumptions concerning the basic geometry, the assumed properties of a lower layer (representing an adiabatic deep fluid interior) and the relative size of the Rossby deformation radius (i.e. the effective Burger number). Under a wide variety of conditions, many of these models are able to exhibit the ability to support stable isolated oval eddies embedded in a zonal flow which reverses sign with latitude. The resulting solution in the reference frame moving with the eddy at zonal velocity c has the form $\mathbf{u}.\nabla q = 0$ which, for horizontally non-divergent flow, reduces to

$$J(\Psi,q) = 0 \qquad (5)$$

where q is the appropriate potential vorticity and Ψ is the Doppler-shifted streamfunction (= ψ +cy). The solution to (5) is $\Psi = \Psi(q)$, so that contours of q follow the streamlines. Such a solution is called a 'free mode', and may be spatially isolated (cf 'modons') or wave-like (and domain-filling) in character.

3. MODELS OF LONG-LIVED EDDIES: BAROTROPIC QG MODELS

The simplest form of such a model assumes no variation in flow structure with height, and can therefore be represented by the barotropic form of (1). Various authors (including [9],[10]) have shown semi-analytically that such a system may admit monopole isolated eddy solutions ('modons'), which numerical experiments show to be stable under typical conditions. Furthermore, pairs or groups of such eddies will readily merge together to form a smaller number of larger features. This property was demonstrated numerically by [10] and (most recently) by Marcus [11] - in which it was suggested that the GRS could form spontaneously and maintain itself by repeated vortex mergers from smaller features (generated convectively on Jupiter?).

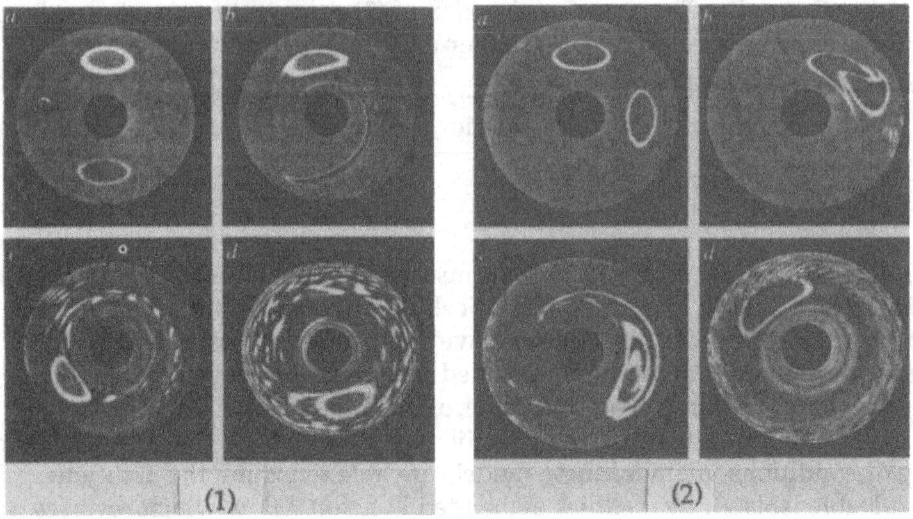

Figure 1. Two sequences taken from QG numerical simulations [11], showing the persistence of a single anticyclonic eddy in an anticyclonic zonal shear zone from two different initial conditions (1) a complementary pair of eddies (cyclone & anticyclone) and (2) merger of two initial anticyclones.

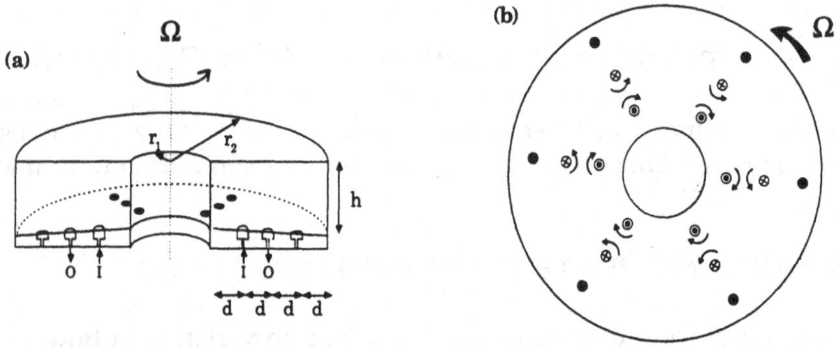

Figure 2. Cross-section (a) and plan view (b) of the tank used [12] to investigate barotropic eddies on a β-plane in a reversing zonal flow.

These numerical experiments prompted a laboratory study by Sommeria et al. [12], in which a reversing zonal flow was driven in a cylindrical apparatus by an arrangement of mass sources and sinks, and the variation of Coriolis parameter with latitude was simulated by a conical bottom. Oval eddies formed in the region of strong shear and were found under some conditions to merge - ultimately forming a single eddy. The patterns obtained were similar in some respects to the Stewartson layer instabilities studied by Niino & Masawa ([13]; see [14]) - although the latter did not observe flows with a single eddy. Sommeria et al. [12] stressed the possible importance of turbulent effects (produced by the point-like mass sources and sinks) in sustaining their isolated eddy, though it is not wholly clear what role these effects might have.

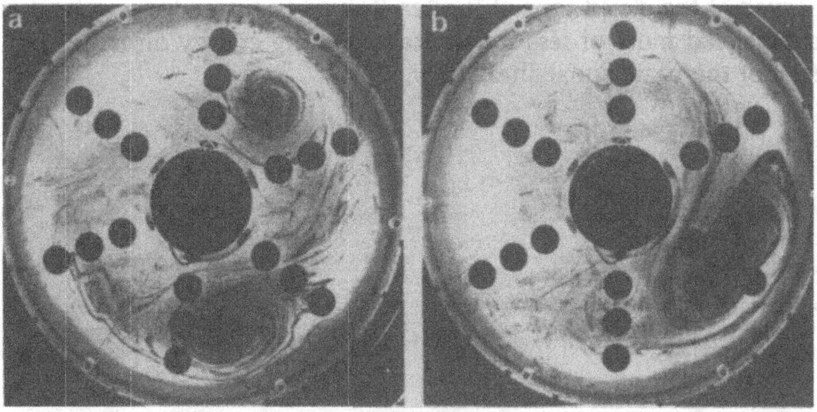

Figure 3. The merger of two cyclonic eddies in the rotating tank experiments of [12]. The ejection of potential vorticity is rendered visible by a dye tracer.

4. MODELS OF LONG-LIVED EDDIES: SHALLOW LAYER MODELS

The observation that Jovian (and Saturnian and Neptunian) stable eddies are predominantly anticyclonic and (apparently) much larger in scale than the Rossby deformation radius prompted some workers to suggest that QG models were not appropriate for this kind of phenomenon, since the QG forms of the equations are symmetrical with respect to the sign of relative vorticity. Williams & Yamagata [15] and Nezlin [16] independently noted that isolated eddies in a shallow water system would tend to favour the formation of anticyclones at large scale, because of the nonlinear divergence term (associated with the ageostrophic part of the term $h\nabla.\mathbf{u}$ from (3), and which becomes significant at large disturbance amplitude as an additional nonlinear effect due to the finite displacement h of the interface) normally omitted in QG dynamics. This term opposes the dispersion produced by the β-effect for anticyclones, but reinforces dispersion for cyclones, rendering the latter more difficult to sustain.

Various workers have carried out numerical experiments using this system, with differing assumptions concerning the role of dissipation and the form of u_2 in the lower layer. Dowling & Ingersoll [17] showed, however, that their qualitative results were not critically sensitive to u_2. The results were able to demonstrate the formation of an isolated vortex via a series of merging events from a perturbed zonal flow state which resembled the zonal wind pattern observed at the cloud-tops of Jupiter. The eddy formed was robust and steady, and persisted more or less indefinitely, gaining energy directly from the kinetic energy of the basic zonal shear flow.

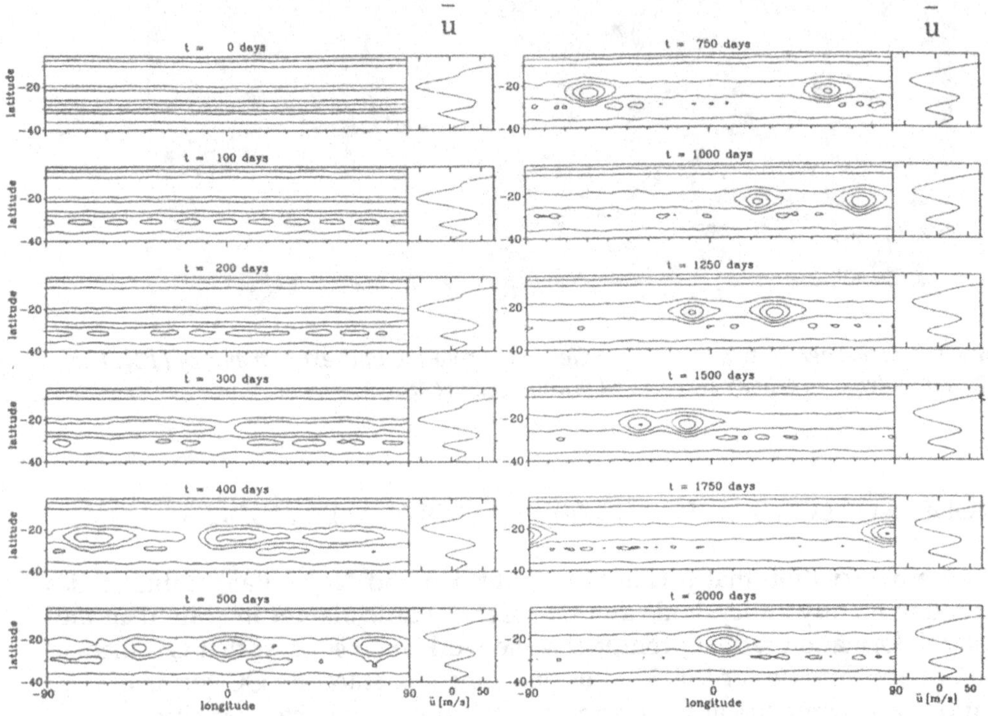

Figure 4. Contours of free surface height ($h + h_2$) during a numerical simulation of shallow layer flow initialised with a sinusoidal wave-like perturbation [17].

An attempt to produce a laboratory analogue of the large-scale shallow-layer vortex has been made by Antipov et al. [16], who used a very rapidly rotating vessel of parabolic cross-section. The action of centripetal accelerations caused fluid to spread over the surface of the vessel in a thin layer of uniform depth when the system was rotated at such a rate that the vessel boundary was effectively a geopotential. The β-effect could be simulated by rotating the vessel at a slightly different rate, producing a non-uniform variation of depth with radius from the rotation axis. A zonal shear flow was driven mechanically by differential rotation of annular rings set into the wall of the vessel.

Like the numerical experiments described above, the zonal shear flow in Antipov et al.'s experiments was barotropically unstable under some circumstances, and would lead to the formation of regular chains of vortices. When the zonal flow was forced most strongly, the shear instability lead to the formation of a single 'solitary Rossby vortex'. The authors noted that anticyclonic eddies (forming in an anticyclonic zonal shear) would develop much more readily than cyclones (in cyclonic shear), which Nezlin [16] interpreted as being due to the asymmetry of the nonlinear divergence noted above, though this may not be the only possible explanation (e.g. see [18]).

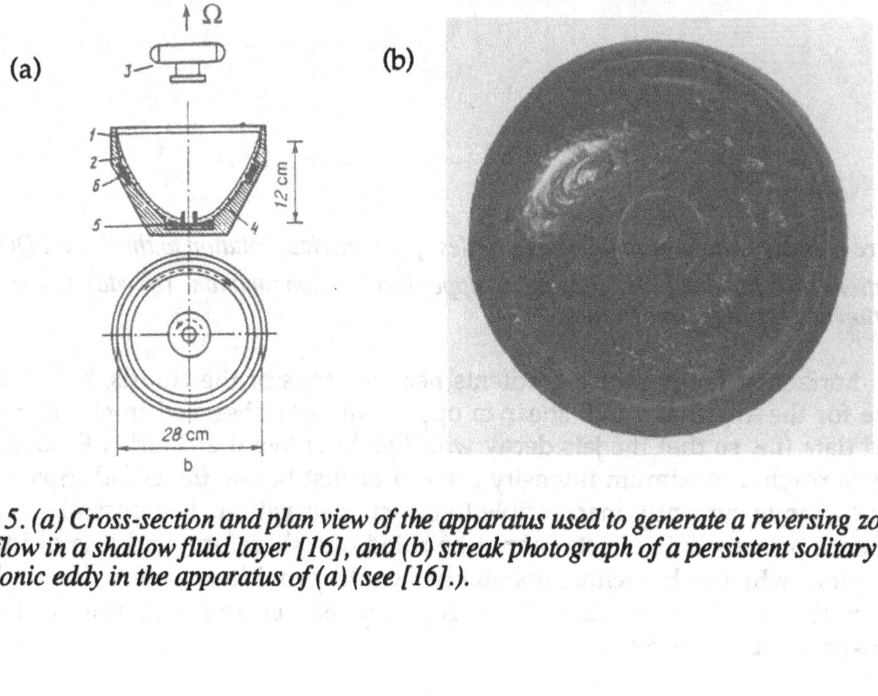

Figure 5. (a) Cross-section and plan view of the apparatus used to generate a reversing zonal shear flow in a shallow fluid layer [16], and (b) streak photograph of a persistent solitary anticylonic eddy in the apparatus of (a) (see [16].).

5. MODELS OF LONG-LIVED EDDIES: BAROCLINIC MODELS

All of the above models make the assumption that variations in the flow structure with altitude are not significant for the dynamics of the large eddies. This would be clearly defensible in a homogeneous fluid (even with compressibility), but is much less secure in a non-homogeneous fluid with spatial contrasts of density and temperature - in which baroclinic instabilities may occur. Observations of the major planets indicate the presence of

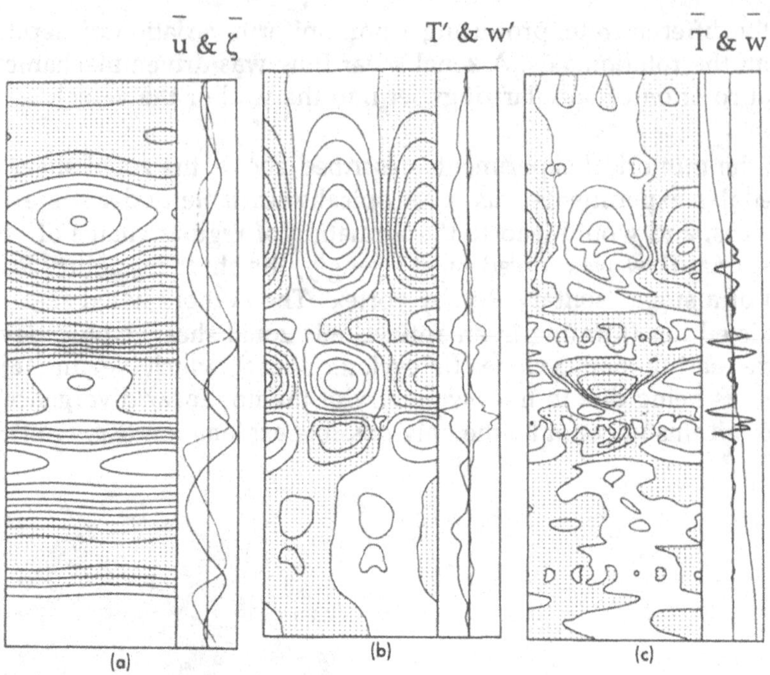

Figure 6. Basic fields and zonal mean profiles of a numerical solution to the 2-level QG equations on a β-plane [19], showing (a) upper level stream function, (b) eddy stream function and (c) eddy temperature.

horizontal temperature gradients near the tops of the clouds, but in the sense for the 'thermal wind' shear to oppose the jets observed in cloud-tracked wind data (i.e. so that the jets decay with height above the clouds). Evidently, the jets reach a maximum intensity around or just below the cloud tops, in an altitude range currently inaccessible to direct observation. The possible influence of baroclinic effects cannot be ruled out, therefore, and it is of interest to explore whether baroclinic instabilities could provide an alternative explanation for the formation of coherent oval eddies and their role in the atmospheric circulation.

Williams [19] studied a 2-level QG model under conditions notionally appropriate to the Jovian atmosphere, with a zonal flow driven by differential heating between equator and pole. His model demonstrated the formation of oval eddies and a set of parallel zonal jet streams, separated by a distance comparable both to $(U/\beta)^{1/2}$ and to the deformation radius L_R. Williams' models did not, however, exhibit the formation of a single compact eddy analogous to the GRS or GDS on Neptune. Other studies of the linear baroclinic instability of flows analogous to Jupiter suggested that the effect of the near-adiabatic convective interior would tend to suppress baroclinic instabilities, though the results were quite strongly model-dependent.

Read & Hide [20] studied a cylindrical laboratory system wherein a zonal flow which changed sign with radius was driven by a distribution of heating and cooling such that the horizontal temperature gradient changed sign about

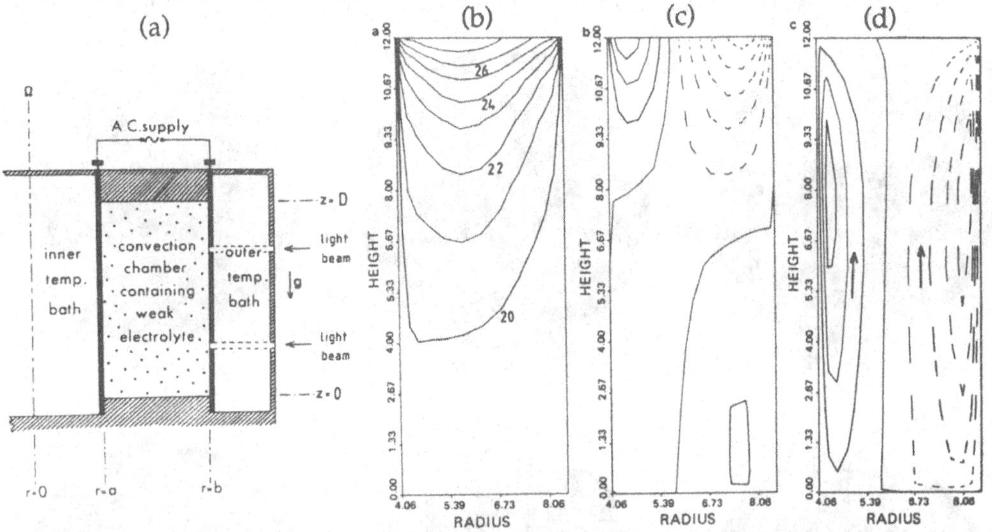

Figure 7. (a) Cross-section of apparatus used [4],[20] to produce a thermally-driven pair of reversing zonal jets; (c)-(f) (r,z) cross-sections of basic axisymmetric fields produced in a rotating annulus heated internally and cooled at both sidewalls. Fields shown are (b) temperature, (c) zonal velocity, and (d) meridional streamfunction.

the middle of an annular channel. Heating was applied electrically to the body of the fluid, and *both* sidewalls of the annulus were cooled. The resulting flow consisted of two pairs of oppositely-directed jets, antisymmetric about the mid-level. The thermally-driven meridional circulation carried heat upwards and towards the cooled sidewalls, establishing a statically stable thermal stratification (cf the wall-heated annulus in [21]).

Experiments showed that this basic axisymmetric flow would become baroclinically unstable under conditions similar to those for the formation of baroclinic waves in the conventional annulus. The form of the fully-developed waves in the internally heated system, however, took the form of chains of compact oval eddies circulating in the same sense as the shear of the zonal flow *at that level* - i.e. anticyclonically at upper levels, and cyclonically at lower levels. The system exhibited a range of regular and irregular wave regimes in a manner similar (though not identical to) the waves in the conventional annulus. Close to the 'upper symmetric transition', it was found that the m=1 state obtained with strong heating was highly asymmetric, forming a very compact stable eddy which was strongly anticyclonic at upper levels and 'shadowed' by a weaker cyclonic feature.

Further experiments showed that (as with some of the barotropic experiments) the baroclinic oval eddies would behave as domain-filling waves, yet individual eddies would trap fluid within them for long periods.

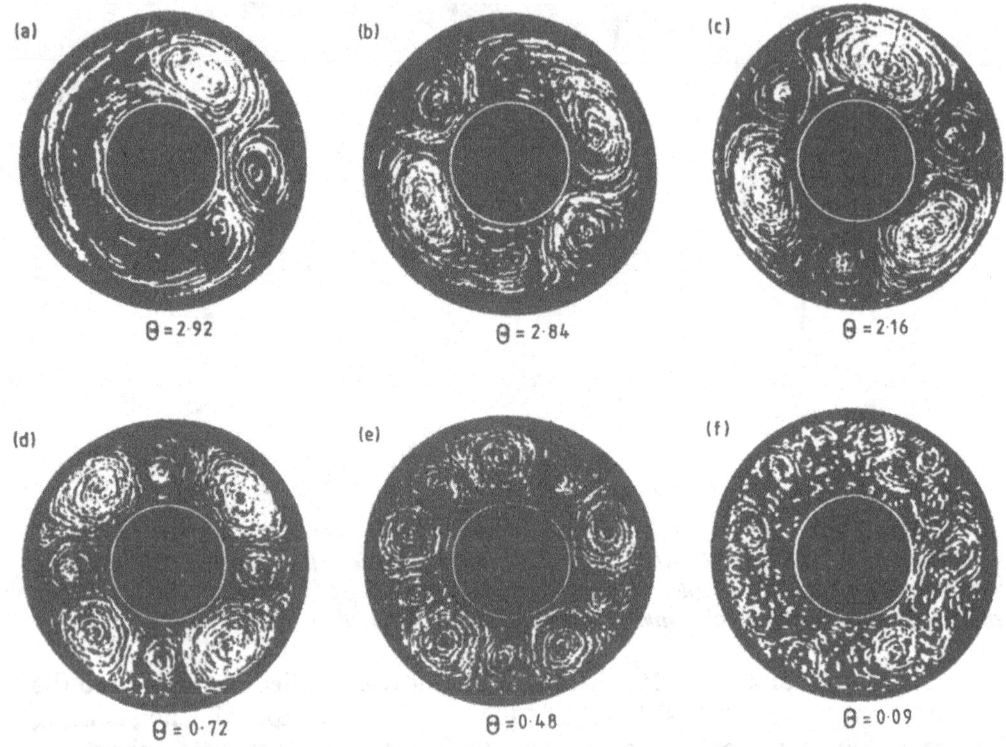

Figure 8. A series of upper level streak photographs of baroclinic eddy flows in an internally-heated rotating annulus with sidewall cooling [4], at a number of different values of thermal Rossby number Θ.

The onset of time-dependence took the form of near-periodic eddy merging and splitting events - as the pattern cyclically changed its dominant wavenumber - leading to mixing of fluid between the interior of each eddy and its surroundings.

The numerical experiments of Lewis [23] showed that these types of eddy merger events do not necessarily act to reinforce the dominant eddy, in contrast to the suggestion made regarding the barotropic eddies, but could act to *erode* the main eddy by entraining potential vorticity of opposite sign (note

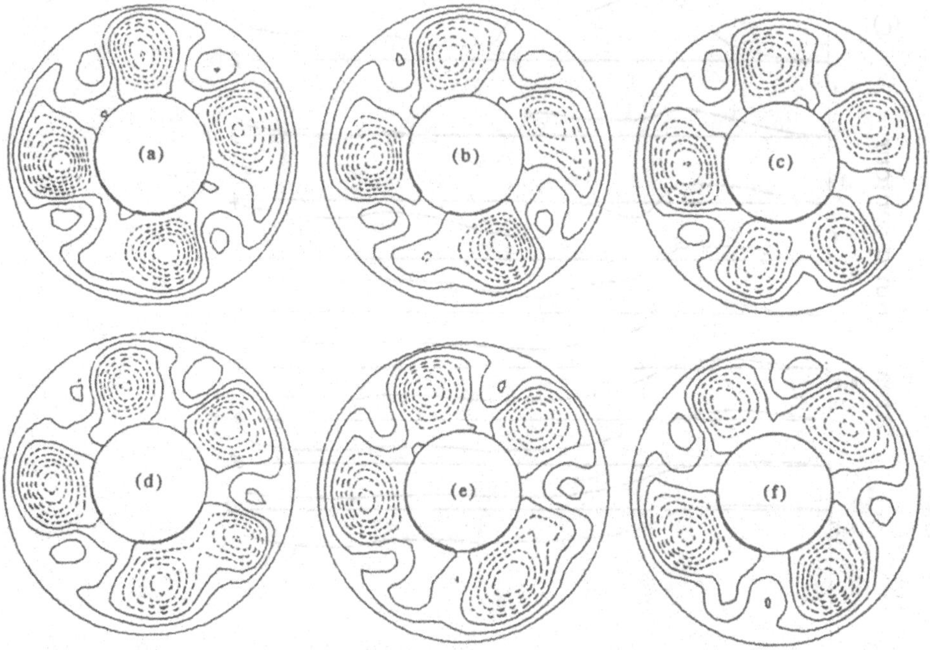

Figure 9. Sequence of horizontal streamfunction patterns obtained from laboratory velocity measurements [22] in an internally heated rotating annulus, during a cycle of a 'wavenumber vacillation'.

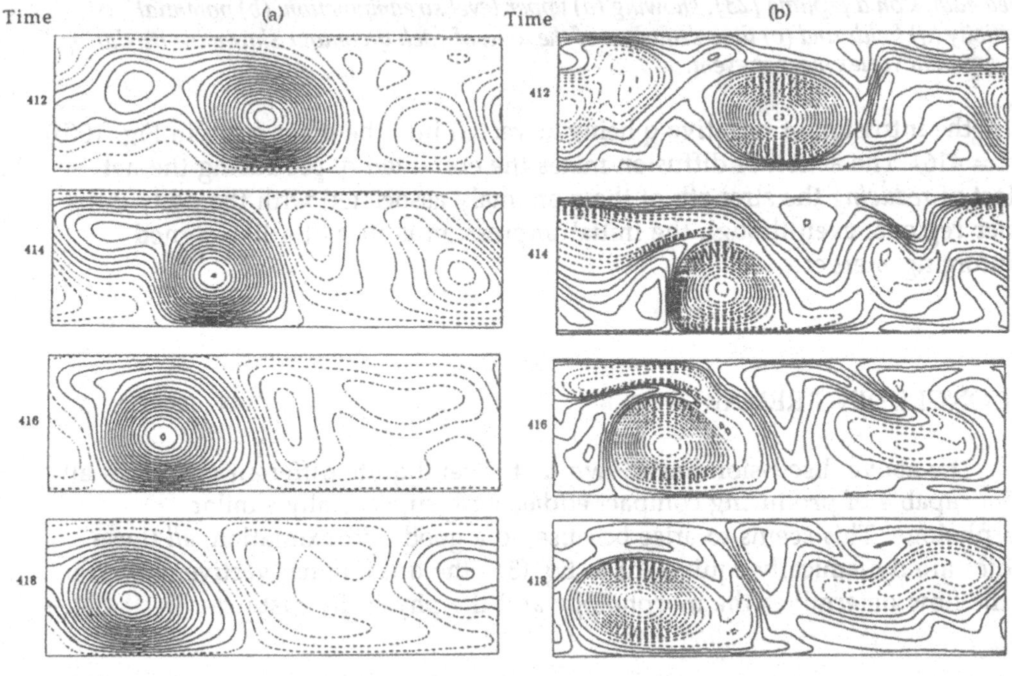

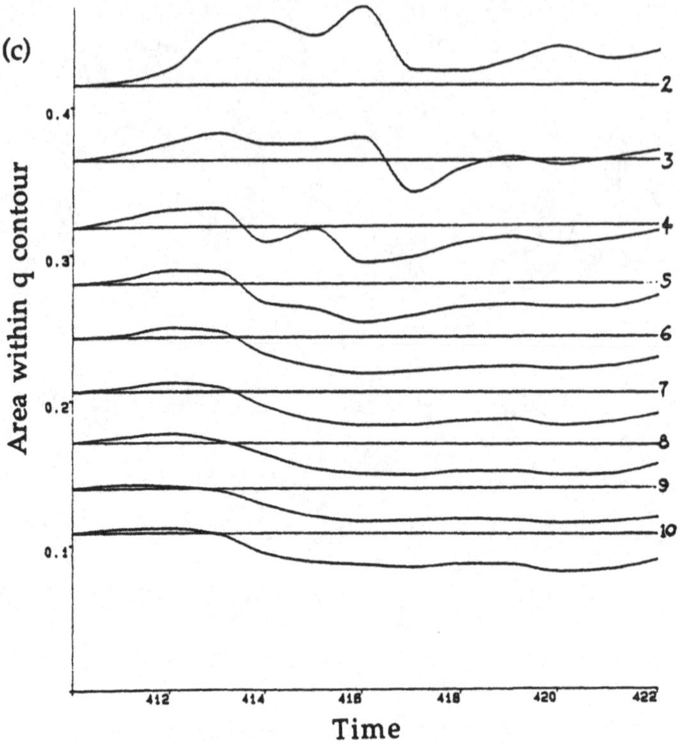

Figure 10. Sequence of flow patterns from a QG two-level numerical simulation of long-lived eddies on a β-plane [23], showing (a) upper level streamfunction, (b) potential vorticity (q) fields and (c) time-sequence of the area of each q-contour shown in (b), during an apparent vortex merger event.

the intrusion of negative q being advected into the main eddy in Fig. 10(b) at $t = 416$). The action of diffusion mixes the entrained q, producing the net effect of *reducing* the strength of the q anomaly associated with the eddy in its outer regions, even though the initial impression is that of a like-signed reinforcing merger.

6. CONCLUDING REMARKS

The above discussion has shown that several quite different processes all seem capable of producing compact eddies with superficially similar morphology. This seems to arise because, to a good approximation, *all* these eddies are dynamically equivalent to Eq. (5) - the quasi-nonadvective character of the flow determines the morphology and the physically distinct characteristics are manifest only as small departures from the zeroth order 'free mode'. Elucidation of these small departures on Jupiter etc. poses enormous challenges to observers making measurements of these planets. It may be a long time before the true nature of the GRS and GDS is revealed. In the meantime, it is to be hoped that our speculations will continue to stimulate investigations which may lead to important advances in our understanding of the basic dynamics of rotating fluids.

ACKNOWLEDGEMENTS

Much of this chapter originated as part of a series of graduate lectures given by the author at the University of Oxford, and I am grateful to many colleagues and students for discussions concerning the material herein. Particular thanks are due to Raymond Hide, for introducing me to the problems of Major Planet dynamics, and to Andy White, Barney Conrath and Peter Gierasch for sharing their insights into various aspects of these problems.

REFERENCES (Reviews denoted *)

1. Stephenson, D. J.: Interiors of the giant planets, *Ann. Rev. Earth Plan. Sci.*, **10** (1982), 257-295.

2. Gierasch, P. J. & Conrath, B. J.: Vertical temperature gradients on Uranus: implications for layered convection, *J. Geophys. Res.*, **92** (1987), 15019-15029.

3. Conrath, B. J. & Gierasch, P.: Global variation of the *para* hydrogen fraction in Jupiter's atmosphere and implications for dynamics on the outer planets, *Icarus*, **57** (1984), 184-204.

4. Read, P. L.: Stable, baroclinic eddies on Jupiter and Saturn: a laboratory analog and some observational tests, *Icarus*, **65** (1986), 304-334.*

5. Achterberg, R. K. & Ingersoll, A. P.: A normal-mode approach to Jovian atmospheric dynamics, *J. Atmos. Sci.*, **46** (1989), 2448-2462.

6. Ingersoll, A. P.: Atmospheric dynamics of the Outer Planets, *Science*, **248** (1990), 308-315.*

7. Polvani, L. M., Wisdom, J., DeJong, E. & Ingersoll, A. P.: Simple models of Neptune's Great Dark Spot, *Science*, **249** (1990), 1393-1398.

8. Hide, R.: Origin of Jupiter's Great red Spot, *Nature*, **190** (1961), 895-896.

9. Ingersoll, A. P.: Jupiter's Great Red Spot: a free atmospheric vortex?, *Science*, **312** (1973), 1346-1348.

10. Ingersoll, A. P. & Cuong,: Numerical model of long-lived Jovian vortices, *J. Atmos. Sci.*, **38** (1981), 2067-2076.

11. Marcus, P. S.: Numerical simulation of Jupiter's Great Red Spot, *Nature*, **331** (1988), 693-696.

12. Sommeria, J., Myers, S. D. & Swinney, H. L.: Laboratory simulation of Jupiter's Great Red Spot, *Nature*, **331** (1988), 689-693; Meyers, S. D., Sommeria, J. & Swinney, H. L.: Laboratory study of the dynamics of Jovian-type vortices, *Physica*, **37D** (1989), 515-530.

13. Niino, H. & Misawa, N.: An experimental and theoretical study of barotropic instability, *J. Atmos. Sci..*, **41** (1984), 1992-2011.

14. Read, P. L.: Dynamics and instabilities of Ekman and Stewarston boundary layers, in: this volume (1992).

15. Williams, G. P. & Yamagata, T., Geostrophic regimes, intermediate solitary vortices and Jovian eddies, *J., Atmos. Sci.*, **41** (1984), 453-478; Williams, G. P. & Wilson, R. J., The stability and genesis of Rossby vortices, *J. Atmos. Sci.*, **45** (1988), 207-241.

16. Antipov et al. 1981-85 - for review see Nezlin, M. V.: Rossby solitons (experimental investigations and laboratory model of natural vortices of the Jovian Great Red Spot type), *Sov. Phys. Usp.*, **29** (1986), 807-842.

17. Dowling, T. E. & Ingersoll, A. P.: Potential vorticity and layer thickness variations in the flow around Jupiter's Great Red Spot and White Oval BC, *J. Atmos. Sci.*, **45** (1988), 1380-1396; Jupiter's Great Red Spot as a shallow water system, *J. Atmos. Sci.*, **46** (1989), 3256-3278.

18. Read, P. L.:Soliton theory and Jupiter's Great Red Spot, *Nature*, **326** (1987), 337-338.

19. Williams, G. P.: Planetary circulations: 2. The Jovian quasi-geostrophic regime, *J. Atmos. Sci.*, **36** (1979), 932-968; Ultra-long baroclinic waves and Jupiter's Great Red Spot, *J. Met. Soc. Japan*, **57** (1979), 196-198.

20. Read, P. L. & Hide, R.: Long-lived eddies in the laboratory and in the atmospheres of Jupiter and Saturn, *Nature*, **302** (1983), 126-129; An isolated baroclinic eddy as a laboratory analogue of the Great Red Spot on Jupiter?, *Nature*, **308** (1984), 45-49.

21. Read, P. L.: Rotating annulus flows and baroclinic waves, in: this volume (1992).

22. Read, P. L.: Coherent baroclinic waves in a rotating, stably-stratified fluid and transitions to disordered flow, in: *Proceedings of IMA Conference "Waves & Turbulence in Stably-Stratified Flows" (ed. King, J. C. & Mobbs, S. D.).* Oxford University Press 1991. (in press).

23. Lewis, S. R.: Long-lived eddies in the atmosphere of Jupiter, *D. Phil. Thesis*, University of Oxford 1988.

24. Kuiper G. P.: Lunar and Planetary Laboratory studies of Jupiter - II., *Sky & Telescope*, **43** (1972), 75-81.

25. Maxworthy, T. & Redekopp, L. G.: A solitary wave theory of the Great Red Spot and other observed features in the Jovian atmosphere, *Icarus*, **29** (1976), 261-271.

Chapter V.4

VORTEX FLOW GENERATED BY A ROTATING DISC

M. Mory, A. Spohn
University J.F. and CNRS, Grenoble Cedex, France

Flows produced by a rotating disc are of considerable interest in many practical applications. In computers, for instance, the flow near the rotating hard disc influences heat convection and therefore the temperature of adjacent electronic components. Centrifugal separators, centrifugal pumps, disc viscometers and methods for the production of crystals used in computer memories are other examples that reveal the importance of rotating disc flows in industrial devices. The idealised configuration of a rotating disc of infinite radius is considered first. The exact solution of the Navier-Stokes equations obtained by von Karman [1] is used for the boundary layer to obtain some physical insight into the flow structure produced by rotating discs. In the second part, the discussion is enlarged to confined flow geometries.

1. MOTION PRODUCED BY AN INFINITE ROTATING DISC

11. Von Karman solution

Von Karman [1] considered the boundary layer near a disk of infinite radius rotating with constant angular velocity Ω inside a fluid of viscosity ν at rest far from the disc. Under the assumptions that the flow is steady and axisymmetric around the axis of rotation Oz, the flow is determined by solving the Navier-Stokes equations, which are simplified in the cylindrical coordinate system (O,r,θ,z):

$$u\frac{\partial u}{\partial r} + w\frac{\partial u}{\partial z} - \frac{v^2}{r} = -\frac{1}{\rho}\frac{\partial p}{\partial r} + \nu\left\{\frac{\partial}{\partial r}\left(\frac{1}{r}\frac{\partial (ru)}{\partial r}\right) + \frac{\partial^2 u}{\partial z^2}\right\} \tag{1}$$

$$\frac{u}{r}\frac{\partial (rv)}{\partial r} + w\frac{\partial v}{\partial z} = \nu\left\{\frac{\partial}{\partial r}\left(\frac{1}{r}\frac{\partial (rv)}{\partial r}\right) + \frac{\partial^2 v}{\partial z^2}\right\} \tag{2}$$

$$u\frac{\partial w}{\partial r} + w\frac{\partial w}{\partial z} = -\frac{1}{\rho}\frac{\partial p}{\partial z} + \nu\left\{\frac{1}{r}\frac{\partial}{\partial r}\left(r\frac{\partial w}{\partial r}\right) + \frac{\partial^2 w}{\partial z^2}\right\} . \tag{3}$$

The continuity equation is

$$\frac{1}{r}\frac{\partial (ru)}{\partial r} + \frac{\partial w}{\partial z} = 0 . \tag{4}$$

The origin O of the cylindrical coordinate system is taken to be the center of the rotating disc. The boundary conditions associated with the problem are

$$u = w = 0 \quad \text{and} \quad v = r\,\Omega \quad \text{for } z = 0 \quad ; \quad u = v = 0 \quad \text{for } z \to \infty . \tag{5}$$

Von Karman found an exact solution of equations (1-5) by introducing the dimensionless vertical coordinate $\zeta = z\,(\Omega/\nu)^{1/2}$. The velocity components u,v,and w were supposed to have the following similarity form

$$u = r\,\Omega\, F\left(z\sqrt{\frac{\Omega}{\nu}}\right) \quad ; \quad v = r\,\Omega\, G\left(z\sqrt{\frac{\Omega}{\nu}}\right) \; ; \quad w = \sqrt{\nu\,\Omega}\; H\left(z\sqrt{\frac{\Omega}{\nu}}\right) . \tag{6}$$

Introducing these particular functions in (1) to (4) leads to the following set of dimensionless ordinary differential equations depending on the single variable ζ, from which pressure has been eliminated:

$$\begin{aligned} F^2 - G^2 + H F' - F'' &= 0 \\ 2GF + H G' - G'' &= 0 \\ 2F + H' &= 0 \end{aligned} \tag{8}$$

The boundary conditions are now

$$F(0) = H(0) = 0 \text{ and } G(0) = 1 \quad ; \quad F(\infty) = G(\infty) = 0 . \tag{9}$$

The approximate analytical solution of (8) found by von Karman [1] was later improved by Cochran [2] by means of a numerical integration. Figure 1 shows the dimensionless velocity profiles and a schematic representation of the streamlines. Viscous forces drag the fluid adjacent to the disc and throw it radially outwards under the action of centrifugal forces. The centrifuged fluid is replaced by new fluid, which is axially sucked into the boundary layer. The rotating disc is thus acting as a centrifugal pump. One of the most striking properties of the flow is that the boundary layer thickness $(\nu/\Omega)^{1/2}$ and the velocity component perpendicular to the disc ($w = -0.886\,(\nu\Omega)^{1/2}$ far from the disc) are independent of the radial position r. The flow described by this solution is laminar. Experiments indicate that it remains laminar when the radial coordinate r is such that the local Reynolds number $Re_r = \Omega r^2/\nu$ is less than $1.9\ 10^5$.

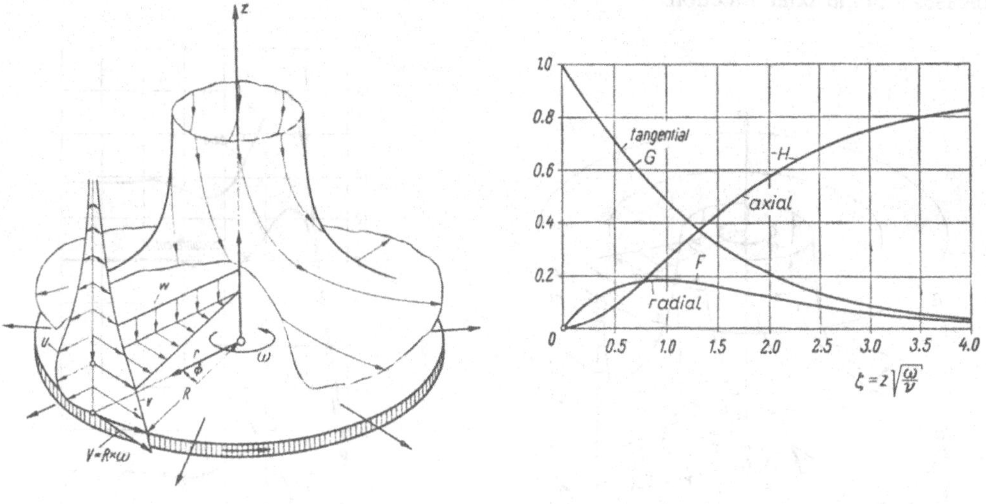

a b

Figure 1: Flow produced by a rotating disc of infinite radius.
a). Schematic representation of the streamlines.
b). Profiles of dimensionless velocity components according to (6).
(Figures reproduced from Schlichting [3]).

12. Rotating flow over a fixed plate

The boundary layer on a fixed plate when the flow far from the plate is in solid body rotation with constant angular velocity Ω around an axis Oz perpendicular to the plate is very similar to the rotating disc case discussed previously. Bödewadt [4] found a solution in the same way as Von Karman. Dimensionless similarity functions (eq. 6) are also used to reduce the steady axisymmetric Navier-Stokes equations to a system of ordinary differential equations, which are solved numerically. We will not present this solution in detail. The reader may refer to Schlichting [3]. Figure 2 shows the profiles of the dimensionless velocity components and a three-dimensional representation of the streamlines. As in the case of the rotating disc, the velocity component w and the boundary layer thickness are independent of the radial position r. While the boundary layer above the rotating disc sucks the fluid in axially and centrifuges it out radially, the boundary layer over the fixed plate transports the fluid radially towards the axis and releases it in the axial direction.

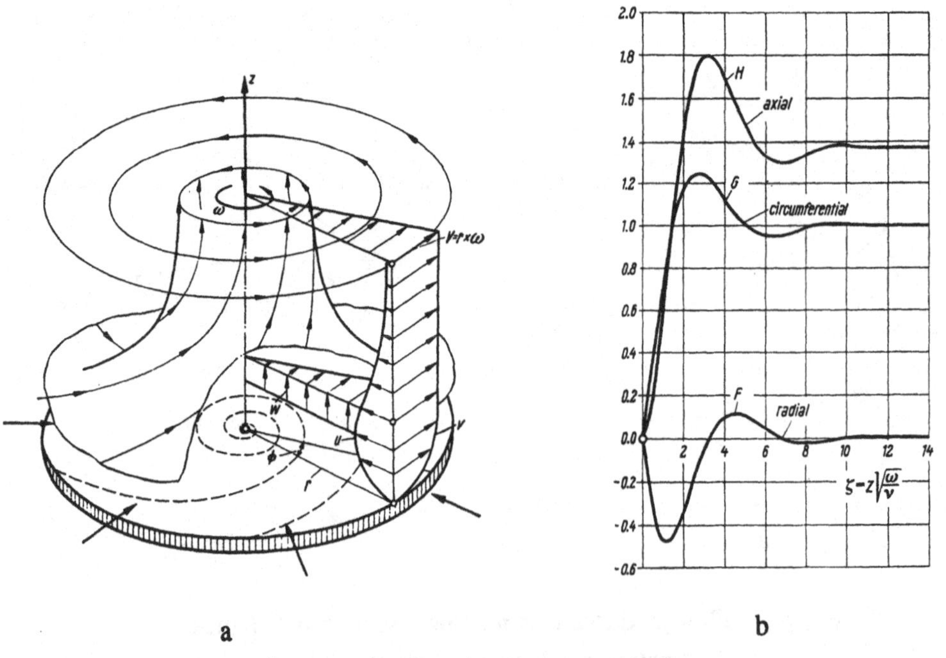

a b

Figure 2: Boundary layer flow between a fixed plate and a fluid in solid body rotation.
a). Schematic representation of the streamlines.
b). Profiles of dimensionless velocity components according to (6).
(Figures reproduced from Schlichting [3]).

13. Combinations of rotating disc flows

Flows studied by Von Karman and Bödewadt are particular cases of the flow between two infinite discs, separated by a finite distance H, which rotate around the same axis Oz, with angular velocities Ω_1 and Ω_2 respectively. Configurations of this type are governed by the Reynolds number $Re=\Omega_1 H^2/\nu$ and the ratio parameter of the angular velocities of the two discs $s=\Omega_1/\Omega_2$.

In the simplest case, one disc is rotating with angular velocity Ω, while the other is at rest. The flow structure can be physically interpreted with the help of the von Karman and Bödewadt solutions presented above. The fluid is pumped axially and centrifuged radially by the boundary layer on the rotating disc, while the fluid adjacent to the fixed disc converges radially inwards and is reinjected axially from the boundary layer. Since the velocity component w is independent of the radial position r, the balance of the flow rates pumped by the rotating disc and injected by the boundary layer on the fixed plate is obtained when the fluid between the two discs is in solid body rotation outside the boundary layers. Batchelor [5] and Grohne [6] have shown that the value of the angular velocity of the fluid between the two discs is 0.3 Ω. When the two discs are corotating in opposite directions but at the same speed, Batchelor [5] conjectured that two cells appear in the main body of the fluid between the two discs. The fluid is counter-rotating in the two cells, which are separated by a transition layer.

This paper will confine itself to the basic features given by the von Karman, Bödewadt and Batchelor solutions, which are essential to understand the processes of pumping and injection caused by rotating and fixed discs, respectively. Nevertheless, determination of the flow between two infinite discs has in the recent years been the subject of intense work. The difficulties encountered have been mostly theoretical and mathematical, because the questions of the existence and non-uniqueness of the solutions have been posed. For further information on this subject, the very complete review by Zandbergen and Dijkstra [7] is recommended.

2. FLOW PRODUCED BY A ROTATING DISC IN CONFINED GEOMETRY

The confined flow domains found in practice often lead to recirculation of the fluid. This means that confinement effects must be taken into account. Figure 3 shows the schematic of an experiment which produces such a recirculating flow. The experiment is conducted inside a circular cylindrical tank. A disc rotates with angular velocity Ω at the bottom of the tank. The upper boundary is either a rigid cover or a free surface.

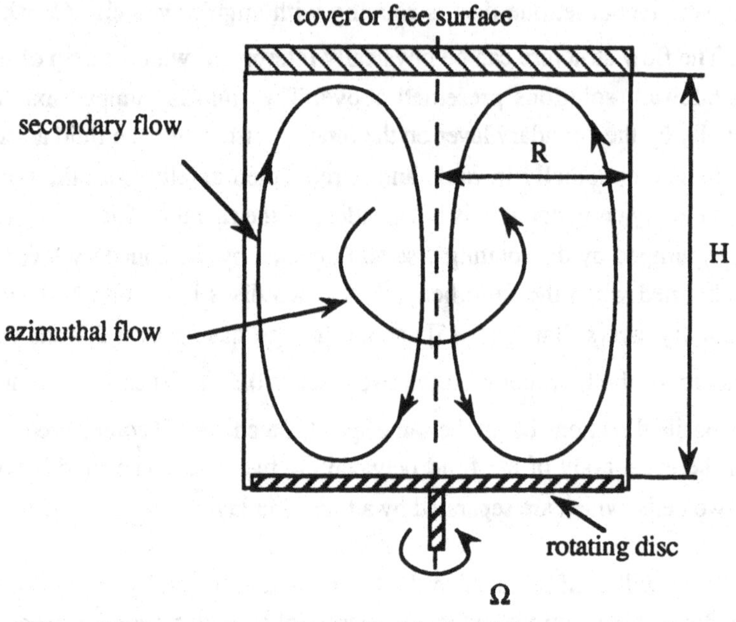

Figure 3: Flow produced in a cylindrical tank by a rotating disc. Schematic representation of an experiment (Spohn, [8]).

The flow is governed by two dimensionless parameters [1]:

$$\text{the Reynolds number} \qquad Re = \frac{\Omega R^2}{\nu} \tag{10}$$

$$\text{the aspect ratio} \qquad \delta = \frac{H}{R} \, . \tag{11}$$

(1). When the upper boundary is a free surface, the flow structure also depends on the chemical composition at the free surface (see Spohn [8]).

The structure of the flow depends very much on the interactions between the different boundary layers. The finite radial extent modifies the flow, but the main features predicted by the models described above are again observed. Above the rotating disc, the boundary layer is again acting as a pump centrifuging the fluid toward the lateral boundary. When the upper boundary is a rigid cover, a boundary layer similar to the one described by Bödewadt is produced on this cover. It makes the fluid converge radially toward the axis and reinjects this fluid inside the central part of the tank. The structure of the boundary layer on the lateral cylindrical wall is less well understood (see Tomlan & Hudson [9]). This boundary layer is rather like a spiralling wall jet, which carries the fluid centrifuged toward the lateral cylinder and feeds the boundary layer on the upper boundary. When the upper boundary is a free surface, the fluid is also reinjected by the upper boundary layer, but the recycling of flow does not concern a central core contained in the fluid. Actually the boundary layer is no longer of the kind described by Bödewadt.

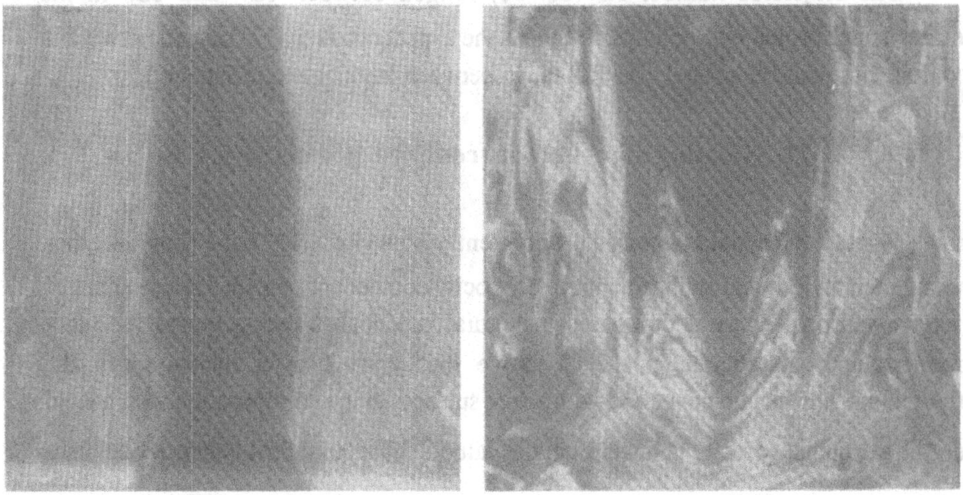

rotating disc — rotating disc

a. Top boundary is a rigid cover — b. Top boundary is a free surface

Figure 4: Visualisations of the transport of fluorescein dye injected at the border of the rotating disc when the flow is in steady state. Photographs are taken a few rotation periods after the dye has been injected (Spohn, [8]; Re = 10^4, H/R = 2).

Figure 4 compares the secondary motions observed in a vertical sheet of light passing through the rotation axis of the disc (4a: configuration with rigid cover; 4b: configuration with a free surface). The pictures were taken a few rotation periods after the fluorescein dye was injected through a hole in the cylinder wall near the border of the rotating disc. In both cases the boundary layer on the lateral cylinder carries the fluid up to the upper boundary. In the configuration with a rigid cover at the top (4a), the boundary layer redistributes and reinjects the flow over the whole surface of the rigid cover. The whole section is coloured with dye after a duration of the order of $H/R\ Re^{1/2}$ rotation periods of the disc, due to advection. In the configuration with a free surface (4b), secondary motions are confined to an annulus bounded by an inner cylinder with a radius about half that of the rotating disc. The time taken for the whole section to be coloured with dye is much longer, depending on the diffusion time scale.

The Reynolds number Re (eq. 10) ranged between 10^3 and 10^5 in the experiments performed by Spohn [8], whereas the aspect ratio (eq. 11) ranged between 1 and 4. It is thus possible to study the full range between laminar and turbulent flows.

21. Rotating disc flow in confined geometry for Re >> 1

When the Reynolds number is sufficiently high ($Re > 10^4$), the flow becomes more and more cylindrical and the azimuthal velocity component becomes independent of the axial position. Figure 5 compares the radial profiles of the azimuthal velocity component at a distance $z/R = 1.45$ above the rotating disc, measured in the configurations with a rigid cover and with a free surface. In the configuration with a rigid cover, the azimuthal velocity is lower than the value $0.3\ \Omega\ r$ obtained between two discs of infinite radius (§ 13). The angular velocity of the flow reaches the value $0{,}3\ \Omega$ only for $r < 0.1\ R$. In the configuration with a free surface, the flow is in solid-body rotation in the central part of the tank ($r < 0.1\ R$) with the angular velocity of the rotating disc. For this reason, the pumping of fluid by the rotating disc vanishes in the central core of the flow, tallying with the observation of figure 4b. The radial profile of the azimuthal velocity component, however, depends on the aspect ratio; increasing the aspect ratio decreases the angular momentum, an effect which is accentuated with increasing values of the radial coordinate. The axial velocity component w has not been measured; the Von Karman model (eq. 6) gives the scaling $w = O(\ (\nu\Omega)^{1/2}\ )$.

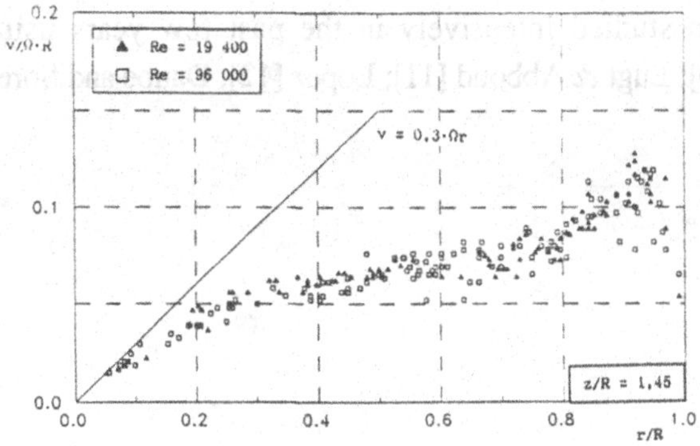

a. Top boundary is a rigid cover

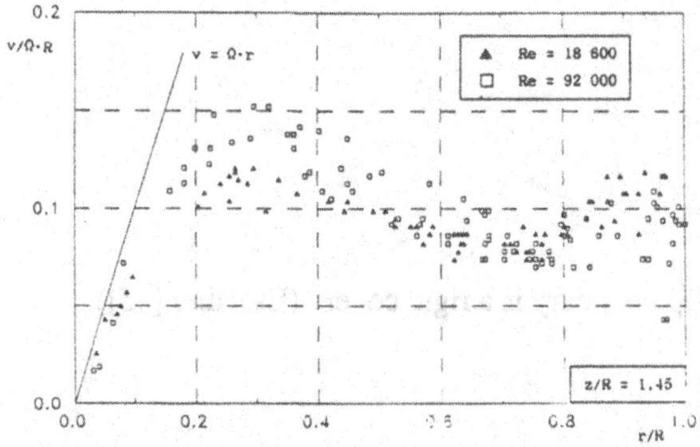

b. Top boundary is a free surface

Figure 5: Radial profiles of the azimuthal velocity component with a large Reynolds number. Aspect ratio is H/R = 2.9. (Spohn, [8]).

22. Vortex breakdown phenomena (Re=O(10^3))

The occurrence of vortex breakdown phenomena is a typical property of the experiment shown in Figure 3. These events appear for low Reynolds numbers and high swirl numbers as compared to the Reynolds numbers and swirl numbers for which breakdowns usually occur in aeronautics and for which they have been studied before. This vortex breakdown occurs in laminar conditions. Flow visualisations are therefore a particularly interesting way of gaining insight into the kinematics of the flow. A further interest of the present configuration is that the geometry is closed so that the occurrence of vortex breakdown is not linked to particular inflow conditions. For this reason, this configuration has been studied intensively in the past few years using numerical simulations (Neitzel [10]; Lugt & Abboud [11]; Lopez [12]; Daube and Sorensen[13]).

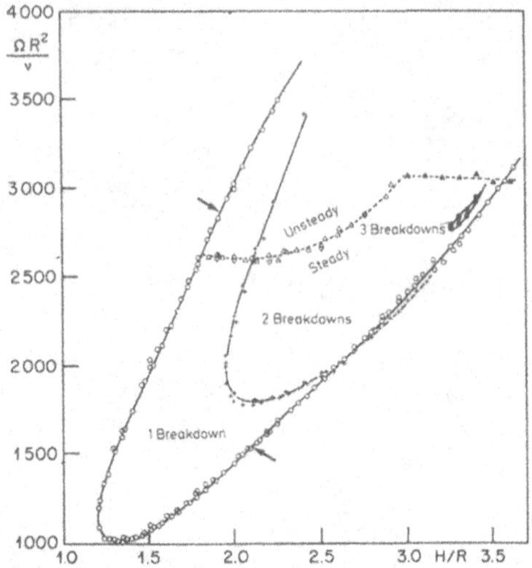

a. Top boundary is a rigid cover (Escudier [15])

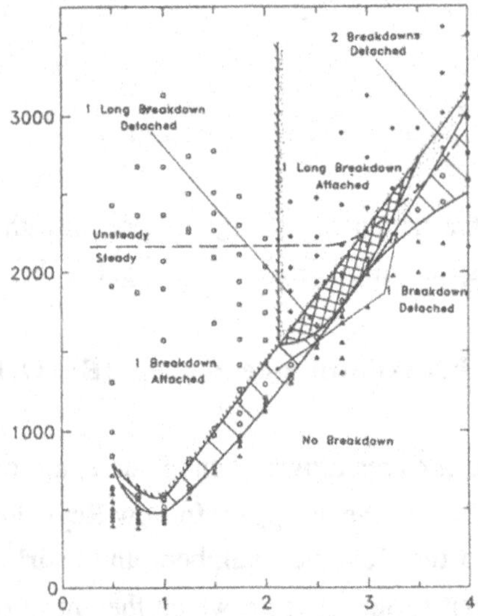

b. Top boundary is a free surface (Spohn [8])

<u>Figure 6:</u> Experimental conditions for the occurrence of vortex breakdown in disc flow with confined geometry.

The occurrence of vortex breakdown phenomena in the configuration with a rigid cover was first noticed by Vogel [14]. More recently, Escudier [15] determined experimentally the conditions under which breakdown occurs in the configuration with rigid cover. The results are reported by plotting a map in the (Re, δ=H/R) coordinate system, on which several domains are identified corresponding to the various regimes observed (no breakdown, one breakdown, two breakdowns, unsteadiness...). This map is shown in Figure 6a. A similar map obtained by Spohn in the configuration with a free surface is shown in Figure 6b. Compared to the configuration with a rigid cover, the domain of occurrence of vortex breakdown is enlarged in the configuration with a free surface. It is observed up to the highest Reynolds numbers and down to the smallest aspect ratios considered. The locations of the recirculating bubbles are also very different. Figure 7, which shows visualisations in a vertical sheet of light, illustrates these differences. In the configuration with a rigid lid (Fig. 7a), the bubble is always detached from the boundary. In the free surface configuration, the recirculation zone is in general an elongated eddy attached to the free surface.

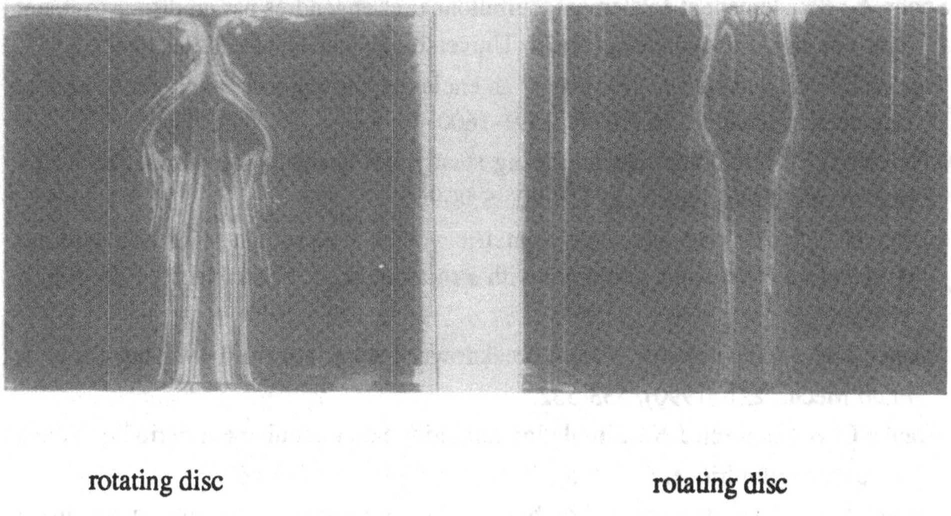

a. Top boundary is a rigid cover   b. Top boundary is a free surface

Figure 7: Visualisation of vortex breakdown bubbles in disc flow with confined geometry [8]. Reynolds number: Re=1850. Aspect ratio: H/R = 1.75.

REFERENCES

1. Von Karman T.: Über laminare and turbulente Reibung, ZAMM, 1 (1921), 233-252.
2. Cochran W.G.: The flow due to a rotating disk, Proc. Cambridge Philos. Soc., 30 (1934),365-375.
3. Schlichting H.: Boundary layer theory, McGraw Hill, 1968.
4. Bödewadt U.T.: Die Drehströmung über festem Grunde, ZAMM, 20 (1940), 241-253.
5. Batchelor G.K.: Note on a class of solutions of the Navier-Stokes equations representing steady rotationally-symmetric flow, Q. J. Mech. Appl. Math., 4 (1951), 29-41.
6. Grohne D.: Über die laminare Strömung in einer kreiszylindrischen Dose mit rotierendem Deckel, Nachr. Akad. Wiss. Göttingen Math.-Phys. Klasse, 12 (1955), 263-282.
7. Zandbergen P.J. and Dijkstra D.: Von Karman swirling flows, Ann. Rev. Fluid Mech., 19 (1987), 465-491.
8. Spohn A.: Ecoulement et éclatement tourbillonnaires engendrés par un disque tournant dans une enceinte cylindrique,Thèse Université J. Fourier, Grenoble, 1991.
9. Tomlan P.F. & Hudson J.L.: Flow near an enclosed rotating disc: analysis, Chemical Engineering Science, 26 (1971), 1591-1600.
10. Neitzel G.P.: Streakline motion during steady and unsteady axisymmetric vortex breakdown, Phys. Fluids, 31 (1988), 958-960.
11. Lugt H.J. & Abboud M.: Axisymmetric vortex breakdown with and without temperature effects in a container with a rotating lid, J. Fluid Mech., 179 (1987), 179-200.
12. Lopez J.M.: Axisymmetric vortex breakdown. Part 1: confined swirling flow, J. Fluid Mech., 221 (1990), 533-552.
13. Daube O. & Sorensen J.N.: Simulation numérique de l'écoulement périodique dans une cavité cylindrique, C.R.A.S., 308 (1989), série II, 463-469.
14. Vogel H.U.: Experimentelle Ergebnisse über die laminare Strömung in einem zylindrischen Gehäuse mit darin rotierender Scheibe, MPI Bericht 6, 1968.
15. Escudier M.P.: Observations of the flow produced in a cylindrical container by a rotating endwall, Exp. Fluids, 2 (1984), 189-196.

PART VI

TURBULENCE

Chapter VI.1

ROTATION EFFECTS ON TURBULENCE

M. Mory
University J.F. and CNRS, Grenoble Cedex, France

We consider the general case of a turbulent flow in a rotating coordinate system. Rotation is known to produce anisotropy in a turbulent flow, revealed by an increase of the turbulent lengthscales in the direction parallel to the rotation axis as compared to those in a plane perpendicular to it. This mechanism, which relies on the dimensional analysis presented earlier in this volume by Hopfinger [1], leads to the evolution of turbulence towards a two-dimensional state. Before this process is described in section 2 of the present paper, the effect of rotation on second order correlations (Reynolds stresses tensor) is considered in section 1.

With respect to geophysical fluid dynamics problems, the subject of turbulence submitted to rotation is also considered in this volume by Maxworthy [2] and Hopfinger [3]. For this reason the focus here is mainly on rotating flows with large aspect ratios, namely when $D/L \gg Ro^{-1}$. Experimental results concerning such flows are discussed at length in section 3 of this paper. Finally section 4 mentions some consequences for the numerical modelling of rotating turbulence.

1. THE ONE POINT CLOSURE APPROACH TO TURBULENCE IN ROTATING FLUIDS

11. The Reynolds equations

As is usual for turbulence, a distinction is made between mean flow (whose characteristic pressure and velocities are written in capital letters) and turbulent flow (written in small letters). Ensemble averaging of the turbulent quantities, denoted $\overline{}$, gives the useful property

$$\overline{u} = \overline{v} = \overline{w} = \overline{p} = 0 \ .$$

However, turbulent second or third order correlations, such as for instance $\overline{u^2}$, $\overline{uv}$, $\overline{pw}$ or $\overline{uvw}$, may not vanish. This is obviously true for the square of turbulent velocity components, which estimate the kinetic energy of turbulence. In shear flows and near walls, other second order or third order correlations may also be non-zero.

The Reynolds equations, which are the averaged Navier-Stokes equations, describe mean flow evolution. These equations are written here in a coordinate system rotating with angular velocity Ω around the Oz axis. The Reynolds equations are similar to the usual Reynolds equations established in a non-rotating coordinate system as far as turbulence terms are considered. The only difference concerns the Coriolis force associated with the mean flow.

$$\frac{DU}{Dt} - 2\Omega V = -\frac{1}{\rho}\frac{\partial P}{\partial x} + \nu \Delta U - \frac{\partial \overline{u^2}}{\partial x} - \frac{\partial \overline{uv}}{\partial y} - \frac{\partial \overline{uw}}{\partial z} \tag{1}$$

$$\frac{DV}{Dt} + 2\Omega U = -\frac{1}{\rho}\frac{\partial P}{\partial y} + \nu \Delta V - \frac{\partial \overline{uv}}{\partial x} - \frac{\partial \overline{v^2}}{\partial y} - \frac{\partial \overline{vw}}{\partial z} \tag{2}$$

$$\frac{DW}{Dt} = -\frac{1}{\rho}\frac{\partial P}{\partial z} + \nu \Delta W - \frac{\partial \overline{uw}}{\partial x} - \frac{\partial \overline{vw}}{\partial y} - \frac{\partial \overline{w^2}}{\partial z} \tag{3}$$

The Reynolds equations are the simplest ones that account for the effect of turbulence on mean flow. Obviously these equations are of no use in the absence of mean flow or when the mean flow is unidirectional and uniform. Nevertheless, an effect of rotation on second order correlations may indirectly modify in its turn the mean flow structure.

21. Evolution equations of second order correlations in a rotating coordinate system

The evolution equations for the six terms of the Reynolds stress tensor,

$$\begin{pmatrix} \overline{u^2} & \overline{uv} & \overline{uw} \\ \overline{uv} & \overline{v^2} & \overline{vw} \\ \overline{uw} & \overline{vw} & \overline{w^2} \end{pmatrix}$$

are rather tedious to obtain. The method is described by Tennekes and Lumley ([4], p. 63). For the second order correlations $\overline{u^2}$, $\overline{v^2}$ and $\overline{w^2}$ corresponding to the kinetic energy of the various turbulent velocity components, one obtains:

$$\frac{1}{2}\frac{D\overline{u^2}}{Dt} - 2\Omega\,\overline{uv} = -\overline{u^2}\frac{\partial U}{\partial x} - \overline{uv}\frac{\partial U}{\partial y} - \overline{uw}\frac{\partial U}{\partial z} - \frac{1}{2}\left(\frac{\partial \overline{u^3}}{\partial x} + \frac{\partial \overline{u^2 v}}{\partial y} + \frac{\partial \overline{u^2 w}}{\partial z}\right) - \frac{1}{\rho}\overline{u\frac{\partial p}{\partial x}} + \nu\,\overline{u.\Delta u} \tag{4}$$

$$\frac{1}{2}\frac{D\overline{v^2}}{Dt} + 2\Omega\,\overline{uv} = -\overline{uv}\frac{\partial V}{\partial x} - \overline{v^2}\frac{\partial V}{\partial y} - \overline{vw}\frac{\partial V}{\partial z} - \frac{1}{2}\left(\frac{\partial \overline{v^2 u}}{\partial x} + \frac{\partial \overline{v^3}}{\partial y} + \frac{\partial \overline{v^2 w}}{\partial z}\right) - \frac{1}{\rho}\overline{v\frac{\partial p}{\partial y}} + \nu\,\overline{v.\Delta v} \tag{5}$$

$$\frac{1}{2}\frac{D\overline{w^2}}{Dt} = -\overline{uw}\frac{\partial W}{\partial x} - \overline{vw}\frac{\partial W}{\partial y} - \overline{w^2}\frac{\partial W}{\partial z} - \frac{1}{2}\left(\frac{\partial \overline{w^2 u}}{\partial x} + \frac{\partial \overline{w^2 v}}{\partial y} + \frac{\partial \overline{w^3}}{\partial z}\right) - \frac{1}{\rho}\overline{w\frac{\partial p}{\partial z}} + \nu\,\overline{w.\Delta w} \tag{6}$$

The linear effect of rotation, due to the Coriolis terms, needs be distinguished from the non-linear effects, namely third order correlations, second order pressure-velocity correlations and viscous terms. The Coriolis terms redistribute kinetic energy between the two components of velocity perpendicular to the rotation axis when the second order correlation $\overline{uv}$ is non-zero. This is not the case when flow is not submitted to a mean shear, and in particular far from a wall in grid turbulence experiments (see § 3). Summing equations (4) and (5) shows that the Coriolis force does not modify the total amount of turbulent kinetic energy. Modelling the effect of rotation on turbulence therefore involves describing the effect of rotation on the third order correlation, pressure-velocity correlation and, in particular, on viscous dissipation.

The evolution equations for the three other second order correlations $\overline{uv}$, $\overline{uw}$ and $\overline{vw}$ are:

$$\frac{D\overline{uv}}{Dt} + 2\Omega\left(\overline{u^2} - \overline{v^2}\right) = \left(- \overline{vu}\frac{\partial U}{\partial x} - \overline{v^2}\frac{\partial U}{\partial y} - \overline{vw}\frac{\partial U}{\partial z} - \overline{u^2}\frac{\partial V}{\partial x} - \overline{uv}\frac{\partial V}{\partial y} - \overline{uw}\frac{\partial V}{\partial z}\right)$$
$$- \left(\frac{\partial \overline{u^2 v}}{\partial x} + \frac{\partial \overline{v^2 u}}{\partial y} + \frac{\partial \overline{uvw}}{\partial z}\right) - \frac{1}{\rho}\left(\overline{u\frac{\partial p}{\partial y}} + \overline{v\frac{\partial p}{\partial x}}\right) + \nu\left(\overline{u.\Delta v + v.\Delta u}\right) \tag{7}$$

$$\frac{D\overline{uw}}{Dt} - 2\Omega\,\overline{vw} = \left(- \overline{wu}\frac{\partial U}{\partial x} - \overline{wv}\frac{\partial U}{\partial y} - \overline{w^2}\frac{\partial U}{\partial z} - \overline{u^2}\frac{\partial W}{\partial x} - \overline{uv}\frac{\partial W}{\partial y} - \overline{uw}\frac{\partial W}{\partial z}\right)$$
$$- \left(\frac{\partial \overline{u^2 w}}{\partial x} + \frac{\partial \overline{uvw}}{\partial y} + \frac{\partial \overline{uw^2}}{\partial z}\right) - \frac{1}{\rho}\left(\overline{u\frac{\partial p}{\partial z}} + \overline{w\frac{\partial p}{\partial x}}\right) + \nu\left(\overline{u.\Delta w + w.\Delta u}\right) \tag{8}$$

$$\frac{D\overline{vw}}{Dt} + 2\Omega\,\overline{uw} = \left(- \overline{wu}\frac{\partial V}{\partial x} - \overline{wv}\frac{\partial V}{\partial y} - \overline{w^2}\frac{\partial V}{\partial z} - \overline{uv}\frac{\partial W}{\partial x} - \overline{v^2}\frac{\partial W}{\partial y} - \overline{vw}\frac{\partial W}{\partial z}\right)$$
$$- \left(\frac{\partial \overline{uvw}}{\partial x} + \frac{\partial \overline{v^2 w}}{\partial y} + \frac{\partial \overline{vw^2}}{\partial z}\right) - \frac{1}{\rho}\left(\overline{v\frac{\partial p}{\partial z}} + \overline{w\frac{\partial p}{\partial y}}\right) + \nu\left(\overline{v.\Delta w + w.\Delta v}\right) \tag{9}$$

Several conclusions are drawn from the set of equations (4) to (9).

When linear Coriolis terms are non-zero, a first effect of rotation on turbulence occurs; this can be understood as a linear wave propagation of the perturbations, in a way similar to the propagation of inertial waves (Mory [5]). Nevertheless, the conditions under which linear Coriolis terms are non-zero are very restrictive, because the linear terms vanish when

$$\overline{uv} = \overline{uw} = \overline{vw} = 0 \quad \text{and} \quad \overline{u^2} = \overline{v^2} \ , \tag{10}$$

and, in particular, when the Reynolds stress tensor is spherical:

$$\overline{uv} = \overline{uw} = \overline{vw} = 0 \quad \text{and} \quad \overline{u^2} = \overline{v^2} = \overline{w^2} \ . \tag{11}$$

Such conditions are usually met in non-rotating homogeneous and isotropic turbulence. In practice, turbulence is often isotropic or very nearly. In rotating flows, numerical simulations by Bardina, Ferziger and Rogallo [6] have shown that, for an initially isotropic turbulence, the Reynolds tensor of turbulence remains spherical. Other numerical studies, starting from initial anisotropic conditions (Itsweire and al [7]; Roy

[8]; Cambon & Jacquin [9]) checked the strong effects of linear Coriolis terms. To summarise, the sphericity or non-sphericity of turbulence depends closely on the initial state of the Reynolds tensor.

Modelling the effect of rotation on turbulence requires far more sophisticated tools than the second order correlations can provide. We may also anticipate that the anisotropy of turbulence will appear through the anisotropy of the turbulent lengthscales parallel and perpendicular to the axis of rotation. This will have significant consequences on the dissipation rate of the turbulent kinetic energy, because the usual scaling of the dissipation rate $\varepsilon = A\,\overline{u^2}^{\,3/2}/l$ will no longer hold.

2. THE 2-DIMENSIONALISATION OF TURBULENCE BY ROTATION

21. Dimensional analysis

In this volume, Hopfinger [1] presented the equation of motion in a coordinate system rotating with angular velocity Ω around the Oz axis. Introducing the lengthscales L and D and velocity scales U and W to denote the orders of magnitude corresponding respectively to the directions perpendicular and parallel to the rotation axis, the equations of motion depend on the following dimensionless parameters:

the Rossby number $$R_o = \frac{U}{2\Omega L}\ , \tag{12}$$

the Ekman number $$E = \frac{\nu}{2\Omega D^2}\ , \tag{13}$$

the aspect ratio $$\delta = D/L\ , \tag{14}$$

and the reduced frequency $$\gamma = \frac{1}{2\Omega T}\ . \tag{15}$$

based on a typical timescale T. The dimensionless Navier-Stokes equations in the rotating coordinate system for the velocity components perpendicular to the rotation axis are:

$$\gamma \frac{\partial \vec{u}}{\partial t} + R_o\,(\vec{u}.\nabla)\vec{u} + \vec{k}\times\vec{u} = -\frac{P}{\rho 2\Omega UL}\nabla_H p + E\,\delta^2\,\nabla_H^2\vec{u} + E\,\frac{\partial^2\vec{u}}{\partial z^2}\ , \tag{16}$$

whereas the equivalent equation for the velocity component parallel to the rotation axis is:

$$\gamma \frac{\partial w}{\partial t} + R_o (\vec{u}.\nabla) w = - \frac{P}{\rho 2\Omega W D} \frac{\partial p}{\partial z} + E \delta^2 \nabla_H^2 w + E \frac{\partial^2 w}{\partial z^2} . \quad (17)$$

The continuity equation is finally:

$$\nabla_H . \vec{u} + \frac{WL}{UD} \frac{\partial w}{\partial z} = 0 . \quad (18)$$

This dimensional analysis still applies for turbulent flows, provided the scales introduced in the dimensionless numbers are the turbulent velocity scales and turbulent lengthscales such as integral lengthscales. The effect of viscosity is negligible when the Ekman number is small ($E << 1$), except in the boundary layers. Flow is dominated by the effect of rotation of the rotating coordinate system when the Rossby number is small ($Ro << 1$) and when the timescale T is larger than the timescale of rotation ($\gamma << 1$). When flow has no mean component the two conditions are equivalent because the timescale T is the turnover time of turbulence ($T = L / U$). However, when turbulent flow is advected by a mean flow, attention should be paid to the fact that observations must be made over a sufficiently long period to distinguish all the effects of rotation on turbulence.

When the two conditions $Ro << 1$ and $\gamma << 1$ are met, the leading order in equation (16) is the geostrophic equilibrium, which expresses the balance between the pressure gradient in the direction perpendicular to the rotation axis and the Coriolis force. The scaling $P = \rho\, 2\Omega\, U\, L$ is deduced (Hopfinger [1]), and therefore

$$- 2\Omega\ v = - \frac{1}{\rho} \frac{\partial p}{\partial x} \quad (19)$$

$$2\Omega\ u = - \frac{1}{\rho} \frac{\partial p}{\partial y} . \quad (20)$$

The continuity equation then reduces to:

$$\frac{\partial u}{\partial x} + \frac{\partial v}{\partial y} + \frac{\partial w}{\partial z} = \frac{\partial w}{\partial z} = 0 . \quad (21)$$

The main consequence of (21) is that $W L / U D << 1$. It is more usual to develop the equations with respect to the small parameter Ro. We then have $W L / U D = O(Ro)$.

In what follows two flows are considered successively, respectively with a small aspect ratio ($\delta = D / L << 1$) or with an aspect ratio of the order of unity or more. The first case is in general relevant to geophysical flows. "Geostrophic turbulence" is addressed to some extent in this volume by Maxworthy [2] and Hopfinger [3], so that the present review concerns mainly the second case, which is more relevant to industrial situations. A prominent part of the paper is devoted to the discussion of two related experiments.

22. Turbulent flows confined by low aspect ratios

In most geophysical flows the Rossby number is small when large-scale motions are considered. In these motions the thickness D of the layer is such that $D/L << 1$, a property from which it may be deduced that $W = UD/L\ O(Ro)$. Shallow water equations are therefore the appropriate ones to describe the motion. The momentum equation for the velocity component parallel to the rotation axis gives the following order of magnitude

$$\frac{\partial p}{\partial z} = \rho\ \frac{U^2}{L}\ \frac{D}{L}\ O(R_0) \quad , \tag{22}$$

which means that the geostrophic pressure field as defined by (19) and (20) is hydrostatic. Hence the flow is two-dimensional at the lowest order since

$$\frac{\partial u}{\partial z} = \frac{U}{L}\ \frac{D}{L}\ O(R_0) \quad \text{and} \quad \frac{\partial v}{\partial z} = \frac{U}{L}\ \frac{D}{L}\ O(R_0) \quad . \tag{23}$$

23. Turbulent flows with O(1) or large aspect ratios

The scaling given below, derived when the flow is not confined in the direction parallel to the rotation axis, will serve to interpret the experiments on turbulence submitted to rotation presented in section 3. The scale D of motion in the direction parallel to the rotation axis is no longer the depth of the layer as in the geophysical case. It is fixed by the dynamics of motion. A major phenomenon in rotating flows is the occurrence of inertial waves. Perturbations with frequency $\omega << 2\Omega$ propagate in the direction parallel to the rotation axis according to the following dispersion law (Mory [5])

$$\frac{k_z^2}{k_x^2 + k_y^2} = \frac{\omega^2}{4\,\Omega^2 - \omega^2} \quad . \tag{24}$$

k_x, k_y and k_z are the components of the wave vector. It is straightforward to verify that turbulent motion with small Rossby numbers produce motions propagating in the direction Oz, because $\omega = U/L \ll 2\Omega$. Since $k_z = O(1/D)$, $k_x = O(1/L)$ and $k_y = O(1/L)$, the lengthscale of the motions produced by turbulence in the direction Oz is therefore deduced from (24)

$$\frac{D}{L} \approx \frac{2\Omega}{U/L} = O(R_0^{-1}) \quad . \tag{25}$$

The latter estimate actually determines the lengthscale of motions in the direction parallel to the rotation axis when the depth of the layer is larger than D. What is meant by the term "large aspect ratio" is more than $D/L \gg 1$; it is $D/L \gg Ro^{-1}$. Several consequences emerging from (25) are worth mentioning:

(i). The scaling $W L / U D = O(Ro)$ deduced from (21) is still valid, but the relative orders of magnitude of L and D imply that the characteristic velocities U and W are of the same orders of magnitude.

(ii). The horizontal components of velocities u and v are in geostrophic equilibrium because the Rossby number is small. Nevertheless, the Navier-Stokes equation for the velocity component w shows that the pressure gradient with respect to z of the geostrophic pressure is of the order of magnitude of the inertial terms. The geostrophic pressure field is therefore not hydrostatic as for the geophysical case considered above.

(iii). The elongation of the lengthscale of motion in the direction parallel to the rotation axis is found. We deduce:

$$\frac{\partial u}{\partial z} \approx \frac{\partial v}{\partial z} = O(\frac{U}{L} R_0) \quad . \tag{26}$$

These gradients are one order of magnitude smaller than the velocity gradients in the directions x and y. Therefore turbulent motion tends to become two-dimensional as the Rossby number decreases. Nevertheless, the degree to which motion becomes two-dimensional is less than in the case of confined motion in the direction z (compare equations (26) and (23)).

The results of section 1 and section 2 have established that a turbulent flow tends to become two-dimensional in a plane perpendicular to the rotation axis when the Rossby number, based on the turbulent velocity U and the integral lengthscale L of motions in a direction perpendicular to Oz, is small. For geophysical flows, flow is confined to a layer of small depth compared to the lengthscale L of motion. The anisotropy of motion occurs though a significant decrease in the velocity component parallel to the rotation axis. In flows with large aspect ratios ($D/L > Ro^{-1}$) all velocity components are of the same order of magnitude. On the other hand, anisotropy appears through an increase in the turbulent lengthscale in the direction parallel to the rotation axis.

24. Some remarks on two-dimensional turbulence

In view of the results established above, it is useful to recall some theoretical results concerning two-dimensional turbulence, in particular that according to which dissipation in two-dimensional turbulence is less than the dissipation occurring in three-dimensional turbulence with a similar lengthscale and kinetic energy. A spectral analysis of turbulence is required to establish this result. Only the major steps of the analysis are mentioned here (the details can be found in Lesieur [10]).

The kinetic energy spectrum E(k) is introduced such that

$$\frac{\overline{u^2} + \overline{v^2} + \overline{w^2}}{2} = \int_0^{\infty} E(k)\, dk \quad , \qquad (27)$$

k is the modulus of the wave vector $\vec{k}$. The spectral density E(k) dk is the kinetic energy of the turbulent motions with lengthscales l in the range

$$\frac{1}{k} < l < \frac{1}{k + dk} \quad .$$

Measurements of the turbulent kinetic energy spectrum determine the range of scales containing turbulent motion and the relative importance of each scale. Integrating the turbulent kinetic spectrum gives the total turbulent kinetic energy. In practice, the turbulent kinetic energy spectrum can be estimated from the Fourier transform of the velocity fluctuation

$$\hat{\vec{u}}(k) = \left(\frac{1}{2\pi}\right)^2 \int_{-\infty}^{\infty} \vec{u}(\vec{x},t)\, e^{-i\vec{k}.\vec{x}}\, d\vec{x} \quad . \qquad (28)$$

(For 2D turbulence, the coordinate vector and the velocity vector have only two components). The turbulent kinetic energy spectrum is thus

$$E(\vec{k}) = \overline{\hat{\vec{u}}(\vec{k}) \,.\, \hat{\vec{u}}(-\vec{k})} \quad . \qquad (29)$$

When turbulence is isotropic in the plane considered, it is usual to integrate the spectrum with respect to angle:

$$E(k) = 2\pi k \, E(\vec{k}) \quad . \tag{30}$$

In two-dimensional turbulent flow with no mean flow component as well as in three-dimensional turbulent flow, the decay of the total turbulent kinetic energy equals dissipation, which takes the following spectral form:

$$\frac{d}{dt} \int_0^\infty E(k,t)\, dk = -\varepsilon = -2\nu \int_0^\infty k^2 E(k,t)\, dk \quad . \tag{31}$$

In the particular case of two-dimensional turbulence the evolution of dissipation is also expressed as

$$\frac{d}{dt} \int_0^\infty k^2 E(k,t)\, dk = -2\nu \int_0^\infty k^4 E(k,t)\, dk \quad . \tag{32}$$

Equation (32) implies that the integral in the right hand side of equation (31) remains finite for two-dimensional turbulence. Hence, the dissipation rate of the turbulent kinetic energy vanishes when viscosity decreases to zero. In this limiting case, the turbulent kinetic energy is therefore conserved. The process is very different for three-dimensional turbulence; in the absence of a constraint equivalent to (32), the integral in the right hand side of (31) is divergent when viscosity is decreased to zero and the dissipation rate of the turbulent kinetic energy is finite and non-zero. The dissipation of turbulent kinetic energy is therefore much reduced in a two-dimensional turbulence as compared to the value it would reach in three-dimensional turbulence with an equal Reynolds number.

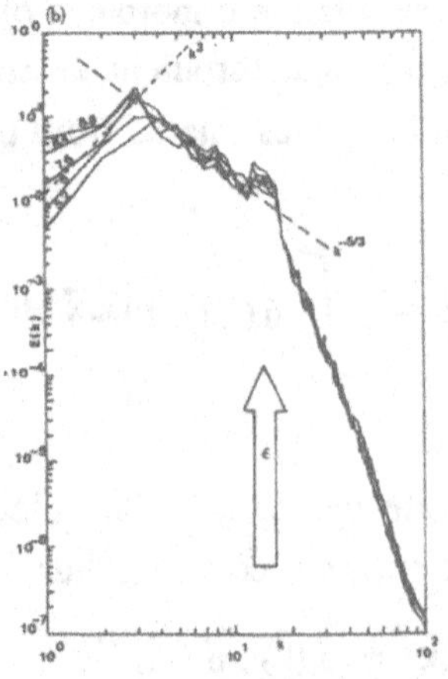

Figure 1: Turbulent kinetic energy spectrum obtained in a numerical simulation of two-dimensional turbulence (Frish and Sulem [11]).

These results have certain consequences on the shape of the turbulent kinetic energy spectrum in the inertial range. For three-dimensional turbulence, the turbulent kinetic energy spectrum usually takes the form

$$E(k) = \varepsilon^{2/3} k^{-5/3} . \tag{33}$$

In two-dimensional turbulence, the turbulent kinetic energy spectrum is in general steeper in the inertial range ($E(k)$ decays like k^{-n} with $n>2$). The larger exponent in the turbulent kinetic energy spectrum is related to the fact that the integral in the right hand side of (31) remains finite even when viscosity approaches zero. Figure 1 shows a turbulent kinetic energy spectrum obtained by Frish and Sulem [11] in a numerical simulation of turbulence. Experimental measurements of turbulent kinetic energy spectra in experiments of turbulence approaching a 2D state do not so convincingly show a k^{-n} dependence with $n=3$, but the property $n>2$ is still observed.

3. ROTATING TURBULENCE WITH LARGE ASPECT RATIO - EXPERIMENTS

The most complete results concerning rotating turbulence with a large aspect ratio have been obtained in the two experimental set-ups sketched in Figure 2. In the first experiment turbulence is produced in a rotating tank by an oscillating grid. Experiments of this type have been carried out by Ibbetson & Tritton [12], Colin de Verdière [13] and Hopfinger, Browand and Gagne [14]. The results obtained by the Grenoble group led by Hopfinger are discussed here. In the second type of experiment turbulence is produced by a mean flow passing through a fixed grid in a rotating air tunnel. The structure of turbulence is studied downstream of the grid. This experiment, first carried out by Wigeland & Nagib [15] was recently redone by Jacquin, Leuchter, Cambon and Mathieu [16]. These recent results are also discussed.

The second type of experiment is well designed to study the evolution in time of turbulence. The turbulent velocity is small as compared to the constant mean velocity U; the properties of turbulence measured at a distance x from the grid in the direction of the mean flow give the state of turbulence at a time $t=x/U$ after turbulent fluctuations have been produced. The effect of rotation on dissipation of turbulent kinetic energy is then directly characterised. In the first type of experiment, the decrease of turbulent kinetic energy is also observed because the turbulent eddies at a distance z from the grid midplane are necessarily produced by the grid; they reach the distance z from the grid midplane after a duration which increases with the distance z. Nevertheless, the relationship between time and distance z is not known on average; it depends on various factors, among which is the structure of turbulence.

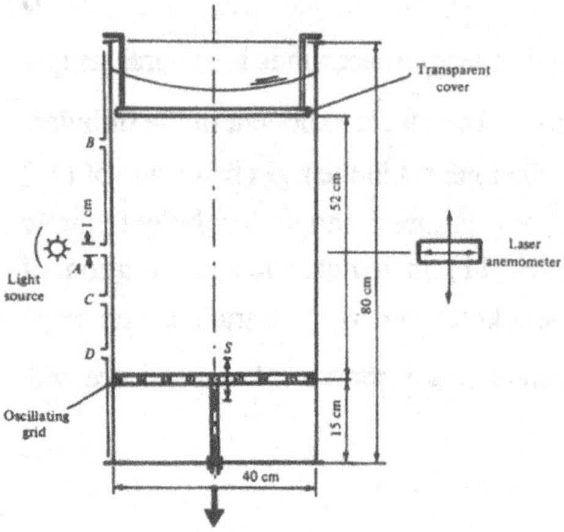

(i) Oscillating grid turbulence in a rotating tank. (Hopfinger *et al* [14])

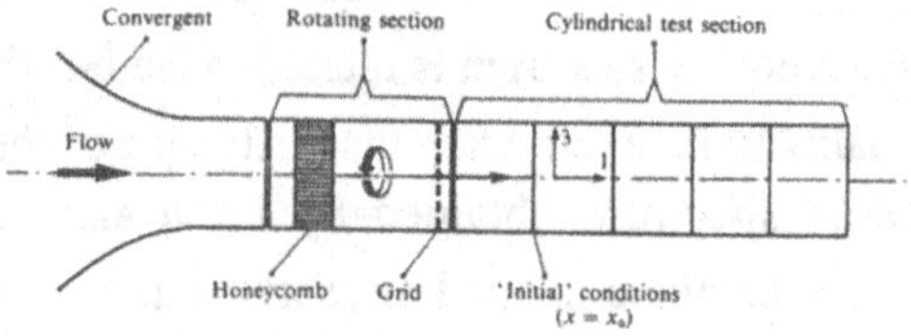

(ii) Mean flow grid turbulence in a rotating channel. (Jacquin *et al* [16])

Figure 2: Laboratory experiments on rotating turbulence with large aspect ratio.

Figure 3 shows the decay of turbulence submitted to rotation measured in the experiments of Jacquin and al [16]. u' and v' are the turbulent velocity components in the directions parallel and perpendicular to the rotation axis, respectively. The decay of turbulence is slower in the presence of rotation as compared to the measurements in non-rotating conditions. The dissipation rate of the turbulent kinetic energy is therefore lower in the presence of rotation. The deviation is stronger as rotation is increased. The structure of the Reynolds tensor is almost unchanged in the presence and in the absence of rotation. The initial structure of turbulence is nearly isotropic and the Reynolds tensor remains approximately spherical in the presence of rotation, as anticipated in section 1. This property is shown in Figure 4, which reproduces the evolution with distance from the grid of the structural coefficient

$$\frac{\overline{v'^2} - \overline{u'^2}}{\overline{v'^2} + \overline{u'^2}} \; . \qquad (34)$$

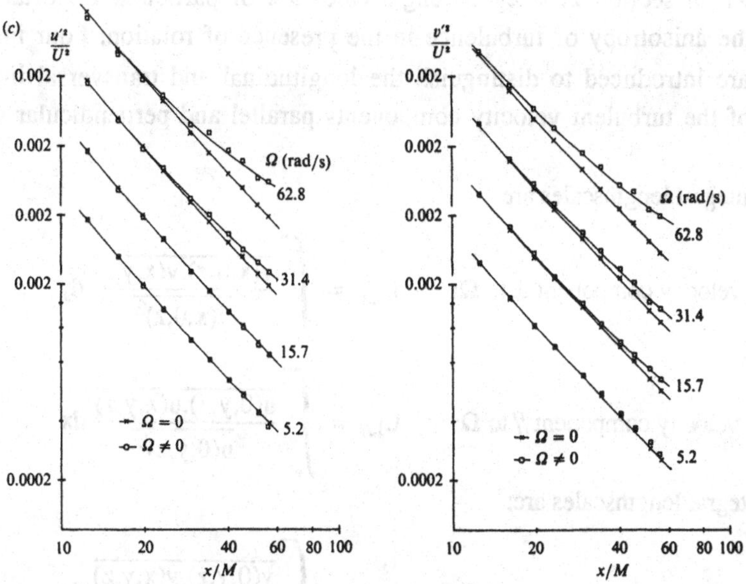

Figure 3: Turbulent velocity decay downstream of a grid in a rotating tunnel (Jacquin, Leuchter, Cambon and Mathieu [16]). Grid mesh is M = 20 mm.

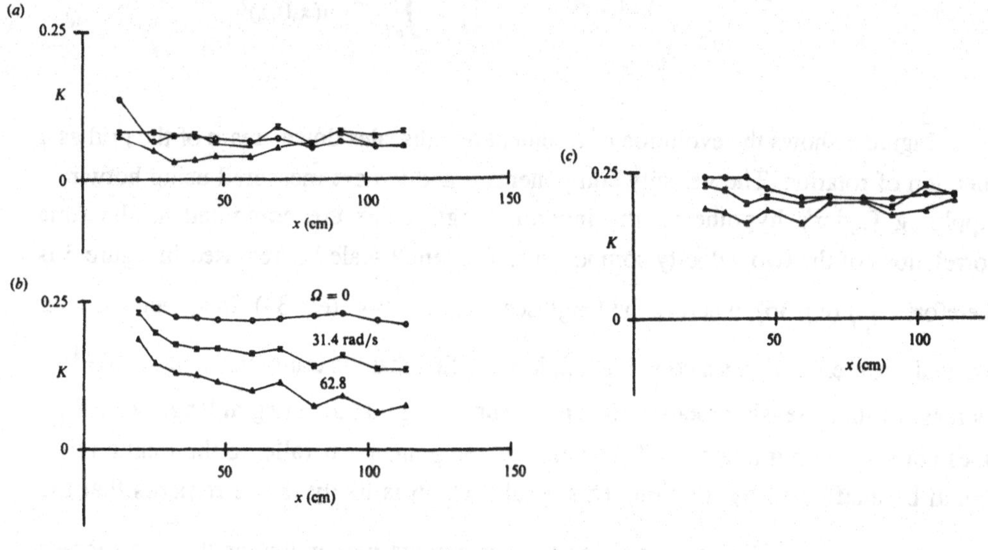

Figure 4: Structural parameter downstream of a grid in a rotating tunnel (Jacquin, Leuchter, Cambon and Mathieu [16]).
(a) M = 10 mm, (b) M = 15 mm, (c) M = 20 mm.

As shown in section 2, integral lengthscales are of particular importance in determining the anisotropy of turbulence in the presence of rotation. Four integral lengthscales are introduced to distinguish the longitudinal and transversal integral lengthscales of the turbulent velocity components parallel and perpendicular to the rotation axis.

Longitudinal integral lengthscales are:

for the velocity component $\perp$ to Ω: $$L_{l,\perp} = \int_0^\infty \frac{\overline{v(x,0,z).v(x,y,z)}}{\overline{v(x,0,z)^2}}\,dy \quad , \qquad (35)$$

for the velocity component // to Ω: $$L_{l,//} = \int_0^\infty \frac{\overline{u(0,y,z).u(x,y,z)}}{\overline{u(0,y,z)^2}}\,dx \quad . \qquad (36)$$

Transversal integral lengthscales are:

for the velocity component $\perp$ to Ω: $$L_{t,\perp} = \int_0^\infty \frac{\overline{v(0,y,z).v(x,y,z)}}{\overline{v(0,y,z)^2}}\,dx \quad , \qquad (37)$$

for the velocity component // to Ω: $$L_{t,//} = \int_0^\infty \frac{\overline{u(x,0,z).u(x,y,z)}}{\overline{u(x,0,z)^2}}\,dy \quad . \qquad (38)$$

Figure 5 shows the evolution of turbulent lengthscales downstream of the grid as a function of rotation. The velocity components u and v were measured using hotwires. Applying Taylor's hypothesis, the integral lengthscales are computed as the time correlations of the two velocity components. The lengthscale L_u reported in figure 5 is therefore $L_{l,//}$ (eq. 36), whereas the lengthscale L_v is $L_{t,\perp}$ (eq. 37). In agreement with the analysis presented in sections 1 and 2, it is verified that the transverse lengthscale $L_{t,\perp}$ increases with increasing rotation. It is more surprising that the integral lengthscale $L_{l,//}$ does not vary with rotation, as if the velocity component u parallel to the rotation axis would be unaffected by rotation. This result, which is likely if one remarks that the Coriolis force is applied to the velocity components perpendicular to $\vec{\Omega}$, disagrees nevertheless with the understanding of the role of inertial waves as described above. According to a linear model of inertial waves, the lengthscale along the direction parallel to the rotation axis is the same for all velocity components.

In figure 6, the turbulent Rossby numbers $R_u = u/2\Omega L_u$ and $R_v = v/2\Omega L_v$ are plotted downstream of the grid as a function of rotation. It can be seen that these numbers decrease with increasing distance from the grid. However, they remain relatively large, reaching values below 1 in a few cases only. We therefore suggest that flow should be studied further from the grid in order to observe a turbulent state approaching a two-dimensional structure. Nevertheless, important effects of rotation on dissipation and on lengthscales are already observed in the region closest to the grid.

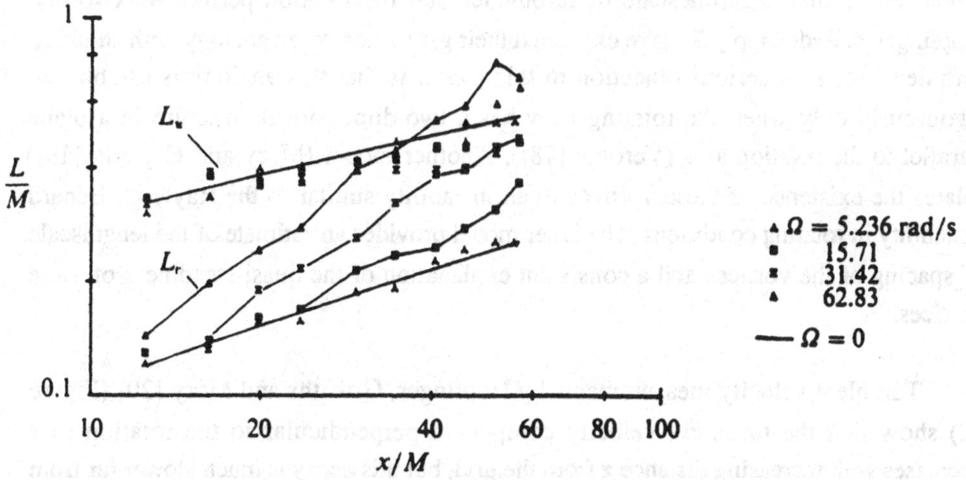

Figure 5: Evolution of integral lengthscales downstream of a grid in a rotating tunnel (Jacquin, Leuchter, Cambon and Mathieu [16]). Grid mesh is M = 20 mm.

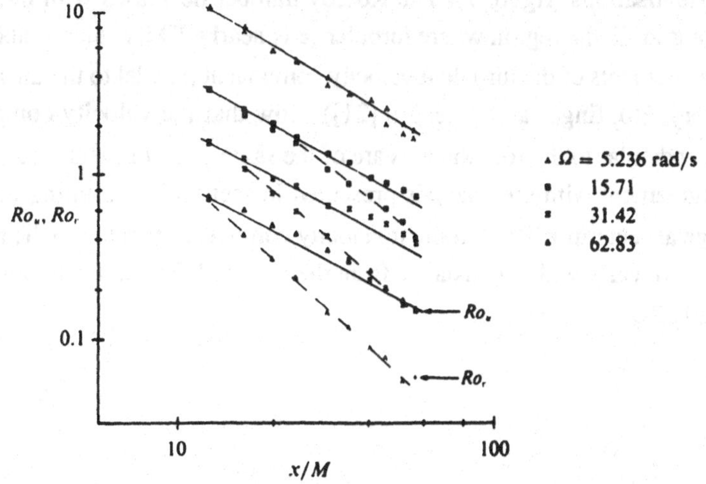

Figure 6: Evolution of Rossby numbers downstream of a grid in a rotating tunnel (Jacquin, Leuchter, Cambon and Mathieu, [16]). Grid mesh is M = 20 mm.

Hopfinger, Browand and Gagne [14] produced a nearly two-dimensional turbulent structure in the oscillation grid experiment in a rotating tank shown in Figure 2. Visualisations of the flow field are shown in figure 7 in planes perpendicular and parallel to the rotation axis. In the plane parallel to the rotation axis, intense vortices parallel to the rotation axis, are visualised. In the vortex cores vorticity reaches up to 50 times the vorticity of the mean rotation of the tank and the radius of the vortex cores is very small compared to the lengthscale of turbulence. These vortices have a duration timescale that is much longer than the timescale of turbulence and the rotation period. Maxworthy, Hopfinger & Redekopp [17] have explained their generation by an analogy with stratified turbulent flows. A serious objection to this model is that this analogy is established rigourously only when the rotating flow has a two-dimensional structure in a plane parallel to the rotation axis (Veronis [18]). Another model (Mory and Capéran [19]) relates the existence of these vortices to an instability similar to the Rayleigh-Bénard instability in rotating conditions. The latter model provides an estimate of the lengthscale of spacing of the vortices and a consistent explanation of the quasi-steadiness of these vortices.

Turbulent velocity measurements by Hopfinger, Griffiths and Mory [20] (Figure 8a) show that the turbulent velocity component perpendicular to the rotation axis decreases with increasing distance z from the grid, but this decay is much slower far from the grid. At a large distance from the grid midplane, the turbulent velocity becomes almost independent of the distance z, and the flow field is nearly two-dimensional as shown by the visualisations (figure 7). The Rossby number decreases with increasing distance from the grid; in the region where turbulence is nearly 2D its value is about 0.2. More recent measurements of the turbulent velocity component parallel to the the rotation axis (Fleury, Mory, Hopfinger and Auchère [21]) show that the velocity components parallel and perpendicular to the rotation axis are of the same order of magnitude (Figure 8b). This is in agreement with the analysis presented in section 2 of rotating turbulent flows with a large aspect ratio. The turbulent velocity component parallel to the rotation axis decays slowly linearly with the distance from the grid midplane in the region where turbulence is nearly 2D.

visualisation of flow in a plane
perpendicular to the rotation axis

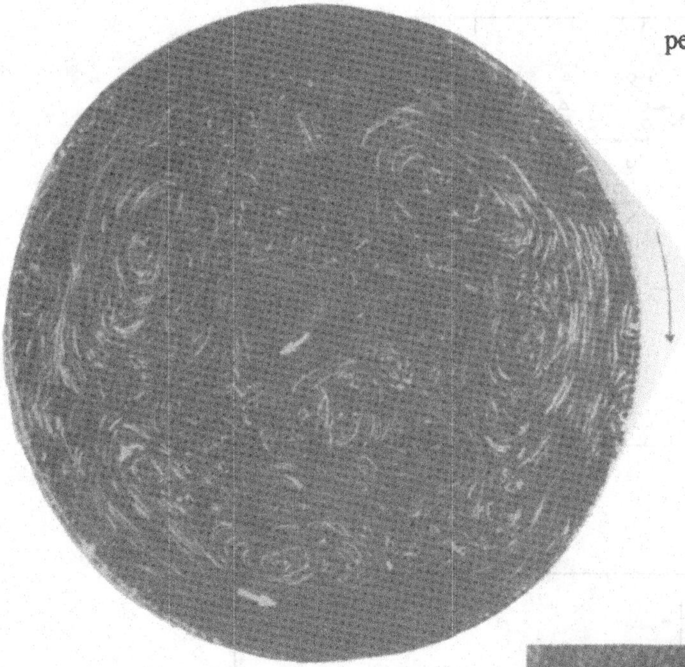

visualisation of flow in a plane
parallel to the rotation axis

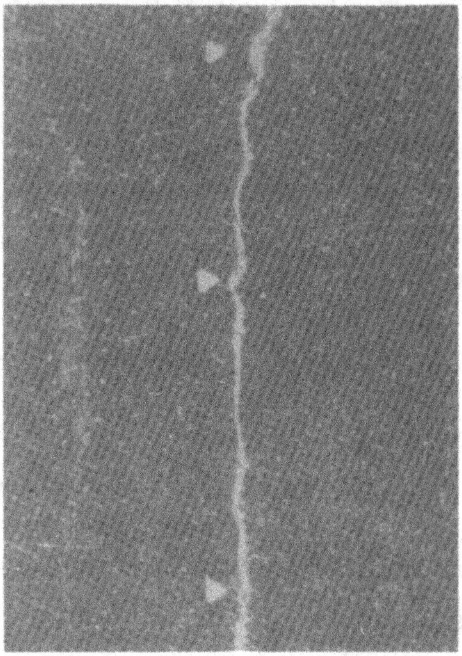

Figure 7: **Visualisations of an oscillating grid turbulent flow in a rotating tank (Hopfinger, Browand and Gagne [14]).**

(a)

(b)

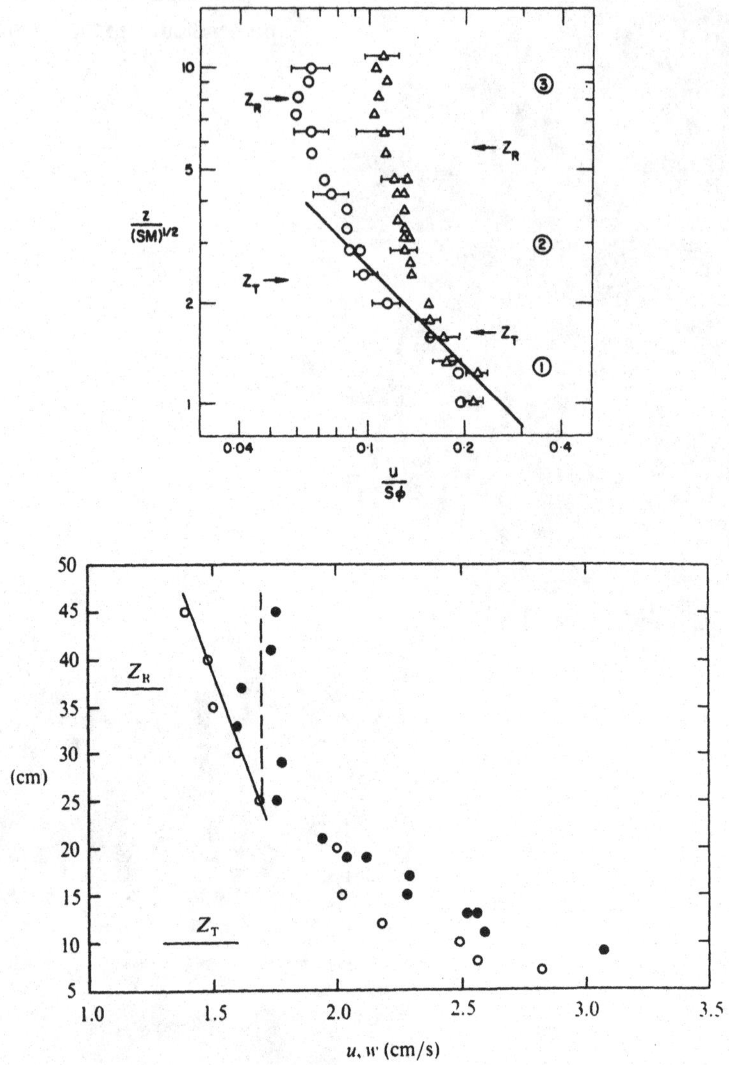

Figure 8: Decay of turbulent rms velocity components produced in a rotating tank by an oscillating grid.

a). turbulent velocity component perpendicular to rotation (Hopfinger *et al* [20]).

S is the stroke (4 cm), M the mesh of the grid (5 cm). The frequency of oscillation ϕ is 3.3 Hz (Δ) or 6.6 Hz (o). Rotation is $\Omega=\pi$ rad.s^{-1}.

b). Comparison of the velocity component parallel to rotation (o) with the velocity component perpendicular to rotation (•) (Fleury *et al* [21]).

We continue the discussion with an estimate of a few characteristic quantities. These estimates provide a quantitative comparison of the experiments by Hopfinger et al [14] and by Jacquin et al [16].

In Jacquin, Leuchter, Cambon and Mathieu's experiment the mean flow velocity across the grid is $U=10\ m.s^{-1}$. Turbulent velocity components are measured downstream of the grid over a distance L_m about 1m long. Turbulence is therefore studied during a period of time T_e of about 0.1 s after it has been produced by the grid. This duration is short compared to the period of rotation of the tunnel $T_r=2\pi/\Omega$, whose value is 1.20 s for the lowest rotation rate ($\Omega=5.236\ rad.s^{-1}$) and 0.1 s for the highest rotation rate ($\Omega=62.83\ rad.s^{-1}$). Returning to Equation (16), it can be seen that the reduced frequency $\gamma = 1/2\Omega T$ is in the range $1 < \gamma < 10$. The reduced frequency is not equal to the Rossby number. Since the time derivative terms are not negligible, the turbulence is certainly not in equilibrium in the sense that the rotation was not applied for a sufficient period in the experiment of Jacquin et al [16] to modify the structure of turbulence as fully as it could.

Calculations of the group velocity of inertial waves in the experiment by Jacquin et al are also most interesting. The group velocity of inertial waves is of the order of $C_g = 2\Omega L_v$. Since the lengthscale is of the order of magnitude of the mesh of the grid (M=1 to 2 cm), the group velocity is $10\ cm.s^{-1}$ for the rotation rate $\Omega=5.236\ rad.s^{-1}$ and $1.25\ m.s^{-1}$ for rotation rate $\Omega=62.83\ rad.s^{-1}$, respectively. These wave speeds are small compared to the mean velocity U. This is a limitation on the role played by inertial waves in this experiment because perturbations are not able to propagate upstream. As explained above, inertial wave propagation is a major mechanism in the process of two-dimensionalisation of turbulence by rotation. In the experiment of Hopfinger, Browand and Gagne [14], on the other hand, inertial waves propagate both ways parallel to the rotation axis. Interaction of all parts of flow is made possible and, in fact, the two-dimensional structure of flow is observed where the Rossby number is sufficiently small (Ro<0,2). The propagation of inertial waves both ways is probably a crucial requirement for generating very intense and concentrated vortices in rotating turbulent flows, as they are observed in the experiment of Hopfinger, Browand & Gagne [14].

4. NUMERICAL MODELLING OF ROTATING TURBULENCE

The numerical modelling of rotating turbulence has also produced important results concerning rotating turbulence, which could not be obtained from experiments. The spectral analysis of turbulence submitted to rotation is of particular interest (Bertoglio [22], Cambon & Jacquin [9]), but is beyond the scope of this paper. We will discuss only the k-ε modelling of rotating turbulence, which is simply related to the physical description of rotating turbulence given in this paper. The presentation by Bardina, Ferziger & Rogallo [6] will be followed here. The subject has been also adressed by Aupoix, Cousteix & Liandrat [23].

In k-ε models, the mean flow is computed by solving the Navier-Stokes equations in which the Reynolds stresses are modelled in the form

$$- \left\{ \overline{u_i u_j} - \delta_{ij} \frac{1}{3} \Sigma_m \overline{u_m^2} \right\} = C_\mu \frac{k^2}{\varepsilon} \left\{ \frac{\partial U_i}{\partial x_j} + \frac{\partial U_j}{\partial x_i} \right\} . \qquad (39)$$

$k = \overline{u_m^2}/2$ is the turbulent kinetic energy and ε is the dissipation rate of the turbulent kinetic energy. In order to solve the Navier-Stokes equations, the evolution equations for k and ε are also written and solved by the program. These equations express the evolution in time of the two quantities under the three main processes to which they are submitted, namely production, dissipation and diffusion. One of the main results established for rotating turbulence is that dissipation tends to be reduced in the presence of rotation. Considering therefore only the dissipation terms in the evolution equations for k and ε, one obtains

$$\frac{\partial k}{\partial t} = - \varepsilon \ . \qquad (40)$$

This equation is written for homogeneous turbulence with no mean flow, in which case production and diffusion terms of turbulence vanish. The evolution equation for the dissipation rate of turbulence also takes the form

$$\frac{\partial \varepsilon}{\partial t} = - D \ . \qquad (41)$$

In the case of non-rotating isotropic turbulence the usual modelling of the dissipation term in (41) is $D = 1{,}92\ \varepsilon^2/k$. The dissipation rate D of ε and the dissipation rate ε of the turbulent kinetic energy k have the same dimensional forms in non-rotating isotropic turbulence because $\varepsilon = k^{3/2}/l$. In rotating turbulence, turbulent lengthscales become non-isotropic and the dissipation rate ε of the turbulent kinetic energy k decreases faster when rotation is applied, starting from any initial value ε_o. Modelling the effect of rotation on the evolution of turbulence therefore enters into the evolution equation for ε: D is increased under the effect of rotation (eq. 41). In this way, the dissipation rate of turbulence will decrease faster in time (eq. 41) and the decrease of turbulent kinetic energy will be slower (eq. 40). Several empirical laws of this type have been proposed in the literature. In particular, Bardina, Ferziger and Rogallo [6] expressed the dissipation of ε in the form

$$D = 1{,}92\ \varepsilon^2/k + 0{,}15\ \Omega\varepsilon\ . \tag{42}$$

The solution of the decay of homogeneous turbulence without mean flow is obtained from the two equations:

$$\frac{\partial k}{\partial t} = -\varepsilon \tag{43}$$

$$\frac{\partial \varepsilon}{\partial t} = -1.92\ \frac{\varepsilon^2}{k} - 0.15\ \Omega\varepsilon\ . \tag{44}$$

For an initial turbulence with turbulent kinetic energy k_o and dissipation rate ε_o, ε decreases faster under the effect of rotation. The asymptotic solution without rotation is:

$$k = k_0 \left\{1 + (1.92 - 1)\ \frac{\varepsilon_0}{k_0}\ t\right\}^{1/(1-1.92)} \tag{45}$$

$$\varepsilon = \varepsilon_0 \left\{1 + (1.92 - 1)\ \frac{\varepsilon_0}{k_0}\ t\right\}^{1.92/(1-1.92)}. \tag{46}$$

Both quantities vanish for infinite time. For comparison, with rotation:

$$k = k_0 \left\{1 + \frac{1 - 1.92}{0.15}\ \frac{\varepsilon_0}{\Omega\, k_0}\ \left(\exp^{-0.15\,\Omega t} - 1\right)\right\}^{1/(1-1.92)}\ . \tag{47}$$

It is worth remarking that the turbulent kinetic energy is constant and non-zero for infinite time whereas the dissipation rate tends to zero.

REFERENCES

1. Hopfinger E.J.: Parameters, Scales, Geostrophic balance, this volume.
2. Maxworthy T.: Convective and Shear flow turbulence with rotation, this volume.
3. Hopfinger E.J.: Baroclinic turbulence, this volume.
4. Tennekes H. & Lumley J.L.: A first course in turbulence, MIT Press, 1972.
5. Mory M.: Inertial waves, this volume.
6. Bardina J., Ferziger J.H. & Rogallo R.S.: Effect of rotation on isotropic turbulence: computation and modelling, J. Fluid Mech., 154 (1985), 321-336.
7. Itsweire E., Chabert L. & Gence J.N.: Action d'une rotation pure sur une turbulence homogène anisotrope, C.R.A.S., 289 B (1979), 197-199.
8. Roy Ph.: Simulation numérique d'un champ turbulent homogène incompressible soumis à des gradients de vitesse moyenne, Thèse d'état, Université de Nice, 1986.
9. Cambon C. & Jacquin L.: Spectral approach to non-isotropic turbulence subjected to rotation, J. Fluid Mech., 202 (1989), 295-317.
10. Lesieur M.: Turbulence in Fluids, Kluwer Academic Publishers, 1990.
11. Frish U. & Sulem P.L.: Numerical simulation of the inverse cascade in two-dimensional turbulence, Phys. Fluids, 27 (1984), 1921-1923.
12. Ibbetson A. & Tritton D.J.: Experiments on turbulence in a rotating fluid, J. Fluid Mech., 68 (1975), 639-672.
13. Colin de Verdière A.: Quasi-geostrophic turbulence in a rotating homogeneous fluid, Geophys. Astrophys. Fluid Dyn., 15 (1980), 213-251.
14. Hopfinger E.J. , Browand F.K. & Gagne Y.: Turbulence and waves in a rotating tank, J. Fluid Mech., 125 (1982), 505-534.
15. Wigeland R.A. & Nagib H.M.: Grid generated turbulence with and without rotation about the streamwise direction, IIT Fluids & Heat Transfer Rep. R. 78-1, Illinois Institute of Technology, 1977.
16. Jacquin L., Leuchter O., Cambon C. & Mathieu J.: Homogeneous turbulence in the presence of rotation, J. Fluid Mech., 220 (1990), 1-52.
17. Maxworthy T., Hopfinger E.J. & Redekopp L.G.: Wave motions on vortex cores, J. Fluid Mech., 151 (1985), 141-165.
18. Veronis G.: The analogy between rotating and stratified fluids, Ann. Rev. Fluid Mech., 2 (1970), 37-66.
19. Mory M. & Capéran Ph.: On the genesis of quasi-steady vortices in a rotating turbulent flow, J. Fluid Mech., 185 (1987), 121-136.

20. Hopfinger E.J. , Griffiths R.W. & Mory M.: The structure of turbulence in homogeneous and stratified rotating fluids, J. Mécanique Théorique and Appliquée, Vol. Spécial (1983), 21-44.
21. Fleury M., Mory M., Hopfinger E.J. & Auchère D.: Effects of rotation on turbulent mixing across a density interface, J. Fluid Mech., 223 (1991), 165-191.
22. Bertoglio J.P.: Homogeneous Turbulent Field within a Rotating Frame, AIAA J., 20 (1982), 1175-1181.
23. Aupoix B., Cousteix J. & Liandrat J.: Effects of rotation on isotropic turbulence, Turbulent Shear Flows, Karlsruhe, 1983.

Chapter VI.2

CONVECTIVE AND SHEAR FLOW TURBULENCE WITH ROTATION

T. Maxworthy
University of Southern California, Los Angeles, CA, USA

1 Turbulent Convection

As an introduction to this section we present an extension of the discussion of Mory [1] to include a case of more general geophysical significance, where the source of the initial turbulent motion is a natural one rather than the rather artificial case of turbulence generated by an oscillating grid. Firstly we give a somewhat different argument to show the effect of rotation on turbulent fluctuations. Referring to figure 1, consider a typical turbulent eddy of scale, l, (the integral scale) and characteristic velocty, u', (the rms velocity fluctuation). When a particle is displaced from A to B a swirling velocity, v', is induced, given in order of magnitude by:

$$\frac{\partial v'}{\partial t} \sim 2\Omega u', \quad \text{where } t \sim \frac{l}{u'} \tag{1}$$

hence

$$v' \sim 2\Omega l.$$

Thus the kinetic energy acquired by the particle, with the swirl velocity (v'), is the same order of magnitude as the original eddy kinetic energy when:

$$\left(\frac{v'}{u'}\right)^2 \sim \left(\frac{2\Omega l}{u'}\right)^2 \sim O(1) \tag{2}$$

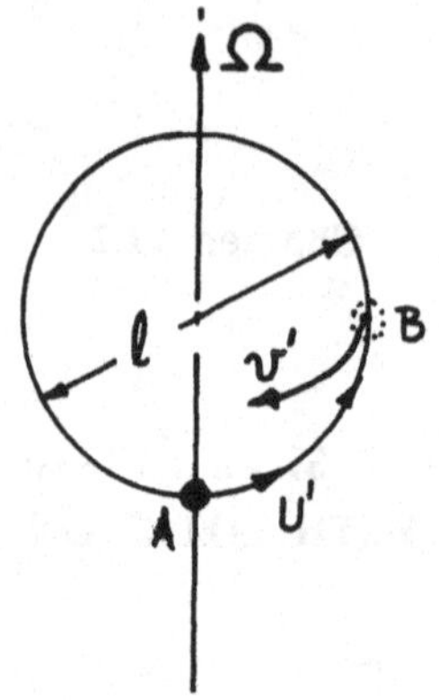

Figure 1: Particle displacement in a rotating fluid, showing the induced swirl velocity.

This argument is, of course, identical to that given for the stabilizing influence of a basic stratification in which case the potential energy gained by the particle replaces the swirl kinetic energy considered above.

In the case considered by Mory [1] the scaling arguments are as follows: An oscillating grid generates turbulence that scales as $u' \sim K/z$ and $l \sim z$ where K is the "grid-action" and z the distance from the grid. Rotation becomes important at a distance z_c given by $(f/K)z_c^2 \sim O(1)$ or $z_c \sim (K/f)^{1/2}$ at which level $u_c' \sim (fK)^{1/2}$ and $l_c \sim (K/f)^{1/2}$, where $f \equiv 2\Omega$. Thereafter fingers of lateral scale l_c propagate along the Ω vector with a velocity of order u_c', generating a strong cyclonic vorticity between them (Maxworthy, Hopfinger and Redekopp [2]).

In the case to be considered now the turbulence is generated by a negative buoyancy flux at the top of the system (Maxworthy and Narimousa [3]). In nature this would result from intense cooling, possibly with ice formation and salt rejection if it occurs in the polar regions. In figure 2 we sketch the sequence of events which occurs after the source of buoyance is "switched-on". In this case $u' \sim (B_0 z)^{1/3}$, $l \sim z$ and the reduced gravity $g' \sim (B_0^2/z)^{1/3}$, where B_0 is the buoyancy flux/unit area. The constraint of rotation again becomes important when (2) is satisfied, so that:

$$z_c \sim \frac{B_0^{1/2}}{(f)^{3/2}} \tag{3}$$

The time to reach this state from the start of the motion is $t_c \sim 1/f$ at which time g_c' and u_c' are given by:

$$g_c' \sim (B_0 f)^{1/2}; \quad u_c' \sim \left(\frac{B_0}{f}\right)^{1/2} \tag{4}$$

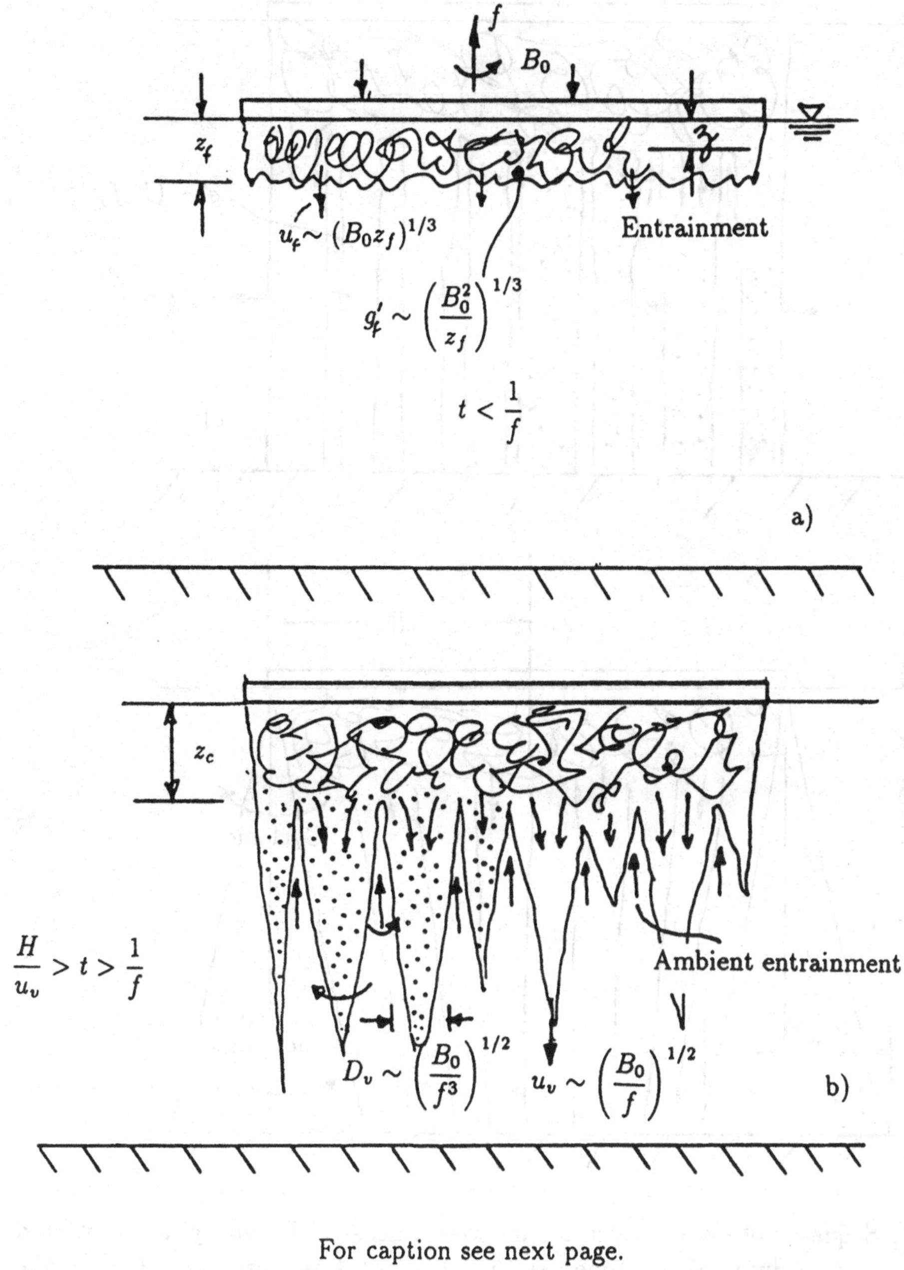

For caption see next page.

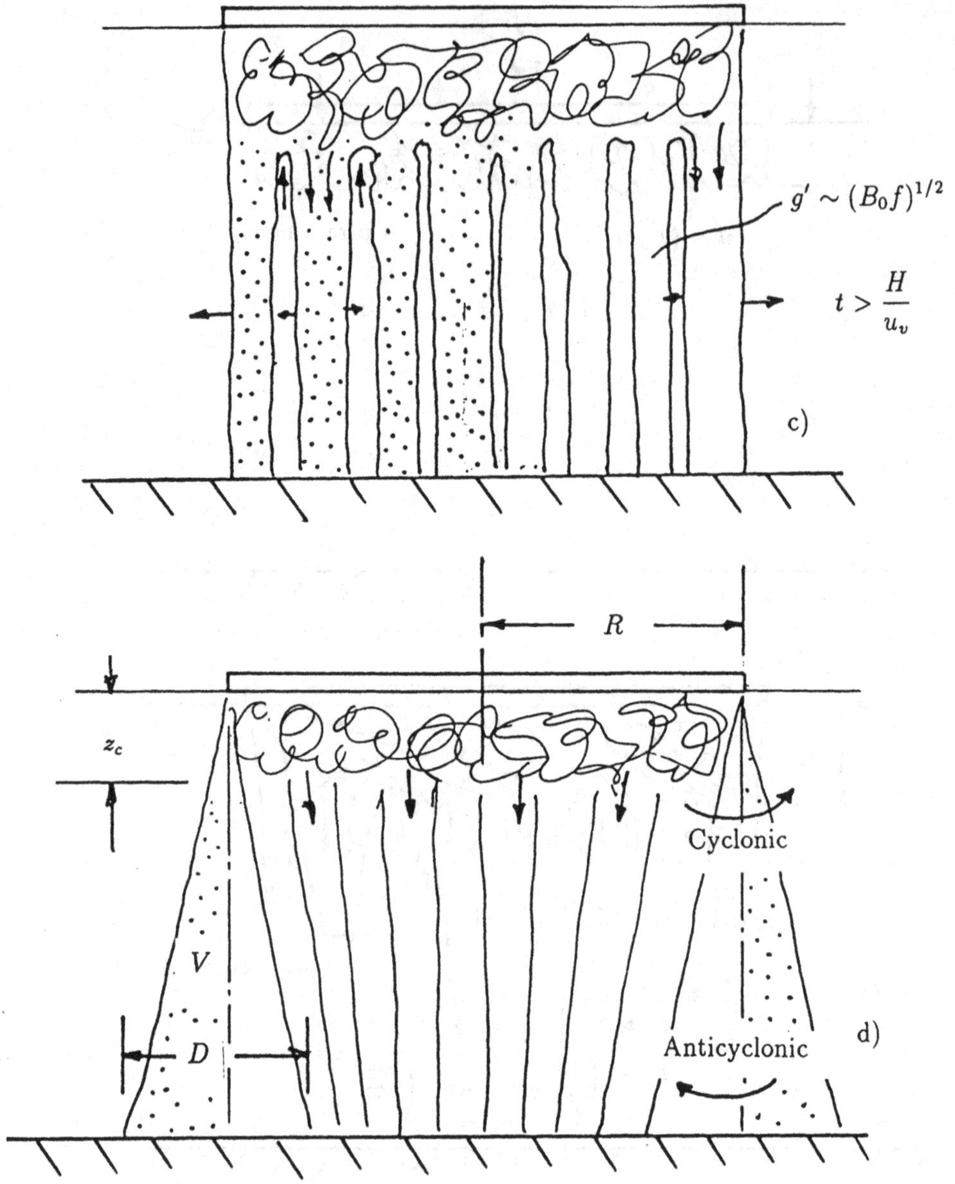

Figure 2: Sequence of events which occurs after a negative buoyancy flux is started in a rotating fluid. a) For time less than $1/f$. b) After rotation influences the turbulence but before it reaches the bottom. c) Filling and tilting of the vortex structures. d) Generation of baroclinic instability.

Equations (3) and (4) also represent the relevant scales of the fingers that propagate away from the mixing region. They interact with the bottom in a time given by $H/(B_0/f)^{1/2}$ at which time information about the existence of the bottom is transmitted upwards and the fingers begin the fill-out (figure 2c) and tilt (figure 2d) until the column becomes unstable to a baroclinic instability of horizontal scale of the order of the Rossby Radius of Deformation:

$$D \sim \frac{[g_c' H]^{1/2}}{f} \sim \frac{[(fB_0)^{1/2}H]^{1/2}}{f} \sim \frac{B_0^{1/4}H^{1/2}}{(f)^{3/4}} \quad \text{or} \quad \frac{D}{H} \sim \left[\frac{B_0}{f^3H^2}\right]^{1/4} \tag{5}$$

The resulting vortices have a heavy core that rotates anticyclonically, while the fluid that has converged above them during their adjustment period is rotating cyclonically.

In the discussion above it has been assumed that the depth, H, is larger than z_c. If this is not so rotation does not have time to act and baroclinic vortices form when the gravity-current outflow has traveled a distance of the order of the Rossby Radius from the periphery of the convecting region. In this case the diameter scales as:

$$D \sim \frac{\left\{\left[\frac{B_0^2}{H}\right]^{1/3} H\right\}^{1/2}}{f} \sim \frac{(B_0H)^{1/3}}{f} \tag{6}$$

or

$$\frac{D}{H} \sim \left[\frac{B_0}{f^3H^2}\right]^{1/3}$$

We have performed a series of experiments [3] in an apparatus identical to that shown in figure 2 and have verifed the scaling given by equations (5) and (6), see figure 3 (where D is now the diameter of the circle of the maximum swirl velocity.) Based on the results of Fernando et al. [4] the transition to the rotationally dominated case occurs when $z_c = 13.5B_0^{1/2}/(f)^{3/2}$, which would place the transition from rotationally to non-rotationally dominated flow i.e. when $z_c = H$, at the location noted on the figure. It appears that the transition actually takes place over a wide range of the parameter Ro^*, the "natural" Rossby Number of the flow rather than at one particular value of this parameter.

Application to natural flows in the Arctic [H = 2000 m; $B_0 = 3 \times 10^{-3}$ cm^2/s^3; $f = 1.25 \times 10^{-4}$/s; $Ro^* = 0.195$; $D/H = 2.3$; $D = 4.6 \pm 90.9$ km] and in the Golfe du Lion [H = 2000 m; $B_0 = 10^{-3}$ cm^2/s^3; $f = 10^{-4}$/s; $Ro^* = 0.50$; $D/H = 2.1$; $D = 4.2 \pm 0.8$ km] gives vortex diameters in good agreement with observations.

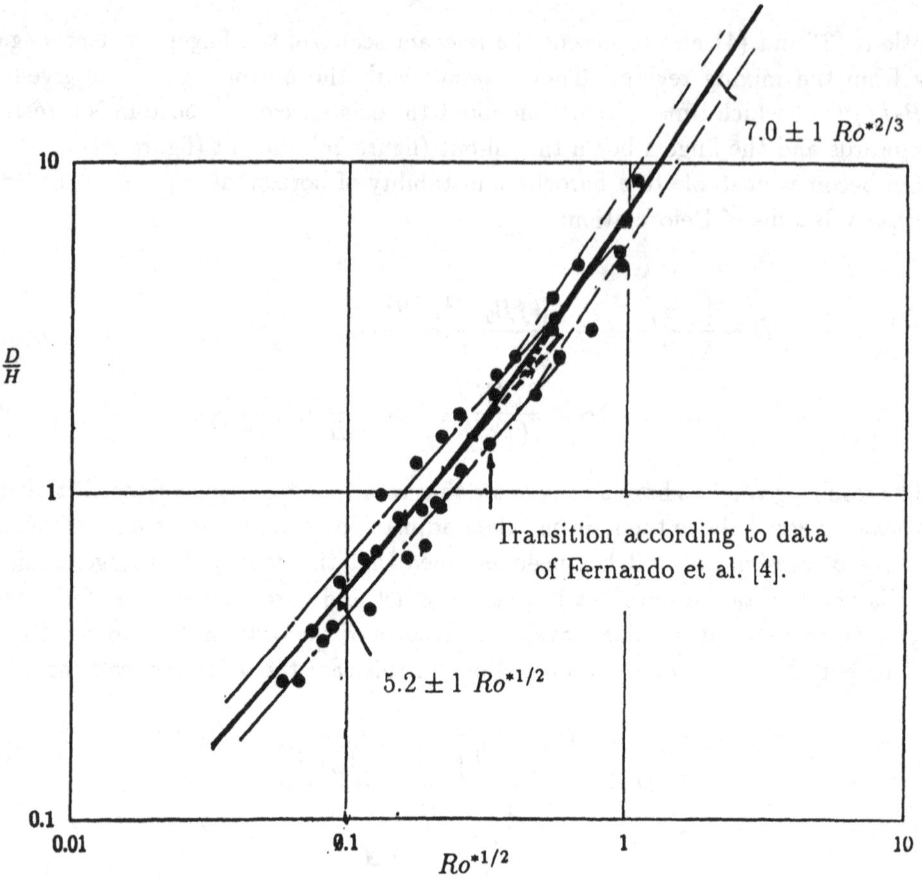

Figure 3: Experimental results relating baroclinic vortex diameter D/H to the natural Rossby Number ($Ro^* = \left(\frac{Bo}{f}\right)^{1/2} \cdot \frac{1}{fH} = \left(\frac{Bo}{f^3H^2}\right)^{1/2}$).

2 Rotating channel flow

In order to focus ideas in such a short section we concentrate on rotating channel flow realizing that a number of the arguments put forward carry over into other case of interest e.g., free shear layers and wall boundary layers in rotating and/or curved channels.

2.1 Instability of the laminar flow

In many of the discussions which have gone before we have shown a strong preference for simple physical models which can explain the underlying dynamics of a particular flow and we continue in the same vein here. We consider a straight, rectangular rotating channel in which a fully developed laminar flow has formed (figure 4). If a particle, A, (figure 5) is displaced from its original location to a new location, B, we can compare the forces that act upon it at B with the force on an undisturbed particle at the same location in order to determine the direction of its further motion, (Tritton and Davies [5]). At position B the particle has a new velocity u_1' given by

$$u_1' - u = \int 2\Omega v dt = 2\Omega\zeta \tag{7}$$

where ζ is the distance the particle has been displaced see eq. (1). Thus at B the particle is being acted upon by a Coriolis force $2\rho\Omega u_1'$ and a pressure gradient force $-2\rho\Omega u_2$, the latter being that required to keep the undisturbed particle C at the level of the velocity u_2. Hence, since $u_2 - u_1 = \zeta(\partial u/\partial y)$, $u_1' < u_2$ only if $(\partial u/\partial y) > 2\Omega$ since

$$u_2' - u_1' = \left(\frac{\partial u}{\partial y} - 2\Omega\right)\zeta \tag{8}$$

Thus in case a) of figure 5 the pressure gradient force $(2\rho\Omega u_2)$ is larger than the Coriolis force $2\rho\Omega u_1'$ and the flow is unstable, unless $u_1' > u_2$ i.e., unless $2\Omega > \partial u/\partial y$. In case b) rotation is always stabilizing since $2\Omega + (\partial u/\partial y)$ is always negative. Thus one has destabilization if the absolute vorticity $2\Omega - (\partial u/\partial y)$ is of opposite sign from the background vorticity, 2Ω. A useful way of parameterizing this effect is to construct an analogue of the Richardson number, namely:

$$Ri_\Omega \text{ or } B = -\frac{2\Omega\left(\frac{\partial u}{\partial y} - 2\Omega\right)}{\left(\frac{\partial u}{\partial y}\right)^2} = S(S+1) \tag{9}$$

where $S = -2\Omega/(\partial u/\partial y)$.

Thus negative B indicates that rotation is destabilizing and if positive stabilizing. It is effectively the same as the Rayleigh criterion for the centrifugal instability of curved flows and points to the similarity to Taylor-Couette and Görtler instability, as will become clear in what follows.

The linear stability of either a rotating plane Couette flow or plane Poisieulle flow was first discussed by Hart [6] and Lezius and Johnston [7] who pointed-out, also, the similarity to Benard convection. Two points are of interest, initially; the theories considered the flow to take place in an infinitely wide duct while, of course, any experimental study must take place in a duct of finite width. The latter has one significant effect. The developed mean flow in the interior has the same form as in the non-rotating case except that a transverse pressure gradient must exist to balance the Coriolis force on a fluid particle. At the top and bottom "side-walls", figure 6, the velocity goes to zero and so the "exterior" pressure gradient can drive an Ekman layer flow which results, ultimately, in secondary vortices that are often difficult to distinguish from the vortices due to the instability of the basic flow.

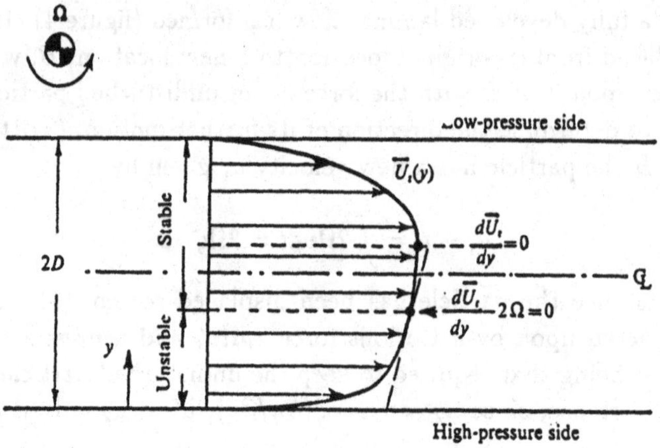

Figure 4: Stable and unstable regions in rotating channel flow, indicating the asymmetry of the turbulent mean velocity profile $U_t(y)$ when $\Omega > 0$.
Reproduced from Lezius et al. [7], with permission.

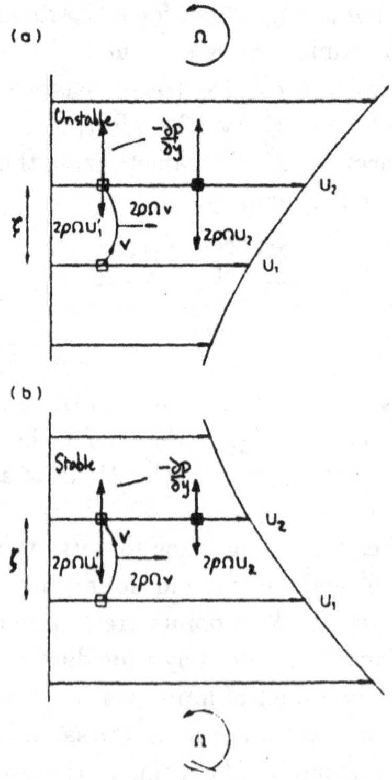

Figure 5: Sketch relating to displaced particle discussion—see text.

Reproduced from Tritton and Davies [5], with permission.

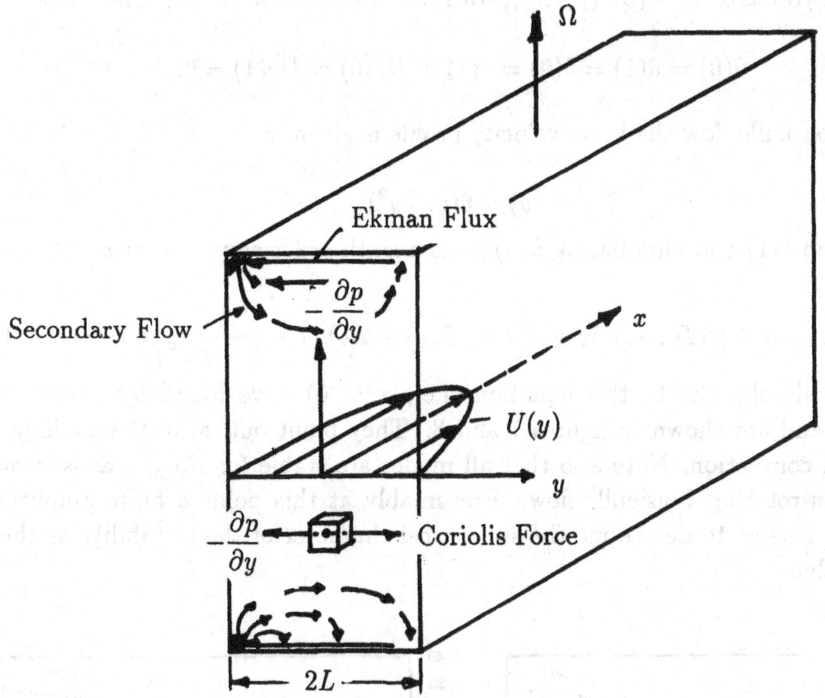

Figure 6: Generation of a secondary flow at the top and bottom walls of a rotating duct.

As discussed by these authors the stability, in an infinite duct, depends on two parameters.

$$Re = \frac{u_m 2L}{\nu} \text{ and } Ro = \frac{2L\Omega}{u_m} \tag{10}$$

where u_m is the average velocity in the duct, and $2L$ is used as a characteristic length scale (figure 6). Substituting a normal-mode type of a disturbance:

$$[\tilde{u}, \tilde{v}, \tilde{p}] = \frac{1}{2}\{\hat{u}(y), \hat{v}(y), \hat{p}(y)]\exp(i\alpha z + \sigma t) + \text{complex conjugate}\}$$

into the linearized equations leads to:

$$[\sigma - Re^{-1}(D^2 - \alpha^2)]\hat{u}(y) + (U' - 2Ro)\hat{v}(y) = 0$$

$$[\sigma - Re^{-1}(D^2 - \alpha^2)](D^2 - \alpha^2)\hat{v}(y) + (U' - 2Ro)\hat{v}(y) - \alpha^2 2Ro\hat{u}(y) = 0 \tag{11}$$

where $D = \partial/\partial y$ and $U' = (\partial U(y)/\partial y)$, with boundary conditions at solid walls:

$$\hat{u}(0) = \hat{u}(1) = \hat{v}(0) = \hat{v}(1) = D\hat{v}(0) = D\hat{v}(1) = 0$$

For plane Poisieulle flow the basic velocity profile is given by

$$U(y) = 6(y - y^2)$$

Substitution in (11) and eliminating $\hat{u}(y)$ gives a sixth order equation for $\hat{v}$:

$$(D^2 - \alpha^2 - \sigma Re)^2(D^2 - \alpha^2)\hat{v} + \alpha^2 4Re^2 Ro(3 - Ro)\left[1 - \frac{6}{3 - Ro}y\right]\hat{v} = 0 \qquad (12)$$

The neutral solutions to this equation (i.e., $\sigma = 0$) were found by Lezius and Johnston [7] and are shown in figures 7 and 8. They point out, also, the analogy to plane Benard convection. Note also that all modes are stable for $Ro \geq 3$ as is known for plane, non-rotating Poisieulle flow. Presumably at this point a finite amplitude calculation is needed to determine the subcritical character of the instability at these parameter values.

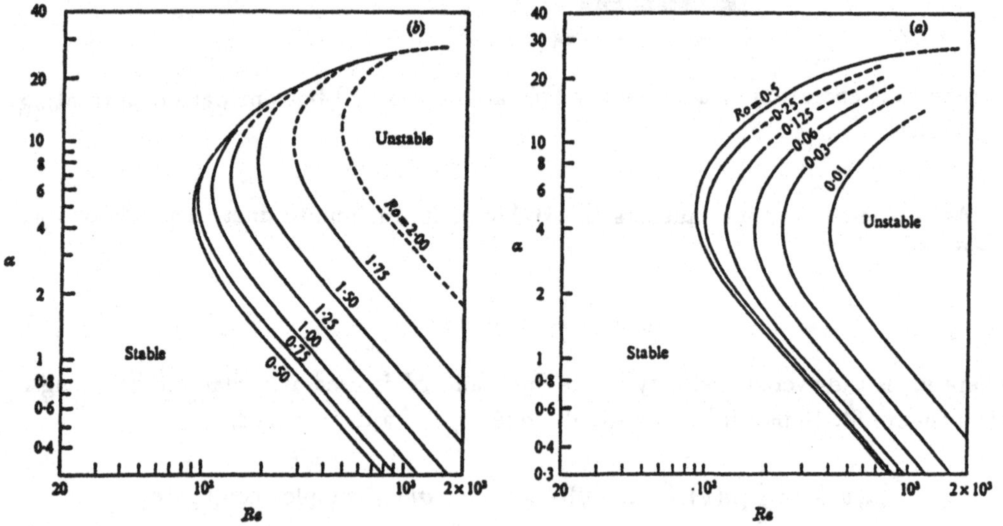

Figure 7: Neutral stability curves for laminar plane Poiseuille flow with rotation for (a) $Ro \leq 0.5$, and (b) $Ro \geq 0.5$.

Reproduced from Lezius et al. [7], with permission.

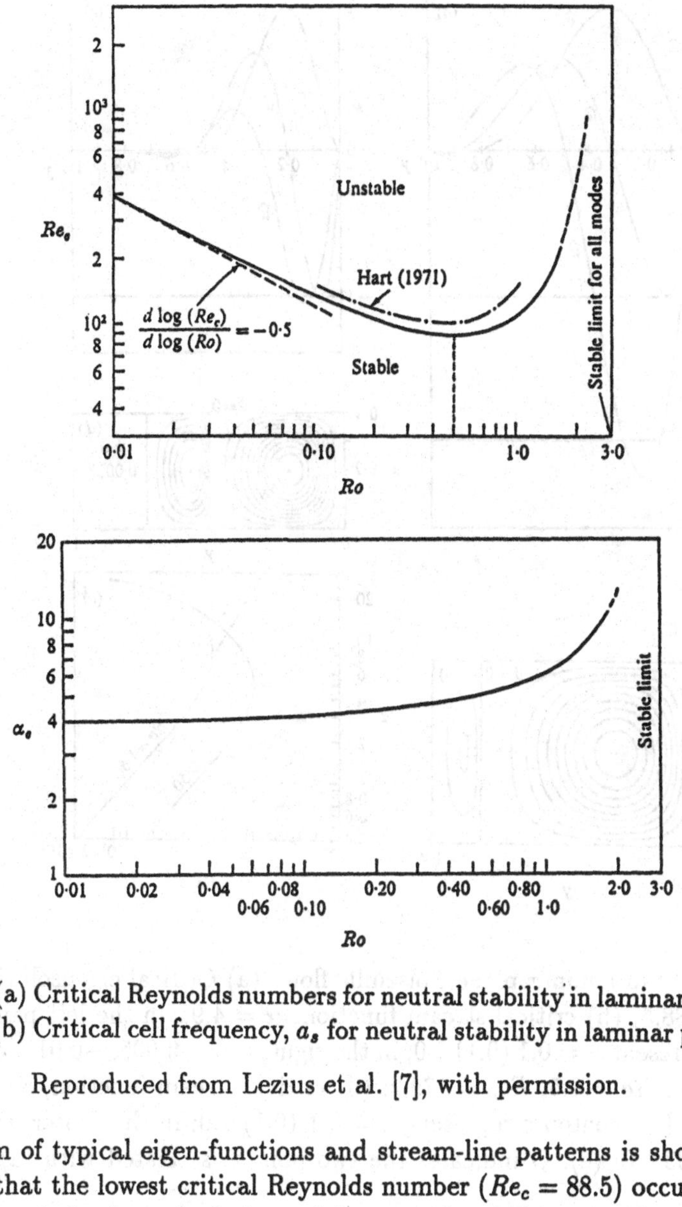

Figure 8: (a) Critical Reynolds numbers for neutral stability in laminar plane Poiseuille flow. (b) Critical cell frequency, a_s for neutral stability in laminar plane Poiseuille flow.
Reproduced from Lezius et al. [7], with permission.

The form of typical eigen-functions and stream-line patterns is shown in figure 9. Note, also, that the lowest critical Reynolds number ($Re_c = 88.5$) occurs at $Ro = 0.5$.

More recently further calculations of the neutral stability curves have been obtained by Alfredsson and Persson [8] who correct the former results at large values of α and calculate curves of constant growth rate (figure 10). The latter authors, have, also performed experiments in a large aspect ratio channel and have verified some of the theoretical predictions. Thus in figure 11 we show the experimental observation and compare in the table theory experiment for the most unstable wave length at an $Re = 167$.

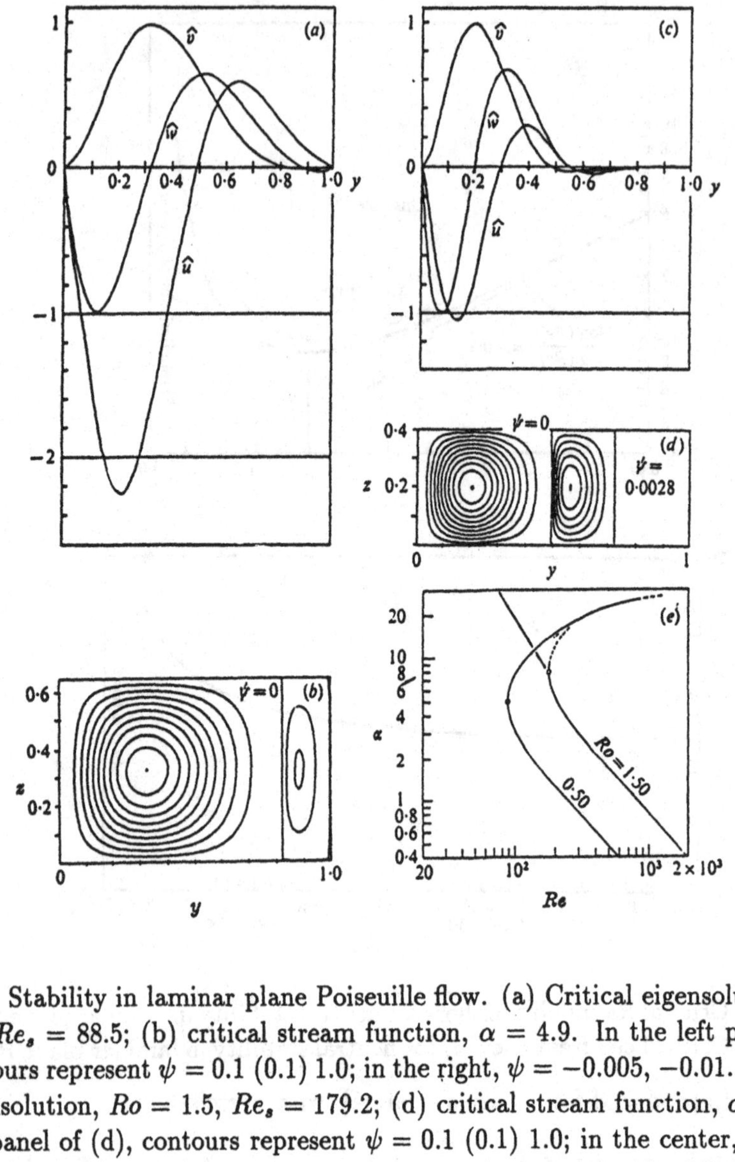

Figure 9: Stability in laminar plane Poiseuille flow. (a) Critical eigensolution, $Ro = 0.5$, $Re_s = 88.5$; (b) critical stream function, $\alpha = 4.9$. In the left panel of (b), contours represent $\psi = 0.1$ (0.1) 1.0; in the right, $\psi = -0.005, -0.01$. (c) Critical eigensolution, $Ro = 1.5$, $Re_s = 179.2$; (d) critical stream function, $\alpha 7.8$. In the left panel of (d), contours represent $\psi = 0.1$ (0.1) 1.0; in the center, $\psi = -0.01$ (0.005) -0.03. In (e), O indicates the two points associated with (a), (b) and (c), (d).

Reproduced from Lezius et al. [7], with permission.

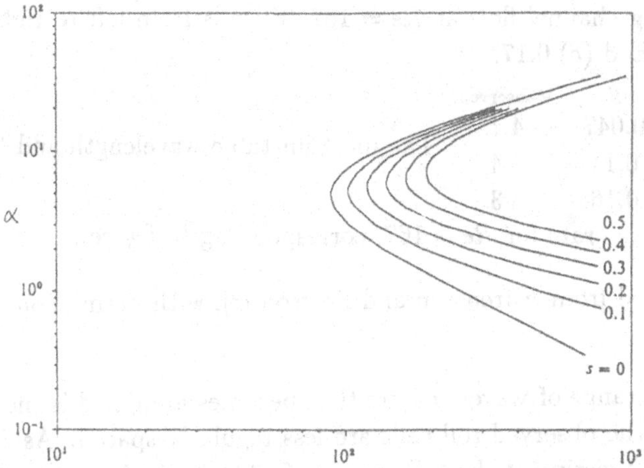

Figure 10: Curves of constant growth rate for $Ro = 0.5$. Reproduced from Alfredsson and Persson [8], with permission.

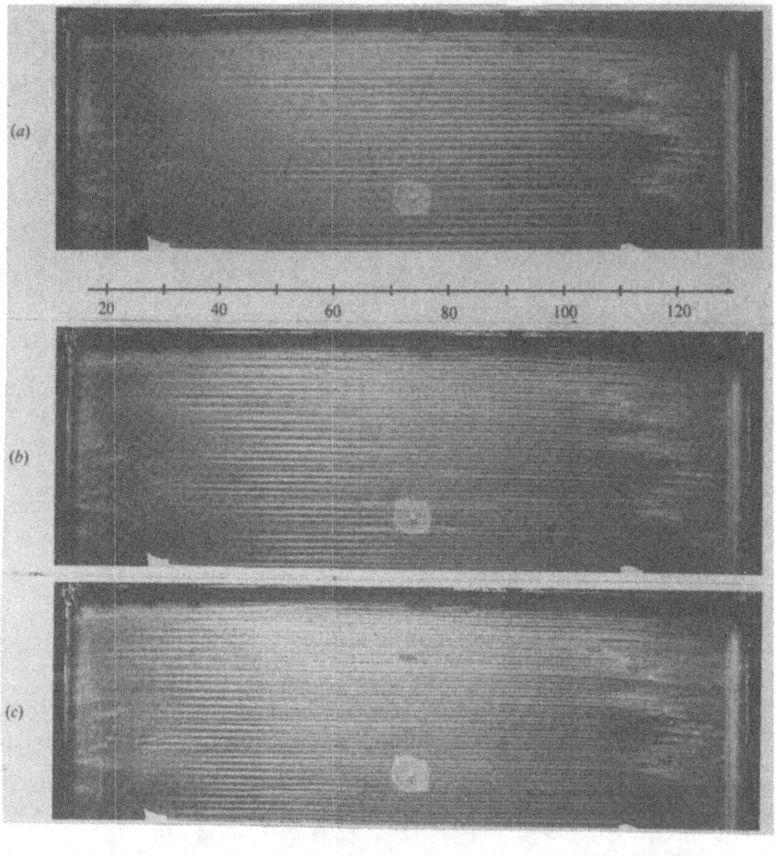

For caption see next page.

Figure 11: Rotating channel flow at $Re = 167$. Flow is from left to right. (a) $Ro = 0.09$. (b) 0.13 and (c) 0.17.

Ro	α	g	α_{observed}
0.09	4.5	0.047	4.7
0.13	4.9	0.11	4.4
0.17	5.2	0.16	5.3

The most unstable wavelength and the growth rate for $Re = 167$, corresponding to figure.

Reproduced from Alfredsson and Persson [8], with permission.

At higher Re the range of wavenumbers that becomes amplified is more extensive with the result that the observed roll-cells are less regularly spaced. As Ro increases at a given large Re a twisting of the cells occurs, first at their downstream ends. The start of twisting moves upstream as Ro increases further (figure 12) and are presumably related to the wavy instability of roll-cells seen in Taylor-Couette flows.

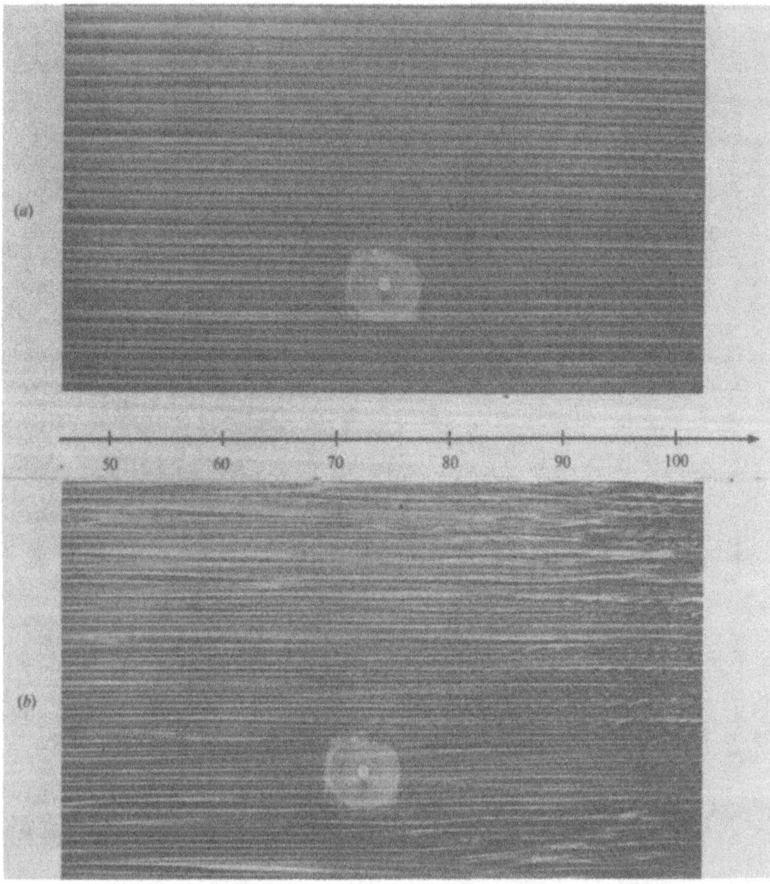

For caption see next page.

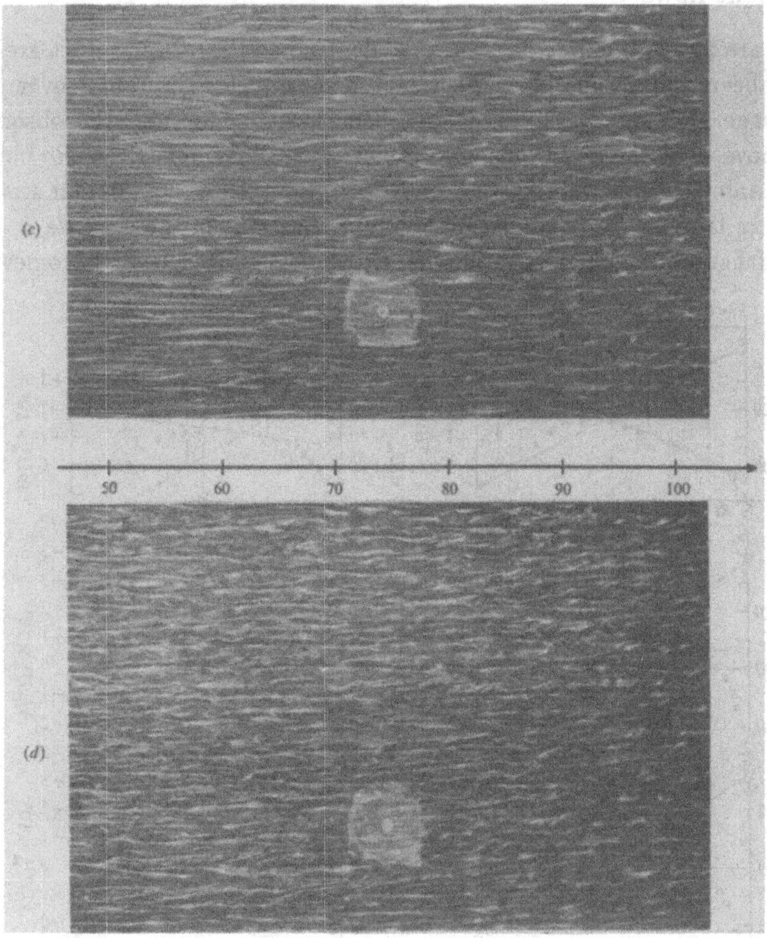

Figure 12: Rotating channel flow at Re = 590 for varying rotation number. (a) $Ro = 0.015$, (b) 0.084, (c) 0.16, (d) 0.26.

Reproduced from Alfredsson and Persson [8], with permission.

2.2 Turbulent flows

As far as we are aware the first data on turbulent flow in a rotating duct are due to Johnston, Halleen and Lezius [9] who measured mean-velocity profiles over a small parameter range and found, as could have been anticipated from the observations mentioned above, that the mean-velocity profiles became asymmetric as Ro increased.

Thus the stabilized side wall layer (B or $Ri > 0$) tends toward a laminar state, and, in fact becomes laminar (as observed by flow visualization) in the case $Re = 11,500$, $Ro = 0.21$ of figure 13. One the other hand the destabilized layer becomes "more

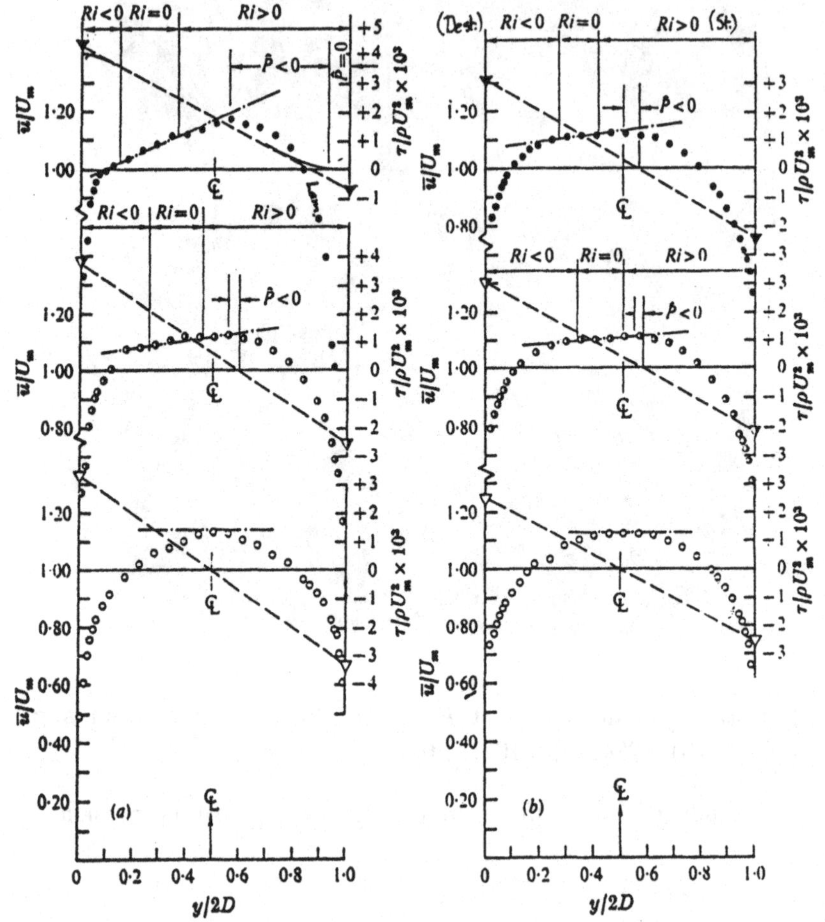

Figure 13: Mean velocity and shear stress profiles at $x/D = 68$. ∇, wall shear stress; o, mean velocity. — · —, slope of $\xi_{abc} = 0$; — — —, total shear stress τ. Open symbols, $Ro = 0$. (a) $Re = 11500$. Half-filled symbols, $Ro = 0.069$; filled symbols, $_o = 0.210$; solid curves for $Ro = 0.21$ are estimate of $-\rho\overline{u'v'}$ where it differs significantly from τ. (b) $Re = 35000$. Half-filled symbols, $Ro = 0.42$, filled symbols, $Ro = 0.068$. Reproduced from Johnston et al. [9], with permission.

turbulent" by generating intense, large-scale, longitudinal vortices, figure 14. These two regions are separated by a layer in which $Ri \approx 0$ and turbulent energy production is zero. Most of the turbulent energy production is taking place at the destabilized wall with a layer of negative production, $\hat{p} < 0$, between it and the stabilized wall. As noted by them "vigorous eddies appeared collide with this 'interface' as if it were a solid wall." Rotation also affected the rate of ejection "bursting" of turbulent fluid from the wall layers thus changing the rate of production of turbulent kinetic energy.

In a later paper Lezius and Johnston [7] proposed a two-layer model of the turbulent flow found in their earlier experiments which although it appears to be somewhat *ad hoc* gives reasonable agreement with experiment.

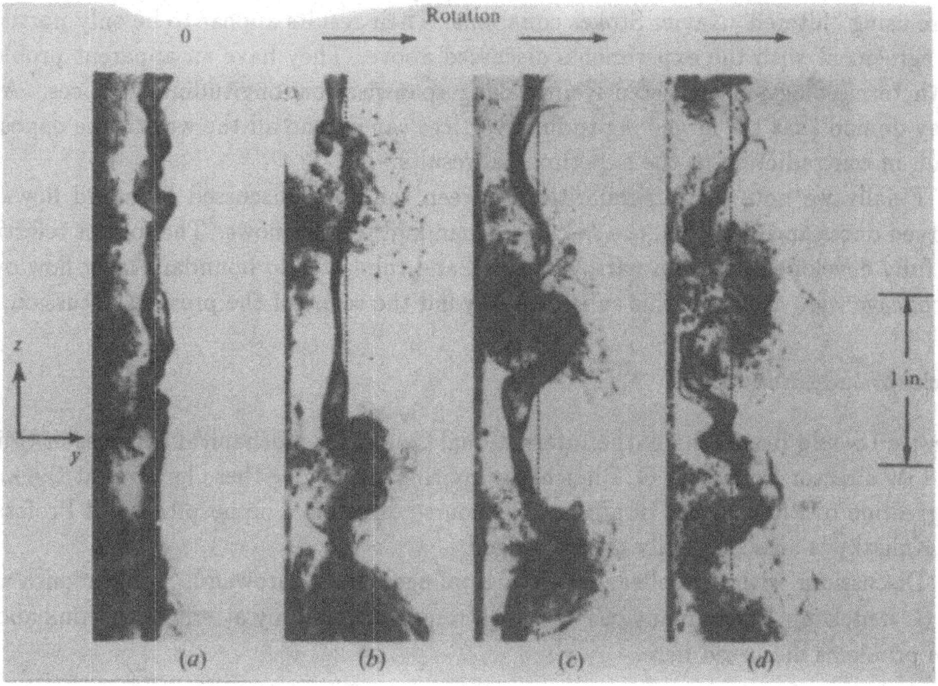

Figure 14: End-view of roll-cell structure on unstable side at $Re_Q = 5900$. Photos taken 0.23 s after relase of bubbles from downstream wire at position shown by dotted line (a) $Ro_Q = 0$. (b) $Ro_Q = 0.22$. (c) $Ro_Q = 0.32$. (d) $Ro_Q = 0.40$.

Reproduced from Johnston et al. [9], with permission.

More recently and almost simultaneously a number of groups have generated numerical solutions to the problems discussed above. In particular Yang and Kim [10] in a direct numerical simulation have verified and clarified the experimental results of Alfredsson and Persson [8]. In all cases longitudinal vortices develop initially with a wavenumber consistent with linear theory. At the later nonlinear stage a further wavenumber selection takes place leading to a different dominant wavenumber. Depending on *Ro* and *Re* these modes are detected at both high and low streamwise wavenumber. The secondary selection process involved the amalgamation and leapfrogging of vortex pairs, as found in the similar case of unsteady Couette instability by Maxworthy [11].

Another recent paper by Tafti and Vanka [12] concerns itself with the fully turbulent case using "filtered" Navier Stokes equations. Their results appear to be only partially in agreement with the experiments discussed above. They have an apparent problem with terminology in consistently mistaking spanwise for longitudinal vortices. Also they do find that the large longitudinal vortices can extend all the way to the opposite wall, in contradiction to the experimental results.

Finally we note many similarities between the flows discussed here and flows in curved ducts and channels, the so-called Dean and Görtler flows. The former referring to fully developed flow in a narrow channel and the latter to boundary layer flow over a concave wall. However this subject is beyond the scope of the present discussion.

Acknowledgments

Firstly I would like to thank the International Center for Mechanical Sciences in Udine and its director Professors G. Bianchi for inviting me to give these lectures at the kind suggestion of Professor E. Hopfinger, the course director. The hospitality of Professor L. Kaliszky is also gratefully acknowledged.

Discussions with my colleagues E. J. Hopfinger, F. K. Browand, S. Narimousa and L. G. Redekopp among many others have strongly affected my attempts to think about the problems discussed here.

References

[1] Mory, M., "Rotation Effects on Turbulence", in this volume, 1992.

[2] Maxworthy, T., Hopfinger, E. J. and Redekopp, L. G., "Waves on vortex cores", *J. Fluid Mech.*, **151**, 141–165, 1985.

[3] Maxworthy, T. and Narimousa, S., "Vortex generation by convection in a rotating fluid", *Ocean Modelling*, **92**, 1991, (unpublished manuscript).

[4] Fernando, H. J. S., Chen, R.-R. and Boyer, D. L., "Effects of rotation on convective turbulence", *J. Fluid Mech.*, **228**, 513–547, 1991.

[5] Tritton, D., and Davies, P. A., "Instabilities in geophysical fluid dynamics", in "Hydrodynamic Instabilities and the Transition to Turbulence", *Topics in Appl. Phys.*, **45**, Springer, 1968.

[6] Hart, J. E., "Instability and secondary motion in rotating channel flow", *J. Fluid Mech.*, **45**, 341, 1971.

[7] Lezius, D. K. and Johnston, J. P., "Roll-cell instabilities in rotating laminar and turbulent channel flows", *J. Fluid Mech.*, **77**, 153–175, 1976.

[8] Alfredsson, P. H. and Persson, H., "Instabilities in channel flow with system rotation", *J. Fluid Mech.*, **202**, 543–557, 1989.

[9] Johnston, J. P., Halleen, R. M. and Lezius, D. K., "Effects of spanwise rotation on the structure of two-dimensional, fully-developed, turbulent channel-flow", *J. Fluid Mech.*, **56**, 533–557, 1972.

[10] Yang, K.-S., and Kim, J., "Numerical investigation of instability and transition in rotating, plane Poisieuille flow", *Phys. Fluids*, **A3**, 633–641, 1991.

[11] Maxworthy, T., "A simple observational technique for the investigation of boundary-layer stability and turbulence" in *Turbulence Measurements in Liquids*, Department of Chemical Engineering, University of Missouri-Rolla, Continuing Education Series, 1971.

[12] Tafti, D. K. and Vanka, S. P., "A numerical study of the effects of spanwise rotation on turbulent channel flow", *Phys. Fluids*, **A3**, 642–656, 1991.

Chapter VI.3

BAROCLINIC TURBULENCE

E.J. Hopfinger
University J.F. and CNRS, Grenoble Cedex, France

1. INTRODUCTION

Baroclinic turbulence is stratified geostrophic turbulence where rotation and stratification are of equal order. A measure of the relative importance is the internal radius of deformation Λ, which compares (stable) density stratification effects with respect to rotation. In a two-layer fluid, when $\Lambda \rightarrow 0$, the problem reduces to a one layer barotropic turbulence and when $\Lambda \rightarrow \infty$, the two layers are uncoupled. In the intermediate case a vertical shear is allowed and the tilting of the planetary vorticity $f\partial u/\partial z$ is compensated by $-(g/\rho_0) \times \nabla\rho$ [1]. The perturbation pressure is essentially hydrostatic $\partial p/\partial z \approx - g\,\rho'$.

The impossibility of a direct energy cascade to smaller scales, known from 2D turbulence is still valid in baroclinic turbulence. But, contrary to 2D turbulence, vortex stretching is made possible due to vertical variability. In a two layer stratified rotating fluid the interface displacement causes vortex lines to stretch in one layer and to contract in the other. Here, only two-layer stratified geostrophic turbulence is discussed. Geostrophic turbulence with continuous stratification is similar but more complex because of its vertical variability which must be taken into account [2].

2. POTENTIAL VORTICITY AND ENERGY CONSERVATION

It is of interest to discuss briefly some aspects of 2D turbulence which is a model study of geostrophic turbulence and also of baroclinic turbulence. In an incompressible, inviscid barotropic fluid layer the equation of potential vorticity equation is:

$$\frac{D}{Dt}\left(\frac{\zeta+f}{H}\right)=0, \tag{1}$$

where $D/Dt = \partial/\partial t + \mathbf{u}\cdot\nabla$. Expanding this equation gives:

$$\frac{\partial \nabla^2\psi}{\partial t} = J(\psi, {}^2\psi) + \beta\frac{\partial\psi}{\partial x} - \frac{f}{H}J(\psi,H) = 0, \tag{2}$$

↑______↑

2D turbulence

↑__________ ____↑

linear Rossby waves

↑______________________↑

topographic waves

where $\mathbf{u} = (-\frac{\partial\psi}{\partial y}, \frac{\partial\psi}{\partial x})$, $\beta=\partial f/\partial y$ and J the Jacobian operator. We see from this equation that 2D turbulence is a special case of geostrophic turbulence. The most advanced numerical simulations of 2D turbulence seem to have been made by Brachet et al [3]. Their results show changes in the energy spectral slope when in the physical space folding of vorticity sheets occurs. The slope for the enstrophy cascade generally lies between k^{-3} and k^{-4}.

In a two or multilayer stratified fluid the potential vorticity equation (1) still holds in each layer in the form

$$\Pi_i = (\nabla^2\psi_i + f)/h_i \tag{3}$$

where $h_i = \eta_{i+1} - \eta_i + H_i$. The pressure is $p_{i+1} - p_i = g\,\rho_i\eta_i$. When viscosity is neglected, energy $\frac{1}{2}q^2 = \frac{1}{2}(\nabla\psi)^2$ is also a conserved quantity. It is transported toward large scales by an inverse cascade process.

In real fluid systems there is dissipation of energy and also of potential vorticity and the equation for energy for instance is

$$D(\frac{1}{2}q^2)/Dt = \text{forcing - dissipation.} \tag{4}$$

The dissipation has two terms in rotating fluids: one is ν times the the vorticity squared and the other is due to the Ekman drag on the end boundaries. The equation of the spatially averaged energy in a one layer fluid without forcing is:

$$\frac{d(<u^2> + <v^2>)}{dt} = -\frac{2}{\tau}(<u^2> + <v^2>) - 2\nu<\zeta'^2>, \tag{5}$$

where $< \;\; > = \int_D (\;\;)ds$ and $<\zeta'^2>$ is the spatially averaged square vorticity of the turbulence. The time scale τ is the Ekman spin-down time which is for one horizontal boundary $\tau = f^{-1}(E)^{-1/2}$, where the Ekman number $E = \nu/fH^2$. The molecular viscosity must be replaced by a turbulent viscosity when the Ekman boundary layer is turbulent. Another point of importance is that Ekman dissipation is scale independent but it sets a limiting upper scale to the eddy size where the eddy turnover time $L/U \approx \tau$. The contribution of $\nu <\zeta'^2>$ is generally negligible in 2D turbulence compared with Ekman drag.

3. TWO-LAYER STRATIFIED GEOSTROPHIC TURBULENCE

For a two-layer stratified fluid the horizontally and depth averaged energy equation is:

$$\frac{d}{dt}(H_1<U_1^2> + H_2<U_2^2> + g'<\eta^2>) = -(\nu f)^{1/2}<U_2^2>, \tag{6}$$

where $<U^2> = <(u^2 + v^2)>$, $g' = g\,\Delta\rho/\rho_0$, $\Delta\rho=(\rho_2-\rho_1)$, η is the interface displacement and Ekman friction was assumed to act on the lower layer 2. We make furthermore the simplification that the two layers are of equal depth $H_1=H_2=H$ and that the turbulence is initially of equal intensity in each layer $U_1^2=U_2^2=U^2$. These conditions were realized in the experiments of Griffiths and Hopfinger [4]. The energy equation (6) then simplifies to:

$$\frac{d}{dt}(H<U^2> + \frac{1}{2}g'<\eta^2>) = -(\nu f)^{1/2}<U^2>. \tag{7}$$

The potential energy term is related with U through the geostrophic balance relation

$$g'\nabla_H \eta \approx fk \times (U_1 - U_2)\,, \tag{8}$$

where $(U_1 - U_2)$ is the velocity difference across the interface at each point. Squaring relation (8) allows to write

$$g' \langle\eta^2\rangle /2H \langle U^2\rangle = \xi^2 F\,, \tag{9}$$

where the Froude number F is

$$F = f^2 (l_e/2)^2 / g'H \tag{10}$$

and

$$\xi^2 = \langle(U_1 - U_2)^2\rangle / \langle U^2\rangle = 2(1 - \langle U_1 U_2\rangle / \langle U^2\rangle) \tag{11}$$

The correlation $\langle U_1 U_2\rangle/\langle U^2\rangle$ rapidly approaches unity ($\xi \rightarrow 0$) in freely evolving baroclinic turbulence without Ekman dissipation [5]. For stationary turbulence ξ must be constant. Griffiths and Hopfinger [4] assumed a large correlation coefficient with $\xi^2 = 1/5$.

The Froude number cannot exceed a critical value F_c because baroclinic instability will occur on the eddy scale l_e. By the inverse energy cascade l_e will increase and this leads to an increase in F by equation (10). When F reaches F_c, the large eddies on scale l_e will become baroclinically unstable and potential energy is converted into kinetic energy on the scale of the radius of deformation. The flow will thus remain in a marginally stable state and slowly decaying, with $F \approx F_c$.

The value of F_c is not well known and Griffiths and Hopfinger used $F_c = 50(H_2/H_1)^{1/2}$ taken from the experiments of Griffiths and Linden [6].

Expressing the potential energy in (7) by relation (9) and replacing F by F_c, the solution of equation (7) is:

$$\langle U^2\rangle = \langle U^2(t_0)\rangle \exp\left\{\frac{-(\nu f)^{1/2}}{(1+\xi^2 F_c)\,H}(t - t_0)\right\} \tag{12}$$

The time scale τ_B of the exponential decay of the turbulent velocity is thus

$$\tau_B = 2(1 + \xi^2 F_c)H/(\nu f)^{1/2} = (1 + \xi^2 F_c)\tau \, , \tag{13}$$

where τ is the exponential decay time scale for velocity in freely evolving geostrophic one layer turbulence [7]. With the values for ξ^2 and F_c indicated above ($\xi^2 = 1/5$ and $F_c = 50$), the ratio $\tau_B/\tau \approx 11$.

4. AN EXPERIMENT OF BAROCLINIC TURBULENCE

Baroclinic turbulence can be produced by external forcing or by unstable density fronts. In the experiment of Griffiths and Hopfinger [4], turbulence was generated by baroclinic instability. A sketch of the experimental arrangement is shown in Fig 1.

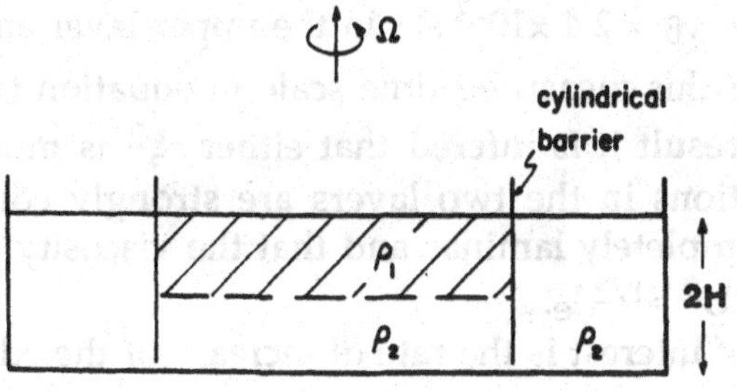

Fig.1 Diagram of the initial state of the experimental arrangement of Griffiths and Hopfinger [4]. The fluid is stationary in the rotating frame.

When the cylindrical barrier (see Fig. 1) is withdrawn, the fluid layer of density $\rho_1 < \rho_2$ moves radially outward under the action of buoancy. By conservation of angular momentum, a differential rotation is established (the lighter layer has an anticyclonic motion) which results in a radially inward acting Coriolis force. Geostrophic equilibrium is reached when the density front has moved through one deformation radius $(g'H)^{1/2}/f$. For the experimental conditions chosen, f=2 rad/s, H = 10 cm, g'= 0.91 cm/s^2,

$(g'H)^{1/2}/f = 1.5$, the density front is baroclinically unstable with typical wavelength $2\pi(g'H)^{1/2}/f$.

Time exposure photographs of neutrally buoyant particles, illuminated by a horizontal light sheet in the mids of the upper layer, are shown in Fig. 2. These are streak photographs showing the direction of particle motion by the dot at the end of each trace. The time exposure was 7.0 s and photo (FIG.2a) was taken 20 rotation periods after the barrier was withdrawn and (Fig. 2 b) after 32 periods. Since initially the horizontal shear component has cyclonic vorticity, the initial vortices are cyclones, but rapidly, vortex pairs are formed and cyclones and anticyclones are equal in number and about equal in intensity. This is shown in Figs.3. At long times, t > 70 periods of rotation, cyclones are rapidly dissipated and only anticyclones remain.

The mean eddy spacing and the turbulent velocities were also measured in this experiment. Their evolutions with time are shown in Fig.4. It is seen that the eddy spacing first decreases as cyclones are split into vortex pairs and then the scale increases by an inverse cascade process. The vortex size reaches a nearly constant value at times > 40 rotation periods when the Froude number approaches F_c and vortex growth and baroclinic instability compete.

From the measured rate of decrease of the turbulent velocity, which follows closely an exponential law as predicted by equation (12), the decay time scale can be evaluated; $\tau_B = 2.4 \times 10^{-3}\ s^{-1}$ in the upper layer and $3.4 \times 10^{-3}\ s^{-1}$ in the lower layer. Using this measured time scale in equation (13) gives $(1 + \xi^2 F_c) \approx 2.5$. From this result it is infered that either ξ^2 is much less than 1/5, meaning that the motions in the two layers are strongly coupled or that the Ekman layer is not completely laminar and that the viscosity must be replaced by an eddy viscosity $<U^2>^{1/2} l_e$.

Another quantity of interest is the rate of increase of the eddy scale which is given by:

$$\frac{dl_e}{dt} = \pi T <U2>^{1/2}, \tag{14}$$

where $l_e = \pi / k_1$, with k_1 the horizontal wave number, was used. The coefficient T must be determined from experiment. From Fig. 4 it is seen that the length scale doubles in the time interval 15 to 50 rotation periods. Solving equation (14) and using (12) and the decay rate of U as indicated in Fig. 4, gives approximately $T \approx 2.5 \cdot 10^{-2}$. This value is close to freely evolving 2D or geostrophic turbulence without stratification [8], [5].

(a)

(b)

Fig. 2. Streak photographs showing the evolution of the turbulence. a), 20 and (b), 32 rotation periods after barrier withdrawal. The exposure time was 7s. and the point at the end of each streak indicates the flow direction.

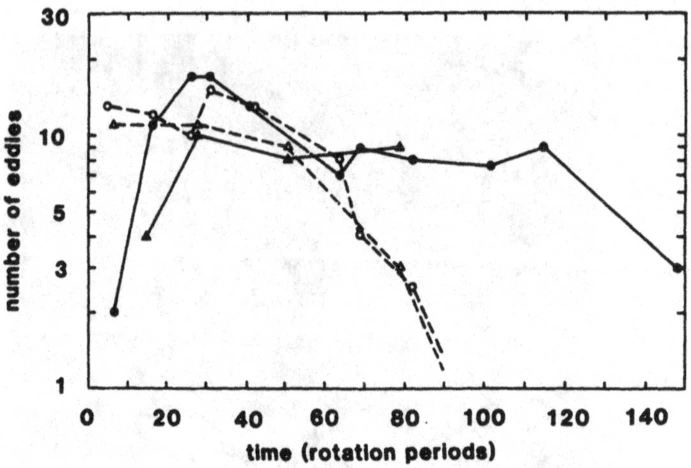

Fig.3 Evolution of cyclones and anticyclones; open symbols cyclones, closed symbols anticyclones; from [4].

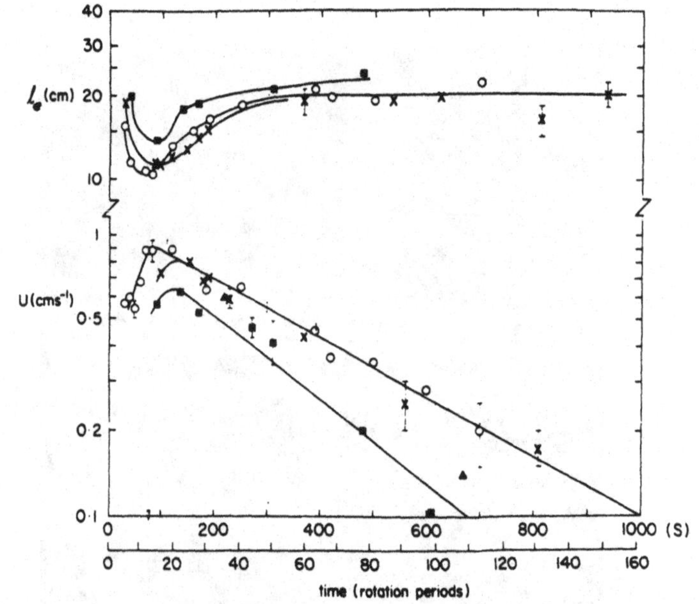

Fig. 4 Evolution of turbulent velocities and length scales in the upper and lower layer. Open symbols upper layer, closed symbols lower; from [4]

4.1 Structure Function and Energy Spectrum:

Energy spectra can be obtained by analysing particle streak images as shown in Fig.2. One possibility is to use the relationship between two-particle dispersion and the one dimensional energy spectrum [9]. Griffiths and Hopfinger [4] used the structure function and also the diffusivity. For unmarked particles the structure function is a well defined quantity and is given by

$$<(\frac{dD}{dt})^2> = <(V_1 - V_2)^2> \ . \tag{15}$$

Two particles already separated by a distance D will separate further due to eddy motions of scale smaller or about equal to the separation. By this assumpion $<(V_1 - V_2)^2> \approx 2<V^2>$; the correllation $V_1 . V_2 \approx 0$. The relation with the energy spectrum is then

$$<(\frac{dD}{dt})^2> \approx 4 \int_{1/D}^{\infty} E(k)\, dk\,, \tag{16}$$

where E(k) is the power spectrum of the kinetic energy. Assuming the existence of a power law $E(k) \sim k^{-\alpha}$, equation (16) gives:

$$<(\frac{dD}{dt})^2> \sim D^{\alpha-1}\,. \tag{17}$$

Experimentally, the structure function (17) is approximated by a finite difference form

$$<(\frac{\delta D}{\delta t})^2> = <\{\frac{(r_2' - r_1') - (r_2 - r_1)}{\delta t}\}^2> \sim D^{\alpha-1}\,, \tag{18}$$

where r_1 and r_2 are the initial and r_1' and r_2' the final positions of two particles 1 and 2 which were moved during the small time interval δt . The structure function shown in Fig. 5, taken from Griffiths and Hopfinger was calculated by averaging over about 500 particle traces separated by $0.3 \leq D \leq 40$ cm. The spectral slope corresponding to Fig.5 is $E \sim k^{2.5}$, which is different from, the k^{-3} to k^{-4} slopes obtained for 2D turbulence.

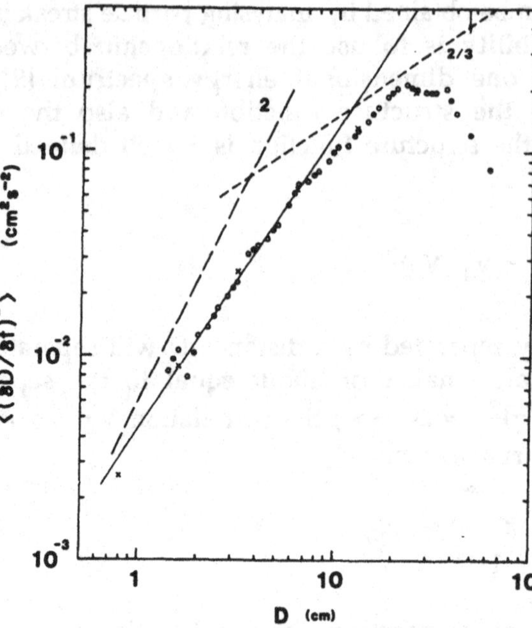

Fig.5 Mean square relative velocity or structure function as a function of D, the particle separation. The slope of the solid line is 1.47; from [4].

Recently, Narimousa et al. [10] determined the spectral slopes in a quasi 2D turbulence by the structure function method discussed above and also by a two-dimensional FFT on the interpolated u'^2 velocity field. The structure function method produced $k^{-2 \pm 0.5}$, whereas the FFT method indicated slopes as large as $k^{-5.5}$. Capéran [11] also obtained steep slopes, k^{-4}, by the FFT method applied to stratified turbulence. The reason for this difference is probably that when $\alpha>3$ in $k^{-\alpha}$, no simple correspondence between the particle structure function and the energy spectrum exists. This has been previously suggested by Babiano et al [12] and is supported by the results obtained in the range where $\alpha<3$; the two methods give in this case identical results.

Acknowledgements:

The support of this work by IFREMER and by the CNRS-INSU is acknowledged. I am particularly indebted to Ross W. Griffiths, whose influence and collaboration is found all along these notes. Many discussions with my close colleagues T. Maxworthy, Ph. Capéran and M. Mory have also greatly contributed to my understanding of the problems discussed here.

REFERENCES:

1. Rhines, P.: Geostrophic turbulence, Ann. Rev. Fluid Mech., 11 (1979), 401.
2. Hua, L.B. & Haidvogel, D.B.: Numerical simulation of the vertical structure of quasi geostrophic turbulence, J. Atmosph. Sci., 43 (1986), 2923.
3. Brachet, M.-E., Meneguzzi, M. and Sulem, P.-L.: Small scale dynamics of high Reynolds number two-dimensional turbulence, Phys. Rev. Lett., 57 (1986), 683.
4. Griffiths, R.W. and Hopfinger, E.J.: The structure of mesoscale turbulence and horizontal spreading at ocean fronts, Deep sea Res., 31 (1984), 245.
5. Rhines, P.: Waves and turbulence on a beta- plane, J.Fluid Mec., 69 (1975), 417.
6. Griffiths, R.W.and Linden, P.: The stability of vortices in rotating stratified fluid, J. Fluid Mech. 195 (1981), 283.
7. Colin de Verdière, A.: Quasi-geostrophic turbulence in a rotating homogeneous fluid, Geophys. astrophys. Fluid Dyn., 15 (1980), 213.
8. Hopfinger, E.J., Griffiths, R.W. and Mory, M.: The structure of turbulence in homogeneous and stratified rotating fluids. J. Méc. Théor. Appl., numéro spéc. 9 (1983), 21.
9. Morel, P. and Larchevèque, M.: Relative dispersion of constant level balloons in the 200 mb general ciculation, J. Atmosph. Sci., 31 (1974), 2189.
10. Narimousa,S., Maxworthy, T. and Spedding, G. R.: Experiments on the structure and dynamics of forced, quasi-two-dimensional turbulence. J. Fluid Mech. 223 (1991), 113.
11. Capéran, P.: Turbulence soumise à des forces extérieures. Thèse d'état, Universite J.F. Grenoble, 1989.
12. Babiano, A., Basdevant, C. and Sadourny, R.: Structure functions and dispersion laws in two-dimensional turbulence, J. Atmos. Sci., 42 (1985), 941.

Chapter VI.4

RESONANT COLLAPSE

P.F. Linden, R. Manasseh
University of Cambridge, Cambridge, UK

When a container full of liquid spins about an unconstrained axis of rotation, the fluid motion within the container may cause an instability. It is observed Pocha [1] that for a cylinder rotating about its symmetry axis, the axis may begin to nutate with the nutation angle growing exponentially in time. Instabilities of this type have serious implications for modern satellites which have fuel tanks filled with liquid fuel. These satellites are often spin stabilized, and spin about their axis of least moment of inertia. Wobble in the spin axis can cause the satellite to point in the wrong direction (or worse to become completely unstable).

There appear to be two basic mechanisms by which the motion of the fluid may cause this instability. There is a 'dynamic instability' which appears to be what prevents a raw egg from spinning when placed on a flat surface. This instability is associated with pressure forces acting on the container walls which may produce torques causing the rotation axis to nutate. In less viscous fluids inertial modes may be responsible for these pressure forces. A second alternative is that energy dissipation within the fluid may cause the container to spin about its axis of greatest inertia. In either case it is clear that an understanding of the fluid motion within a precessing, rapidly rotating cylinder is essential to predicting the behaviour.

The simplest case to consider is the forced response of the fluid within a completely filled, rotating cylinder in which the rotation axis nutates with a given nutation angle θ. Several theoretical studies, beginning with Kelvin [2], have been made of the inviscid forced oscillations. Inertia modes are generated and it is found that the resonant frequencies are dense in parameter space. A careful experimental study of the modes set up by a rotating end cap, placed at an angle to the bottom of the cylinder was carried out by McEwan [3]. He found that linear theory was accurate in predicting the resonances he observed.

It has been observed in a number of related experiments (Johnson, [4]; Malkus, [5]; McEwan [3]; Whiting [6]) that, as the forcing increased, instabilities occurred and turbulence was generated. The most comprehensive study was by McEwan [3] who noted that after forced flow had been established, instabilities appeared (after 10 to 100 rotation periods) and the motion developed disorder and fine-scale turbulence was observed. This phenomenon he called *resonant collapse*. McEwan found that collapses occurred over a wide parameter range and that in many cases the degree of disorder was time dependent.

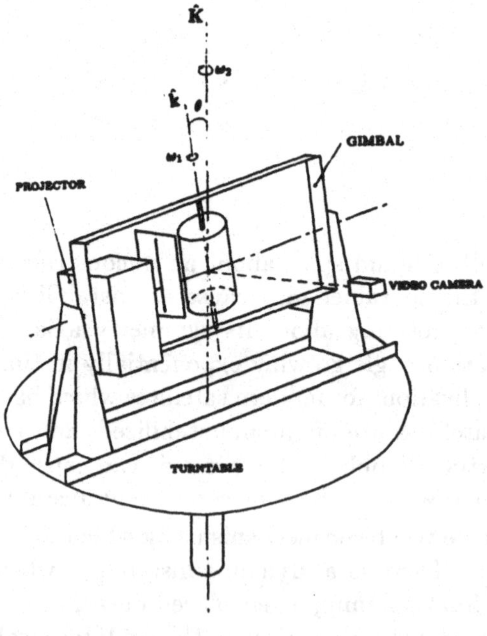

Fig. 1 Sketch of the experimental apparatus for the study of forced precession in a rotating cylinder.

In a recent series of experiments, Manasseh [7] has studied the response of a rapidly rotating cylinder whose axis of rotation precesses with a given nutation angle θ. The experimental set up is shown in figure 1. With the notation shown in figure 1, the equations describing the linearized response to the precessional forcing are

$$\mathbf{u}_t + \omega\hat{\mathbf{k}} \times \mathbf{u} + \nabla p = -(\omega - 2)(r\cos(\phi + t))\hat{\mathbf{k}} \quad , \tag{1}$$

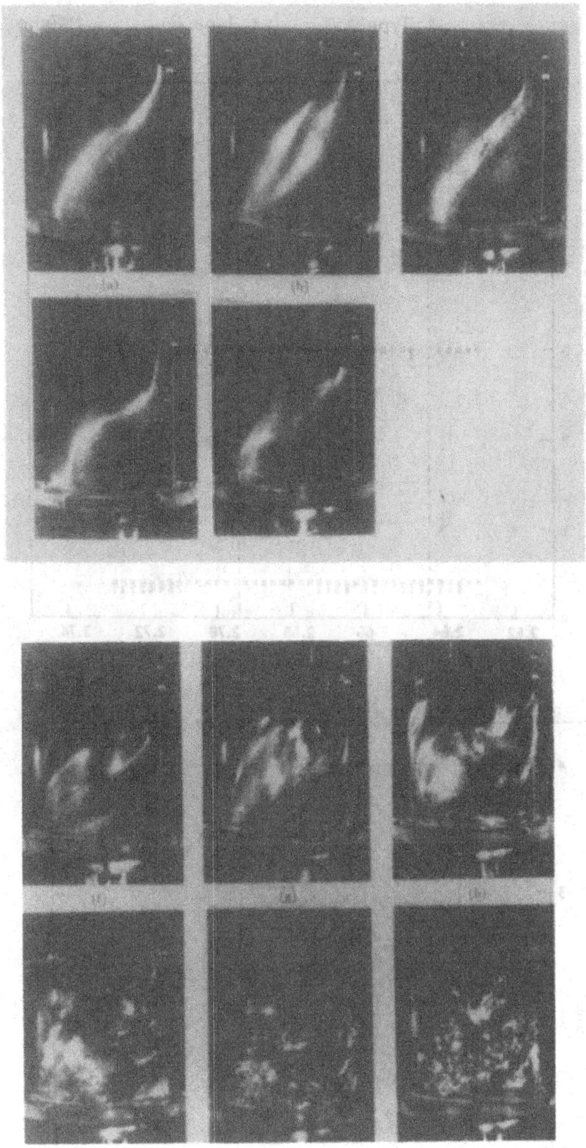

Fig. 2 A time sequence showing the Type A collapse of the (1, 1, 1) mode.

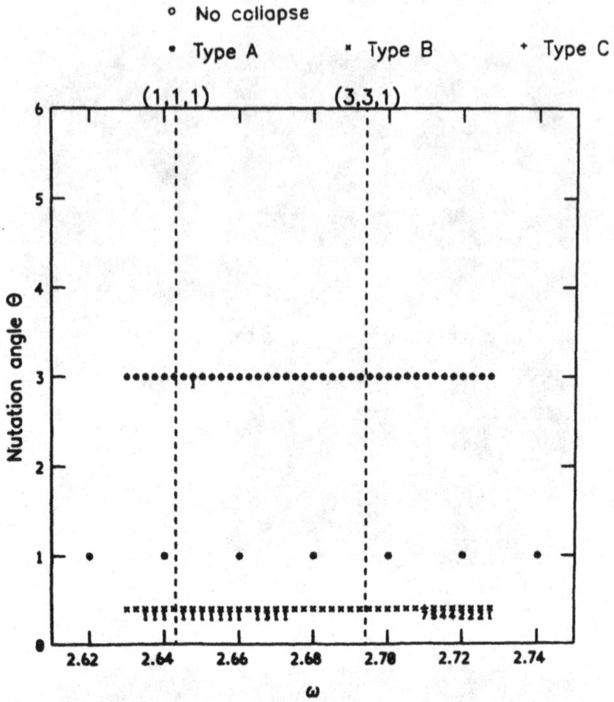

Fig. 3 Régime diagram for the collapse of the (1, 1, 1) mode.

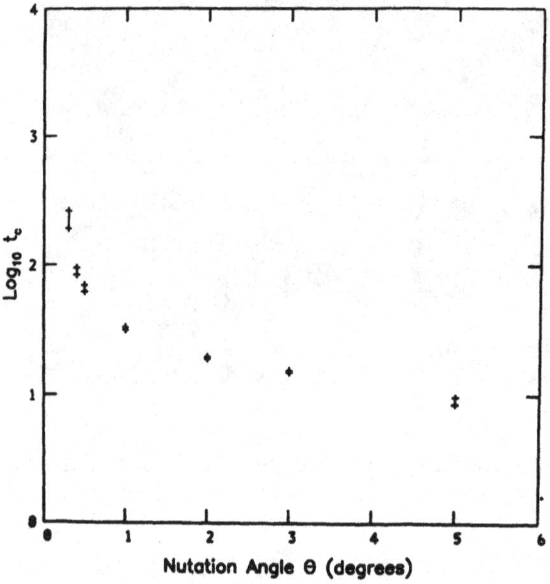

Fig. 4 Collapse times for the (1, 1, 1) mode as a function of the nutation angle θ.

where the dimensionless excitation frequency $\omega = 2\left(1+\frac{\omega_2}{\omega_1}\right)$. Eliminating the velocity components in favour of the pressure p in (1) and assuming an expansion of the form

$$p = \sum Q_n(r\ ,\phi\ ,z)e^{it}\ , \tag{2}$$

the Q_n satisfy Poincaré's equation

$$\frac{1}{r}(rQ_{nr})_r + \frac{1}{r^2}Q_{n\phi\phi} + (1-\omega^2)Q_{nzz} = 0\ . \tag{3}$$

This equation is hyperbolic for $|\omega| > 1$. Application of zero normal velocity boundary conditions on the walls of the cylinder gives the dispersion relation for the free modes of the system, and hence the frequencies at which the forced solution will resonate.

Manasseh [7] describes the behaviour of the system near the resonant frequencies for the low order modes of the system. A very complex range of behaviour is observed, depending on both the form of the low order mode which is near resonance and on the strength of the forcing (the size of the nutation angle). Manasseh's [7] study is a very comprehensive investigation of the fluid flow as revealed by flow visualisation techniques. The reader is referred to Manasseh [7] and [8] for further descriptions of the phenomena outlined below.

We shall concentrate attention on the lowest mode (1, 1, 1) for which Manasseh identifes three types of collapse, identified by A, B and C. Type A collapse (figure 2) in the most common form, occurring for all but the weakest values of the forcing. The waveform figure 2(a) appears to split near the inflexion point, and then the flow develops a modulation and small scale turbulence rapidly develops. At weaker forcing Type B collapses occur in which the eventual state of disorder is weaker and has a more columnar structure than the Type A collapse. These disordered states may undergo further transitions and a Type C collapse was identified. The increased disorder associated with Type C collapse may reduce with time and as subsequent collapses occur.

Figure 3 shows a régime diagram for the collapse types plotted against the nutation θ and the forcing frequency ω. The resonant frequencies for the (1, 1, 1) and the (3, 3, 1) mode are shown as the broken vertical lines. The initial collapse type seems to be dependent only on θ over the range of ω considered, with Type B only occurring for angles $\leq 0.3^o$. The time taken to collapse decreases with increasing θ (see figure 4), but is only weakly dependent on the forcing frequency ω (see figure 5).

The motion within the cylinder was also revealed using Thymol Blue pH indicator and an example for the (1, 1, 1) mode is shown in figure 6. While the initial behaviour shown in figures 6(a) and 6(b) has the same shape and the velocities are in approximate agreement with what would be expected from linear theory, subsequent motion shows high spatial gradients becoming significant in the flow fields even at very early times (figures 6(c) - 6(h)). Similar high wavenumber features were observed on the basic (3, 1, 1) mode (figure 7).

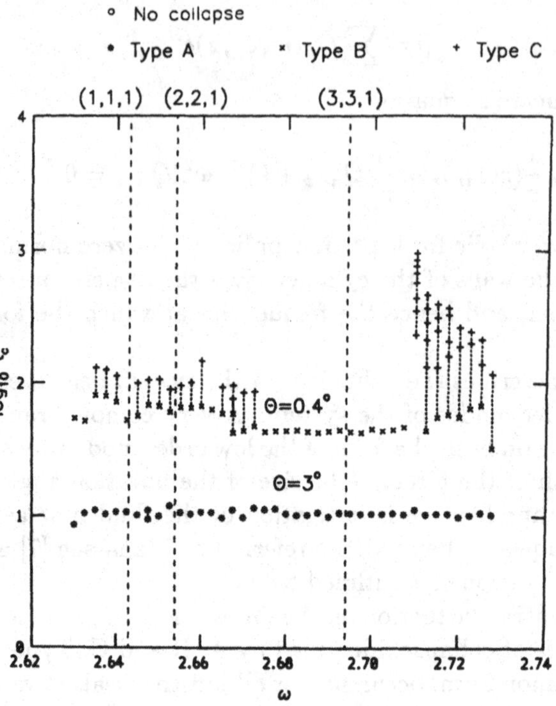

Fig. 5 Collapse times for the (1, 1, 1) mode as a funtion of the nutation frequency ω.

A rich variety of complex structures were observed in these experiments even in the limited range of parameter space that was explored. Linear theory is successful in predicting the resonant frequencies of the low order modes and, at small θ, gives a reasonable quantitative description of the flow fields at early times. However, it is clear from the dyeline experiments that even at early times the flow field is significantly different from that expected on the basis of the linear response of the lowest order mode near resonance. It must be emphasised that this departure occurs well before the collapses discussed above occur.

At this stage an explanation of these complex and intriguing phenomena awaits further study.

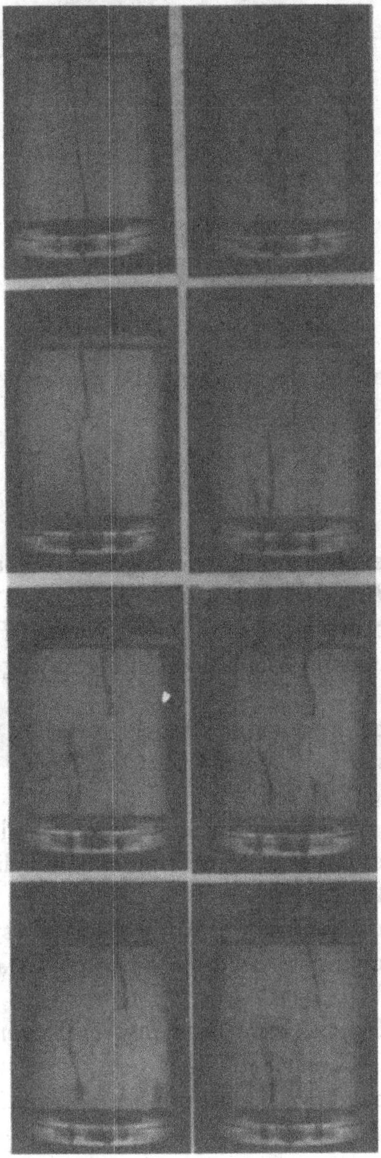

Fig. 6 Dyeline patterns produced by forcing near the resonant frequency of the (1, 1, 1) mode.

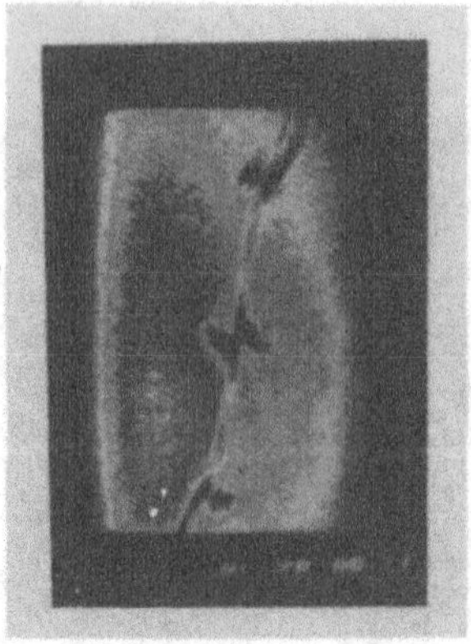

Fig. 7 False colour image of the dyeline near the resonant frequency for the (3, 1, 1) mode.

REFERENCES

1. Pocha, J.J. 1987. An experimental investigation of spacecraft sloshing. *Space Communications and Broadcasting* **5**, 323-332.
2. Kelvin, Lord 1880. Vibrations of a columnar vortex. *Phil. Mag.* **10**, 155-168.
3. McEwan, A.D. 1970. Inertial oscillations in a rotating fluid cylinder. *J. Fluid Mech.* **40**, 603-640.
4. Johnson, L.E. 1967. The precessing cylinder. *Notes on the 1967 Summer Study Program in Geophysical Fluid Dynamics at the Woods Hole Oceanographic Institution, Ref. No. 67-54*, 85-108.
5. Malkus, W.V.R. 1989. An experimental study of the global instabilities due to the tidal (elliptical) distortion of a rotating elastic cylinder. *Geophys. Astrophys Fluid Dynamics* **48**, 123-134.
6. Whiting, R.D. 1981. An experimental study of forced asymmetric oscillations in a rotating liquid filled cylinder. *Ballistic Research Labs.* Report No. ARBRL-TR-02376.
7. Manasseh, R., 1991. Inertia wave breakdown. Experiments in a precessing cylinder. *PhD thesis U. Cambridge.*
8. Manasseh, R. 1992. Breakdown regime of inertia waves in a precessing cylinder. Submitted to *J. Fluid Mech.*

Printed in the United States
By Bookmasters